AF358818

Smart Manufacturing & Automation Control Systems for Industry 4.0/5.0

Smart Manufacturing & Automation Control Systems for Industry 4.0/5.0

Editor

Sergey Y. Yurish

Basel • Beijing • Wuhan • Barcelona • Belgrade • Novi Sad • Cluj • Manchester

Editor
Sergey Y. Yurish
International Frequency
Sensor Association (IFSA)
Castelldefels
Spain

Editorial Office
MDPI AG
Grosspeteranlage 5
4052 Basel, Switzerland

This is a reprint of articles from the Special Issue published online in the open access journal *Processes* (ISSN 2227-9717) (available at: https://www.mdpi.com/journal/processes/special_issues/ 4749139690).

For citation purposes, cite each article independently as indicated on the article page online and as indicated below:

Lastname, A.A.; Lastname, B.B. Article Title. *Journal Name* **Year**, *Volume Number*, Page Range.

ISBN 978-3-7258-1871-6 (Hbk)
ISBN 978-3-7258-1872-3 (PDF)
doi.org/10.3390/books978-3-7258-1872-3

Contents

About the Editor . . . . **vii**

Preface . . . . **ix**

Yongmao Xiao, Jincheng Zhou, Shixiong Xing and Xiaoyong Zhu
Research on Assembly Sequence Optimization Classification Method of Remanufacturing Parts
Based on Different Precision Levels
Reprinted from: *Processes* **2023**, *11*, 383, doi:10.3390/pr11020383 . . . . **1**

Xiaoyong Zhu, Yu Liang, Yongmao Xiao, Gongwei Xiao and Xiaojuan Deng
Identification of Key Brittleness Factors for the Lean–Green Manufacturing System
in a Manufacturing Company in the Context of Industry 4.0, Based on the
DEMATEL-ISM-MICMAC Method
Reprinted from: *Processes* **2023**, *11*, 499, doi:10.3390/pr11020499 . . . . **13**

Xuefei Zhang, Ning Duan, Linhua Jiang, Fuyuan Xu, Zhaosheng Yu, Wen Cheng, et al.
Application of PLC-Based Spectrophotometric System Nitrogen Protection Device to
Automated Direct Measurement of Target Substances in Zinc Hydrometallurgy
Reprinted from: *Processes* **2023**, *11*, 672, doi:10.3390/pr11030672 . . . . **36**

Baisong Hu, Shuo Liu, Chuanzhi Wang and Bingjun Gao
Structural Optimization of High-Pressure Polyethylene Cyclone Separator Based on Energy
Efficiency Parameters
Reprinted from: *Processes* **2023**, *11*, 691, doi:10.3390/pr11030691 . . . . **52**

Yani Zhang, Haoshu Xu, Jun Huang and Yongmao Xiao
Research on Multiple Constraints Intelligent Production Line Scheduling Problem Based on
Beetle Antennae Search (BAS) Algorithm
Reprinted from: *Processes* **2023**, *11*, 904, doi:10.3390/pr11030904 . . . . **78**

Yongmao Xiao, Guohua Chen, Hao Zhang and Xiaoyong Zhu
Optimization of Low-Carbon and Highly Efficient Turning Production Equipment Selection
Based on Beetle Antennae Search Algorithm (BAS)
Reprinted from: *Processes* **2023**, *11*, 911, doi:10.3390/pr11030911 . . . . **94**

Xiaoyong Zhu, Lili Jiang and Yongmao Xiao
Study on the Optimization of the Material Distribution Path in an Electronic Assembly
Manufacturing Company Workshop Based on a Genetic Algorithm Considering Carbon
Emissions
Reprinted from: *Processes* **2023**, *11*, 1500, doi:10.3390/pr11051500 . . . . **111**

Ping Yuan, Tianhong Zhou and Luping Zhao
A Novel Model Prediction and Migration Method for Multi-Mode Nonlinear Time-Delay
Processes
Reprinted from: *Processes* **2023**, *11*, 1699, doi:10.3390/pr11061699 . . . . **129**

Huiyong Liu and Qing Zhao
Numerical Experiments on Performance Comparisons of Conical Type Direct-Acting Relief
Valve—With or without Conical Angle in Valve Element and Valve Seat
Reprinted from: *Processes* **2023**, *11*, 1792, doi:10.3390/pr11061792 . . . . **145**

Yongmao Xiao, Hao Zhang and Ruping Wang
Low-Carbon and Energy-Saving Path Optimization Scheduling of Material Distribution in
Machining Shop Based on Business Compass Model
Reprinted from: *Processes* **2023**, *11*, 1960, doi:10.3390/pr11071960 . . . . **176**

Iuon-Chang Lin, Pai-Ching Tseng, Yu-Sung Chang and Tzu-Ching Weng
IOTA Data Preservation Implementation for Industrial Automation and Control Systems
Reprinted from: *Processes* **2023**, *11*, 2160, doi:10.3390/pr11072160 . . . . **191**

**Alejandro Rubio-Rico, Fernando Mengod-Bautista, Andrés Lluna-Arriaga, Belén
Arroyo-Torres and Vicente Fuster-Roig**
The Industrial Digital Energy Twin as a Tool for the Comprehensive Optimization of Industrial
Processes
Reprinted from: *Processes* **2023**, *11*, 2353, doi:10.3390/pr11082353 . . . . **208**

About the Editor

Sergey Y. Yurish

Sergey Y. Yurish has served as the president of the International Frequency Sensor Association (IFSA)—one of the major professional associations serving the sensor industry and academia—for more than 25 years. Dr. Yurish is a founder of three companies. He is the editor-in-chief of the international peer-reviewed journal *Sensors & Transducers* and editor of several open access multivolume book series. Dr. Yurish obtained his PhD degree in 1996 from the National University Lviv Polytechnic (UA). He has published more than 180 articles and papers in international peer-reviewed journals and conference proceedings. Dr. Yurish holds nine patents and is an author and co-author of 12 books. He has delivered more than 90 speeches, tutorials, and keynote presentations at industries, peer institutions, and professional conferences in 30 countries. Dr. Yurish was a Marie Curie Chairs Excellence Investigator at the Technical University of Catalonia (UPC, Barcelona, Spain) from 2006 to 2009, where he led and developed one of the most successful projects in the UPC on Smart Sensors Systems Design (SMARTSES), totaling EUR 425,000. His professional accomplishments also include a Senior Research Fellowship at the Open University of Catalonia (UOC, Barcelona, Spain) where he spent a year in 2009–2010. Dr. Yurish has 30 years of research and academic experience, during which he has developed numerous projects on an international level in frames of various programmers, including NATO, FP6, and FP7.

Preface

According to the NIST's definition, smart manufacturing is fully integrated, collaborative manufacturing systems that respond in real time to meet changing demands and conditions in the factory, in the supply network, and in customer needs. The global smart manufacturing market was valued at USD 254.24 billion in 2022 and is anticipated to grow at a CAGR of 14.9% from 2023 to 2030. The market is expanding at a faster rate due to factors such as rising Industry 4.0 adoption, more government engagement in supporting industrial automation, increased emphasis on industrial automation in manufacturing processes, and surging demand for software systems that save time and cost.

The special issue entitled "Smart Manufacturing & Automation Control Systems for Industry 4.0/5.0" contains the extended papers selected from the 3rd IFSA Winter Conference on Automation, Robotics & Communications for Industry 4.0/5.0 (ARCI 2023), 22–24 February 2023, Chamonix-Mont-Blanc, France on the following topics: process control, monitoring, smart manufacturing and technologies.

The special issue contains 12 chapters written by 49 authors from China, Spain, and Taiwan. This special issue will inform readers about cutting-edge developments in the field and provide effective starting points and a road map for further research and development. All chapters follow a similar structure: firstly, an introduction to the specific topic under study, and secondly, a description of the field, including applications. Each chapter ends with a list of references, including special issues, journals, conference proceedings, and web sites.

The special issue is intended for researchers and scientists from academia and industry, as well as for graduate and postgraduate students.

Sergey Y. Yurish
Editor

Article

Research on Assembly Sequence Optimization Classification Method of Remanufacturing Parts Based on Different Precision Levels

Yongmao Xiao [1,2,3], **Jincheng Zhou** [1,2,3,*], **Shixiong Xing** [4,5] and **Xiaoyong Zhu** [6]

1   School of Computer and Information, Qiannan Normal University for Nationalities, Duyun 558000, China
2   Key Laboratory of Complex Systems and Intelligent Optimization of Guizhou Province, Duyun 558000, China
3   Key Laboratory of Complex Systems and Intelligent Optimization of Qiannan, Duyun 558000, China
4   School of Electromechanical and Automobile Engineering, Huanggang Normal University,
    Huanggang 438000, China
5   Hubei Zhongke Research Institute of Industrial Technology, Huanggang Normal University,
    Huanggang 438000, China
6   School of Economics & Management, Shaoyang University, Shaoyang 422099, China
*   Correspondence: guideaaa@126.com; Tel.: +86-13016410611

Abstract: Aiming at resolving the problem of low assembly accuracy and the difficulty of guaranteeing assembly quality of remanufactured parts, an optimization classification method for the assembly sequence of remanufactured parts based on different accuracy levels is proposed. By studying the characteristics of recycled parts, based on the requirement that the quality of remanufactured products not be lower than that of the assembly quality of new products, the classification selection matching constraints of remanufactured parts are determined, and the classification selection matching optimization models of remanufactured parts with different precision levels is established. An algorithm combining particle swarm optimization and a genetic algorithm is proposed to solve the model and obtain the optimal assembly sequence. Taking the remanufacturing assembling of a 1.4 TGDI engine crank and a connecting rod mechanism as an example, the comparison of quality data shows that this method can effectively improve the qualified rate of assembly, reduce the cost of after-sale claims, provide new theories and methods for remanufacturing enterprises that need hierarchical assembly, and provide effective guidance for the development of the remanufacturing industry.

Keywords: remanufacturing; assembly sequence; selection matching; combinatorial algorithm; engine

Citation: Xiao, Y.; Zhou, J.; Xing, S.; Zhu, X. Research on Assembly Sequence Optimization Classification Method of Remanufacturing Parts Based on Different Precision Levels. *Processes* **2023**, *11*, 383. https://doi.org/10.3390/pr11020383

Academic Editor: Sergey Y. Yurish

Received: 27 December 2022
Revised: 20 January 2023
Accepted: 22 January 2023
Published: 26 January 2023

1. Introduction

Remanufacturing assembly is a critical step to ensure product quality. Compared with the original new-part manufacturing and assembly, remanufacturing assembly has higher uncertainty, low production efficiency, unstable product quality, frequent abnormal production accidents, and high repair rates [1,2]. Therefore, ensuring the assembly quality of remanufactured products, improving the success rate of product matching, and optimizing the assembly sequence during the remanufacturing assembly process have become key issues that remanufacturers need to solve urgently.

At present, a large quantity of research has been carried out on the optimization of remanufacturing assembly sequences. Yang et al. [3] built a product assembly priority correlation matrix for complex mechanical products and used an improved genetic algorithm to obtain the optimal assembly sequence. Su et al. [4,5] proposed a matching-oriented remanufacturing assembly sequence optimization method and established an assembly sequence optimization model based on the lowest cost of mass loss. According to the assembly characteristics of lithium battery modules, by analyzing the failure condition of used parts, Jiang et al. [6] set up an optimization model of remanufacturing repair scheme of used parts based on failure characteristics with the objective of remanufacturing cost and time. A genetic algorithm was used to optimize the model. Geda et al. [7] studied the remanufacturing assembly combination matching method and established a matching model aiming at minimizing quality loss in product assembly. Xiao et al. [8] studied the remanufacturing assembly inventory cost, proposed an assembly sequence optimization model based on the smallest assembly cost and the highest resource utilization, and used the network flow graph and mathematical linear programming method to solve the model. Zhu et al. [9] developed a remanufacturing cost prediction model based on an improved BP neural network. This model can effectively and accurately predict the remanufacturing cost. Marcin et al. [10,11] carried out assembly sequence planning using artificial neural networks for mechanical parts based on given conditions.

In order to improve the quality of remanufactured products, these studies are mainly conducted by introducing a new classification or calibration method. However, there are few studies on the coupling law of remanufacturing product quality and the quality control of remanufacturing product assembly process in the environment. The existing research methods cannot meet the requirements for the assembly quality of remanufactured products. Based on the above research, this paper proposes an optimization classification method for the assembly sequence of remanufactured parts based on different precision levels to raise assembly accuracy and guarantee assembly quality. By studying the characteristics of recycled parts, and based on the requirement that the quality of remanufactured products not be lower than the quality of newly assembled products, the constraints on remanufactured part classification and selection are determined in the dimension chain constraints and the optimization model of remanufactured parts classification selection with different precision is established. Designing a combined algorithm to solve the target model and obtain the optimal assembly sequence serves to improve the remanufacturing assembly success rate and assembly accuracy.

2. Optimization Model of Remanufactured Part Grading Selection with Different Precision Levels

The assembly error of remanufactured parts fluctuates widely. In order to ensure the quality of remanufactured assembly, this paper establishes different precision standards for remanufactured parts and creates an optimization model for the classification and selection of remanufactured parts under different precision conditions.

2.1. Remanufactured Part Grading Matching Constraints

Assuming that the critical dimension of the original part is $(x^{\alpha}_{-\alpha}, \sigma^2)$, which satisfies the normal distribution, the ideal value is x, the tolerance distribution is $[-\alpha, \alpha]$, and the variance is δ^2. Then, the critical dimension of the remanufactured part is $(x^{\beta}_{-\beta}, \delta^2)$, which also satisfies the normal distribution. The ideal value is x, the tolerance distribution is $[-\beta, \beta]$, and the variance is δ^2. In the actual remanufacturing process, $\alpha \leq \beta, \sigma \leq \delta$, assuming that a certain dimension chain consists of N part sizes, the calculation formula of the dimension chain accuracy of the assembly of new parts can be obtained, as shown in Formula (1):

$$fm = -\sum_{j=1}^{N} \sigma_j^{2} \tag{1}$$

In the formula, σ_j represents the dimensional variance of the j remanufactured part in the dimension chain.

The formula for calculating the accuracy of the assembly dimension chain of remanufactured parts is shown in Formula (2):

$$fr = -\sum_{j=1}^{N} \delta_j^{2} \tag{2}$$

In the formula, δ_j represents the dimension variance of the j remanufactured part in the dimension chain. Obviously, $fm > fr$. According to this model assumption, the parts are divided into two levels in order to ensure that the assembly accuracy of remanufactured parts is not lower than that of new products. That is, it is necessary to meet the tolerances of the two $\delta_i^2 \leq \sigma_i^2$; that is, $\beta \leq \sqrt{2}\alpha$, σ_i represents the dimensional variance of the i remanufactured part in the dimension chain, and δ_i represents the dimension variance of the i remanufactured part in the dimension chain. When the remanufactured part X of the first type is assembled in combination with the remanufactured part Y of the second type, the tolerance of the remanufactured part X of the first type can be relaxed, the selection range is enlarged, and the relaxation ratio is 1.41. According to the above formula, reasoning, and the literature [12], different accuracy standards are established, as shown in Table 1.

Table 1. Classification criteria for different grades.

Classification	Variance Wide Scaling Coefficient	Tolerance Width Casting Coefficient
2	1.8	1.3
3	2.1	1.4
4	2.4	1.5

Mechanical products are assembled using a finite size chain, which is the basis for product assembly accuracy. Assuming that there are multiple size chains in the assembly process, in order to ensure that the quality of remanufactured products is not lower than the assembly quality of newly manufactured products, the product size chain constraints are as follows [13–15]:

$$ES(A_t) = \sum_{i=1}^{m} ES(\overrightarrow{A_i}) - \sum_{i=m+1}^{n-1} EI(\overleftarrow{A_i}) \tag{3}$$

$$EI(A_t) = \sum_{i=1}^{m} EI(\overrightarrow{A_i}) - \sum_{i=m+1}^{n-1} ES(\overleftarrow{A_i}) \tag{4}$$

$$T(A_t) = \sum_{i=1}^{n} T(\overrightarrow{A_i}) + \sum_{i=m+1}^{n-1} T(\overleftarrow{A_i}) \tag{5}$$

In the formula, $\overrightarrow{A_i}$ is the subtraction loop, m is the number of added loops, n is the total loop number, $ES(A_t)$ is the upper deviation of the closed loop, $EI(A_t)$ is the lower deviation of the closed loop, and $T(A_t)$ is the tolerance of the closed loop.

2.2. Comprehensive Model of Remanufactured Part Classification and Selection

Assuming that the remanufactured parts have t dimension chains, that the error coefficient of different precision classifications of the remanufactured parts is $p(t)ij = \delta(t)_{ij}^2 / \sigma(t)_i^2$, and that the cost of mass loss is $c(t)_{ij} = f(p(t)_{ij})$, the remanufacturing assembly cost under different precisions is shown in Formula (6):

$$\min F = \sum_{t=1}^{T} \sum_{i=1}^{n} c(t)_{ij} x(t)_{ij} \tag{6}$$

With the restriction that the product quality of the remanufacturing assembly process of mechanical products not be lower than the assembly quality of newly manufactured products, and the goal of optimizing the remanufacturing assembly cost, a comprehensive selection model of the remanufacturing assembly process under different precision conditions is established, as shown in Formula (7):

$$\min F = \sum_{t=1}^{T} \sum_{i=1}^{n} c(t)_{ij} x(t)_{ij}$$

$$s.t \begin{cases} ES(A_t) = \sum_{i=1}^{m} ES(\overrightarrow{A_i}) - \sum_{i=m+1}^{n-1} EI(\overleftarrow{A_i}) \\[2mm] EI(A_t) = \sum_{i=1}^{m} EI(\overrightarrow{A_i}) - \sum_{i=m+1}^{n-1} ES(\overleftarrow{A_i}) \\[2mm] T(A_t) = \sum_{i=1}^{n} T(\overrightarrow{A_i}) + \sum_{i=m+1}^{n-1} T(\overleftarrow{A_i}) \\[2mm] \sum_{j=1}^{m} x(t)_{ij} = 1 \\[2mm] x(t)_{ij} = 0,1 \\ i \in \{1,2,\ldots,n\} \\ j \in \{1,2,\ldots,m\} \\ t \in \{1,2,\ldots,T\} \end{cases} \tag{7}$$

In the formula, $x(t)_{ij}$ represents the j level of the i attribute of the $x(t)$ component, $c(t)_{ij}$ represents the quality loss cost of the i remanufactured component j level of the t dimension chain, $\delta(t)_{ij}{}^2$ is the variance of the j dimension chain of the i remanufactured component, and $\sigma(t)_i{}^2$ is the variance of the t dimension chain of the i original part.

3. Model Solution

Particle swarm optimization is inspired by the behavior of bird species and is used to optimize the balance of production and assembly lines. In an n-dimensional search range, each particle can be regarded as a search individual. The position of each particle is a candidate solution of the optimal value, and the motion of the particle is the process of searching for the individual. The optimal historical location of individual particles and populations can dynamically adjust the flight speed of particles. Velocity and position are two attributes in particle search, where velocity represents the speed of motion and position represents the direction of motion. The optimal solution found by each particle in motion is called the individual extreme value, and the optimal individual extreme value in the whole particle swarm is regarded as the global optimal solution. Speed and position are updated iteratively. Finally, the optimal solution must meet the termination condition.

The assembly line studied in this paper is a discrete problem. The particle swarm optimization algorithm is used to analyze the assembly line problem, and the following formula is established:

$$X_{id}(t+1) = V_{id}(t+1) + X_{id}(t) \tag{8}$$

$$V_{id}(t+1) = \omega V_{id}(t) + c_1 r_1 (p_{id}(t) - X_{id}(t)) + c_2 r_2 (g_{gd}(t) - X_{id}(t)) \tag{9}$$

Here, X is the position of the particle in the global learning process, representing a feasible solution; V is the particle running speed, representing the global or local optimal learning process of particles; p is the historical optimal location of the current local; g is the historical optimal position of all particles in the global learning process; i is the serial number of the particle, $i = 1,2 \ldots$ n; d is the global dimension of the particle's position or velocity; t is the evolutionary algebra of particles; w is the inertia weight; r_1 and r_2 are any number between [0, 1]; c_1 and c_2 are acceleration factors.

The particle swarm optimization algorithm transforms the optimal search process into the solution of the regular motion of particles in space, in which the motion of particles is affected by three main aspects: the first one is the inertia of particle motion (the first item of Formula (9)); the second is the judgment of the current particle position (i.e., the second term of Formula (9)) and the selection of the historical optimal solution of the particle. $c_1 * r_1$ represents the degree to which the particle judges the historical optimal solution. The third is the cognition of the particle to the global particle and the search for the global optimal solution (the third term of Formula (9)). $c_2 * r_2$ represents the degree to which particles design the current global optimal solution.

Considering that the hierarchical matching model has high requirements for the optimization accuracy of the global search, if a single intelligent optimization algorithm is used, it is easy to fall into the local optimal situation [16–21]. Based on this, this paper proposes an optimization algorithm that combines particle swarm optimization and a genetic algorithm, and integrates the crossover and mutation operators in the genetic algorithm into the particle swarm optimization algorithm to improve the global optimization ability [22].

(a) Crossover operation

Assuming that the two individuals, x_k^m, x_k^n ($m \neq n$) at a certain time k are inherited and crossed, the new individuals generated at the time $n + 1$ after the crossover can be expressed as follows [23,24]:

$$x_{k+1}^m = ax_k^m + (1 - a)x_k^n \tag{10}$$

$$x_{k+1}^n = ax_k^n + (1 - a)x_k^m \tag{11}$$

(b) Mutation operation

Mutation operation is a process of simulating gene mutation in the biological world. Its main function is to increase the local search ability of the new algorithm, which can converge to the optimal solution more quickly. Finally, the position/speed update formula of the mutation operator is introduced as follows:

$$\Delta v_{id}^{k+1} = \Delta v_{id}^k + c_1\xi(\Delta p_{id}^k - \Delta x_{id}^k) + c_2\eta(\Delta p_{gd}^k - \Delta x_{id}^k) \tag{12}$$

$$x_{id}^{k+1} = x_{id}^k + \Delta v_{id}^{k+1} \tag{13}$$

In the formulas, c_1 and c_2 are the weight of the particle tracking itself and the group optimal value, usually set to 2; ξ and η are the $[0, 1]$ random number with uniform interval distribution; γ is the constraint factor [25,26] on the speed, which is the solution step of the algorithm as shown in Figure 1.

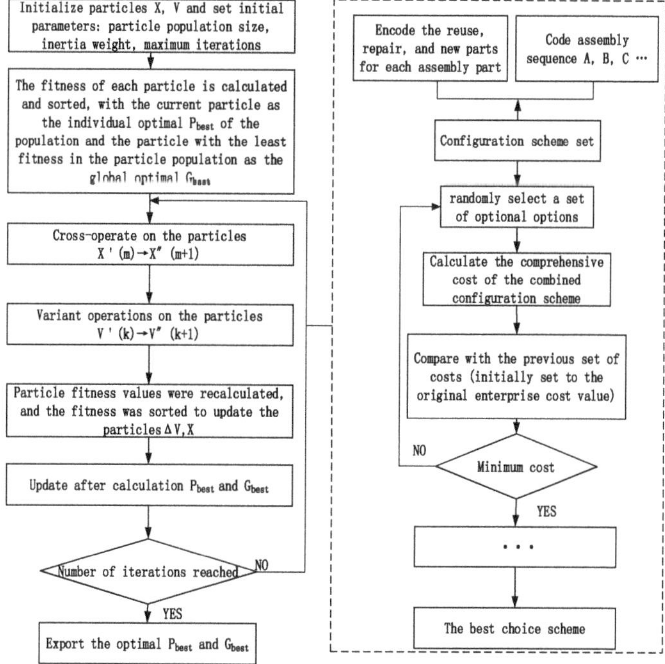

Figure 1. Combined algorithm solution step.

The genetic algorithm (GA) is a typical swarm intelligence algorithm. It obtains the good properties of chromosomes through selection, crossover, and variation, and finally finds the optimal solution through convergence. The genetic algorithm has the advantages of good convergence performance and strong global search ability. Particle swarm optimization and the genetic algorithm are integrated to improve the global search ability of particles to solve some specific problems and obtain better results. This paper adopts the idea of bringing the excellent solutions selected by the genetic algorithm into the next generation. The method involves selecting the best individuals in each generation of particle population and replacing the individuals with weak adaptability in the next iteration process, so as to improve the ability of the algorithm in searching for the best individuals. The specific process is as follows:

(1) Initialize. Set the initial velocity and initial position of the particle, and set cross factor k_o and variation factor T_{st}.

(2) Calculate the fitness index of each particle using the fitness function. *pbest* is the best position of individual particle, and *gbest* is the best position within the whole world.

(3) Constantly update the velocity and position of particles in the global according to the above crossover and mutation operations.

(4) Calculate the particle adaptation value $f(x_i)$ according to the respective positions of particles in the global x_i, and replace the individual optimal position and the global optimal position in a timely manner.

(5) According to the global order of particle fitness values, particles with good fitness values replace particles with bad fitness values, leaving particles with good speed and position.

(6) If the search results meet the set conditions, stop the search and output the best global value at this time; if the stop condition is not met, the next update iteration continues.

4. Example Verification

We applied the optimization model of remanufactured parts classification and selection to a domestic automobile engine remanufacturing enterprise, which has more than 100 waste engine recycling centers across the country. The recycling time and recycling location of each recycling point are different and the quality of different recycled engines is also different. A total of 80% of the remanufactured parts are used in a remanufactured engine, but the assembly success rate is only 84.41% and customer complaints are frequent. Compensation expenses were as high as CNY 9.57 million in 2019, accounting for 7.81% of sales [27–32]. Figure 2 shows the process flow chart of the company's remanufactured engine.

4.1. Method Implementation

Remanufactured engine companies recycle engines from recycling stations with different quality levels. In this case, a 1.4 TGDI gasoline engine crank-connecting rod mechanism is used as an example. Figure 3 is a three-dimensional solid model of the crank-connecting rod mechanism, wherein 1, 2, 36, and 37 are pistons; 3, 6, 33, and 35 are bushings; 4, 5, 32, and 34 are piston pins; 7, 8, 30, and 31 are connecting rod bodies; 9, 10, 28, and 29 are upper half bearings; 11 are crankshafts; 12, 16, 23, and 27 are lower half bearings; 13, 19, 22, and 26 are connecting rod covers; 14, 15, 17, 18, 20, 21, 24, and 15 are connecting rod bolts; 38, 39, 40, and 41 are cylinders.

The remanufactured parts are classified according to the quality grade standards set in Table 1 above. The size standard of the main bearing hole diameter of the new part of the 1.4 TGDI engine is $L_1 = 48^{+0.015}_{0}$ mm; the diameter of the new crankshaft main shaft is $L_2 = 44^{0}_{-0.016}$ mm. Five sets of crank assembly parts were randomly selected, which included the dimensions of five remanufactured crankshafts and five remanufactured cylinder block parts. Tables 2 and 3 show the main bearing hole diameter and crankshaft main journal diameter of some remanufactured cylinder blocks.

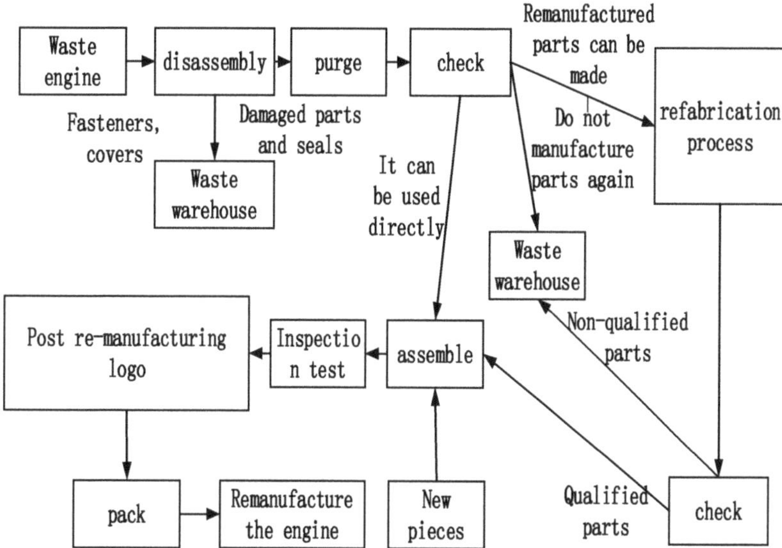

Figure 2. Process flow chart.

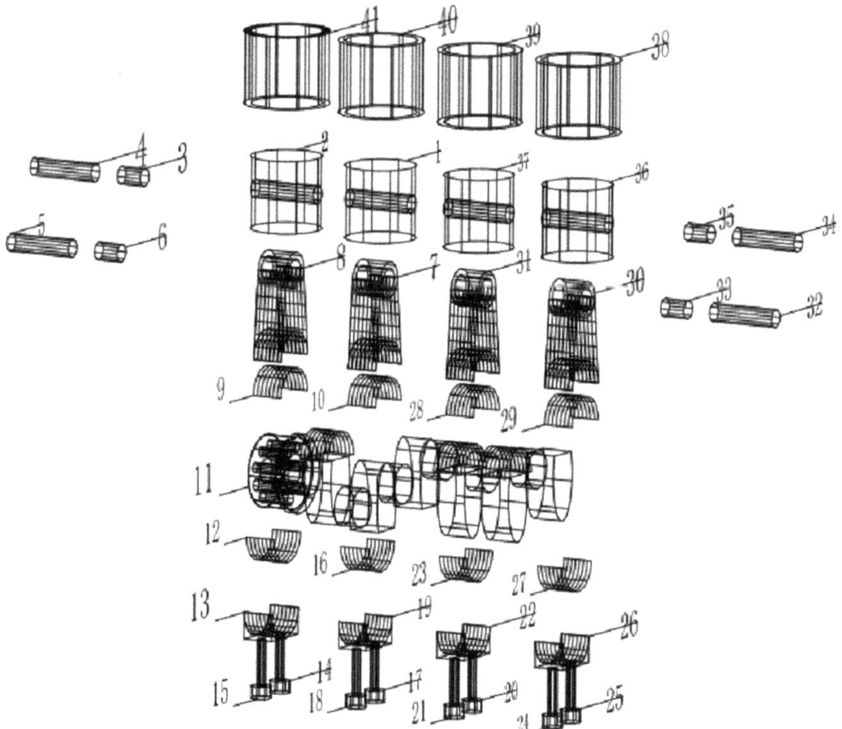

Figure 3. Three-dimensional structural model.

Table 2. Main bearing bore diameter of cylinder block (mm).

Cylinder Code	1 Neck	2 Neck	3 Neck	4 Neck	5 Neck
1-01	48.013	48.005	48.009	48.012	48.006
1-02	48.014	48.012	48.012	48.013	48.013
1-03	48.019	48.014	48.013	48.012	48.002
1-04	48.012	48.012	48.005	48.013	48.005
1-05	48.007	48.011	48.004	48.002	48.013

Table 3. Crankshaft main journal diameter (mm).

Crankshaft Code	1 Neck	2 Neck	3 Neck	4 Neck	5 Neck
2-01	43.987	43.992	43.995	43.986	43.992
2-02	43.982	43.904	43.994	43.987	43.996
2-03	43.986	43.981	43.984	43.985	43.997
2-04	43.988	43.992	43.988	43.986	43.988
2-05	43.996	43.996	43.992	43.989	43.986

Coding the engine block and crankshaft, it can be seen in Tables 2 and 3 that the cylinder block code 1-03 first neck, the crankshaft code 2-02 first neck, and the crankshaft code 2-03 s neck size do not meet the standard of the new part size. This can result in scrap or rework during assembly. The 1.4 TGDI remanufactured engine crank assembly closed-ring constraints are the same as new assembly constraints. The cylinder block of the engine is matched with the crankshaft during assembly; through the calculation results of arrangement and combination, a total of 120 combination schemes can be obtained. The cylinder block and the crankshaft are regarded as a whole. When selecting the upper and lower bearing shells, the commonly used greedy algorithm is used [33–36]. The bearing bush, the lower bearing bush, and 120 kinds of combinations are optimized one by one so as to complete the optimization of an assembly scheme. The structure diagram of the matching process after optimization is shown in Figure 4.

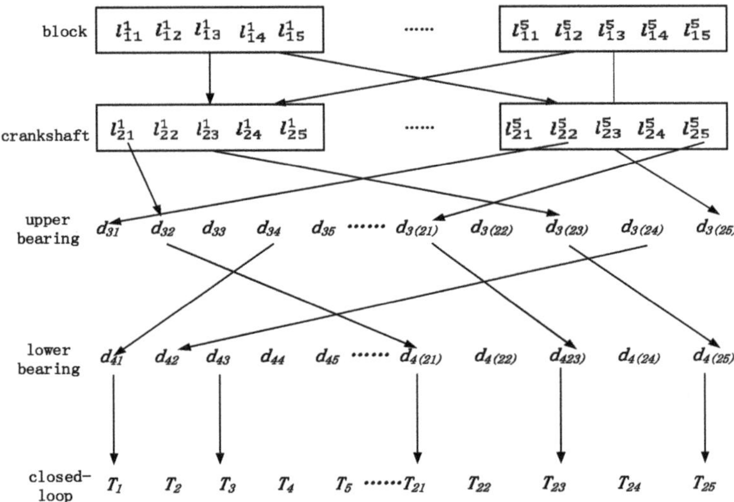

Figure 4. Crank assembly selection process structure.

Through the above analysis, the crankshaft matching scheme is preliminarily generated. Based on this, the assembly sequence of the crankshaft connecting rod mechanism of the remanufactured engine is optimized. The assembly attributes of the crankshaft connecting rod mechanism are shown in Table 4. The serial number, installation direction, and assembly tools used for the parts are described.

Table 4. Assembly properties of the crank linkage.

Serial Number	Part Name	Installation Direction	Assembling Tool
1,2,36,37	plunger	−Z	handwork
3,6,33,35	bush	+X −X	handwork
4,5,32,34	gudgeon pin	+X −X	hacksaw, chassis, heavy hammer
7,8,30,31	shank of connecting rod	−Z	handwork
9,10,28,29	upper half bearing	−Z	handwork
11	bent axle	—	—
12,16,23,27	lower half bearing	+Z	handwork
13,19,22,26	connecting rod cap	+Z	handwork
14,15,17,18,20,21,24,15	screw bolt	+Z	screwdriver, wrench
38,39,40,41	Air cylinder	−Z	workbench, pistol, pliers

Based on the above optimized crank assembly selection scheme, the relevant programs of the remanufactured engine crank-connecting rod mechanism classification selection model are written in Matlab. The reused parts, re-repaired parts, and new parts of each assembly part were coded, and the types of parts in the assembly process were coded according to the assembly sequence. According to the assembly constraints and the classification scheme selection, the particle swarm genetic combination algorithm proposed in this study was used to optimize the model. The specific combination algorithm parameters were set as follows: the initial population size was 50; the maximum number of iterations was 1000; the cross factor k_o was set to 0.9; the variation factor T_{st} was set to 0.4; the learning factor was $c_1 = c_2 = 2$; and Matlab was used to optimize the simulation. The operation results are shown in Figure 5.

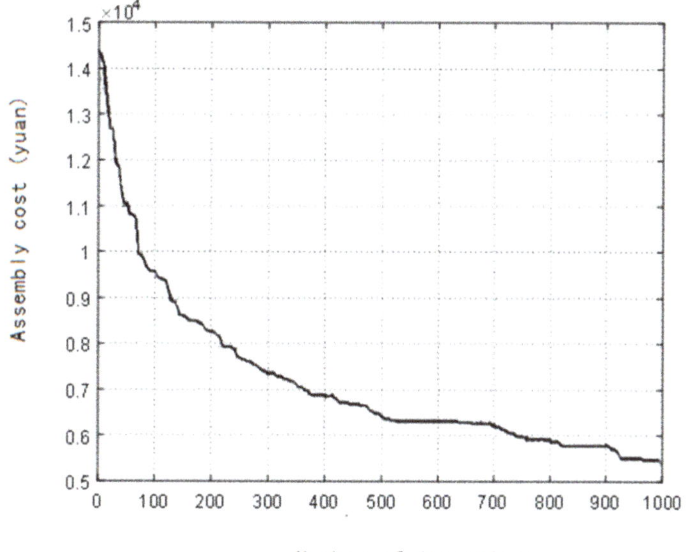

Figure 5. The calculation optimization diagram of the particle swarm genetic combination algorithm for classification selection.

The optimized assembly sequence was optimized by Matlab. The top two optimal assembly sequences are shown in Table 5.

Table 5. Optimal assembly sequence.

Rank	Crank Linkage Assembly Sequence
1	11→9→8→3→2→4→10→7→6→1→5→28→31→35→37→34→29→30→33→36→32→12→16→23→27→13→14→15→19→17→18→22→20→21→26→24→25→38→39→40→41
2	11→10→7→6→1→5→28→31→35→37→34→9→8→3→2→4→29→30→33→36→32→16→19→17→18→23→22→20→21→12→13→14→15→27→26→24→25→38→39→40→41

4.2. Results Comparison

The company began to use the remanufactured part grading and matching method in January 2020, and compared the data for 2019 with those for 2020. According to statistics, the success rate of remanufacturing assembly increased from 84.41% to 90.17%, with an average increase of 5.76%. The company's compensation expenses decreased by CNY 3.21 million, a decrease of 33.5% compared with 2019. The specific comparison is shown in Table 6.

Table 6. Comparison of success rate and compensation cost data for 2019 and 2020.

Project	January	February	March	April	May	June	July	August	September	October	November	December
Success rate in 2019/%	83.27	83.67	84.56	85.23	83.49	84.53	86.19	85.27	84.13	82.94	84.52	85.17
Success rate in 2020/%	89.52	90.34	90.73	88.78	90.36	89.95	88.57	92.31	90.47	90.28	90.32	90.46
Compensation expenses in 2019 (10,000)	84.21	83.46	84.53	78.46	78.42	80.31	9.85	98.34	97.18	88.43	89.56	84.59
Compensation expenses in 2020 (10,000)	62.34	48.57	46.82	47.32	59.15	52.34	58.29	48.73	58.49	46.54	59.48	48.23

4.3. Comparison with the Previous Literature

In the introduction, references [3–11] provide many ideas and methods for remanufacturing and assembly, effectively achieving the goal. In reference [4], an assembly sequence optimization model based on the lowest quality loss cost was established and solved by an ant colony algorithm. A single intelligent optimization algorithm easily falls into the local optimal situation. In reference [6], an assembly sequence optimization model based on the failure feature was established and solved by an ant genetic algorithm. However, few consider the different precision levels of remanufactured parts. The assembly error of remanufactured parts fluctuates widely, so this paper sets up different precision standards for remanufactured parts and establishes the optimization model of classification and selection of remanufactured parts under different precision conditions. In addition, the combination of the particle swarm optimization algorithm and genetic algorithm, which has a wider range and better accuracy, is used to solve the problem.

5. Conclusions

Because the assembly accuracy of remanufactured parts is low and it is difficult to guarantee assembly quality, an optimization classification method of the assembly sequence of remanufactured parts based on different precisions is proposed. The main conclusions are as follows:

1. Remanufactured assembly control is difficult to standardize. Under the condition that the assembly accuracy of remanufactured parts not be lower than that of new products, the classification accuracy standard of remanufactured parts was calculated through mathematical formulas, the optimization model of classification selection under different accuracy conditions was established, and a combinatorial optimization algorithm to solve the model was proposed;

2. This article took the remanufacturing assembly of an engine crank-connecting rod mechanism as an example. The data comparison showed that the optimal assembly

sequence obtained by the hierarchical matching model proposed in this study can effectively ensure different remanufacturing assembly accuracy requirements and improve remanufacturing. The success rate guarantees an improvement in assembly quality and a reduction in after-sale claim costs. The best assembly sequence provides the best assembly quality and the lowest claim cost. The concept of optimality refers to the best assembly time and quality of remanufactured parts with different precisions. The success rate of assembly and the reduction in after-sale claim costs provided new theories and methods for remanufacturing enterprises, which should adopt hierarchical assembly.

Author Contributions: Conceptualization, Y.X., J.Z., and X.Z. ; Methodology, Y.X., and S.X.; Software, Y.X., J.Z., and S.X. ; Formal analysis, Y.X.; visualization, J.Z. and X.Z.; supervision, J.Z. and X.Z.; project administration, S.X.; review and editing, Y.X., and S.X. All authors have read and agreed to the published version of the manuscript.

Funding: The authors are grateful for National Science Foundation, China (No. 61862051); the Natural Science Foundation of Hunan Province, China (No. 2022JJ50244); the China Education Department of Hunan Province (No. 21B0695); the project of the Hunan Social Science Achievement Evaluation Committee in 2022 (No. XSP22YBC081); the Science and Technology Foundation of Guizhou Province under Grant (No. [2019]1299); the Top-notch Talent Program of Guizhou Province under Grant (No. KY [2018]080); the program of Qiannan Normal University for Nationalities under Grant (Nos. QNSY2018JS013, QNSYRC201715).

Data Availability Statement: Not applicable.

Conflicts of Interest: The authors declare no conflict of interest.

References

1. Wang, L.; Guo, Y.; Zhang, Z.; Xia, X. Extensive Concept, State-of-the Art Developing Trends of Remanufacturing Service. *J. Mech. Eng.* **2021**, *57*, 138–153.
2. Yin, L.; Yang, F. Study on selective assembly of remanufacturing parts to be assembled based on NSGA-III. *J. Mach. Des.* **2022**, *39*, 53–60.
3. Yang, W.; Meng, C.; Cao, W.; Xie, K.; Yu, P. Assembly Sequence Optimization based on Improved Genetic Algorithm. *J. Mech. Transm.* **2016**, *40*, 67–70.
4. Su, B.; Huang, X.; Ren, Y.; Wang, F.; Xiao, H.; Zheng, B. Research on Selective Assembly Method Optimization for Construction Machinery Remanufacturing Based on Ant Colony Algorithm. *J. Mech. Eng.* **2017**, *53*, 60–68. [CrossRef]
5. Xiao, Y.; Zhang, H.; Jiang, Z. An approach for blank dimension design considering energy consumption. *Int. J. Adv. Manuf. Technol.* **2016**, *87*, 1229–1235. [CrossRef]
6. Jiang, Y.; Jiang, Z.; Zhang, H.; Cheng, H. Optimization Study of Remanufacturing Reconditioning Scheme for Used Parts Based on Failure Features—Research on optimization of waste parts based on failure feature. *Mach. Tool Hydraul. Press.* **2016**, *44*, 168–172.
7. Geda, M.W.; Kwong, C.K.; Jiang, H.M. Fastening method selection with simultaneous consideration of product assembly and disassembly from a remanufacturing perspective. *Int. J. Adv. Manuf. Technol.* **2019**, *101*, 1481–1493. [CrossRef]
8. Xiao, Y.; Yan, W.; Wang, R. Research on Blank Optimization Design Based on Low-Carbon and Low-Cost Blank Process Route Optimization Mode. *Sustainability* **2021**, *13*, 1929. [CrossRef]
9. Zhu, H.; Ding, Z.; Chen, B.; Jiang, Z. Cost Prediction of End-of-life Products Remanufacturing Based on Improved BP Neural Network. *Mach. Tool Hydraul.* **2020**, *48*, 34–38.
10. Suszyński, M.; Peta, K. Assembly Sequence Planning Using Artificial Neural Networks for Mechanical Parts Based on Selected Criteria. *Appl. Sci.* **2021**, *11*, 10414. [CrossRef]
11. Suszyński, M.; Peta, K.; Černohlávek, V.; Svoboda, M. Mechanical Assembly Sequence Determination Using Artificial Neural Networks Based on Selected DFA Rating Factors. *Symmetry* **2022**, *14*, 1013. [CrossRef]
12. Liu, M.; Liu, C. The invention relates to an assembly size classification method for remanufactured parts. U.S. Patent CN103413024B, 8 February 2017.
13. Oh, Y.; Behdad, S. Simultaneous reassembly and procurement planning in assemble-to-order remanufacturing systems. *Int. J. Prod. Econ.* **2017**, *184*, 168–178. [CrossRef]
14. Wang, Z.; Jiang, X.; Liu, W.; Shi, M.; Yang, S.; Yang, G. Precision prediction and error propagation model of remanufacturing machine tool assembly process. *Comput. Integr. Manuf. Syst.* **2021**, *27*, 1300–1308.
15. Yenipazarli, A. Managing new and remanufactured products to mitigate environmental damage under emissions regulation. *Eur. J. Oper. Res.* **2016**, *249*, 117–130. [CrossRef]

16. Kiranyaz, S.; Pulkkinen, J.; Gabbouj, M. Particle Swarm Optimization. *Expert Syst. Appl.* **2014**, *38*, 2212–2223. [CrossRef]
17. Kiani, A.T.; Nadeem, M.F.; Ahmed, A.; Khan, I.A.; Alkhammash, H.I.; Sajjad, I.A.; Hussain, B. An Improved Particle Swarm Optimization with Chaotic Inertia Weight and Acceleration Coefficients for Optimal Extraction of PV Models Parameters. *Energies* **2021**, *14*, 2980. [CrossRef]
18. Xian, X.; Zhou, Z.; Huang, G.; Nong, J.; Liu, B.; Xie, L. Optimal Sensor Placement for Estimation of Center of Plantar Pressure Based on the Improved Genetic Algorithms. *IEEE Sens. J.* **2021**, *21*, 28077–28086. [CrossRef]
19. Kang, M.; Li, Y.; Jiao, L.; Wang, M. Differential Analysis of ARX Block Ciphers Based on an Improved Genetic Algorithm. *Chin. J. Electron.* **2022**, *32*, 225–236.
20. Vanderstar, G.; Musilek, P. Optimal Design of Distribution Overhead Powerlines using Genetic Algorithms. *IEEE Trans. Power Deliv.* **2021**, *37*, 1803–1812. [CrossRef]
21. Fernandes, J.; Arsenio, A.J.; Arnaud, J. Optimization of a Horizontal Axis HTS ZFC Levitation Bearing Using Genetic Decision Algorithms Over Finite Element Results. *IEEE Trans. Appl. Supercond.* **2020**, *30*, 3601308. [CrossRef]
22. Chen, J. The Research on Critical Assembly Process Quality Control Method and Supporting System for Remanufacturing Engine. Ph.D. Thesis, Hefei University of Technology, Hefei, China, 2016.
23. Xiao, Y.M.; Zhang, H. Multiobjective optimization of machining center process route: Tradeoffs between energy and cost. *J. Clean. Prod.* **2021**, *28*, 21–26. [CrossRef]
24. Zhang, Y. Method and Key Technology of Assembly Quality Control for Remanufactured Engine. Ph.D. Thesis, Hefei University of Technology, Hefei, China, 2017.
25. Zhang, J.; Li, Y. Motor Rolling Bearing Compound Fault Diagnosis Based on Particle Swarm Optimization & Blind Source Separation. *Mach. Tool Hydraul.* **2019**, *47*, 167–172.
26. Xiao, Y.; Wang, R.; Yan, W.; Ma, L. Optimum Design of Blank Dimensions Guided by a Business Compass in the Machining Process. *Processes* **2021**, *9*, 1286. [CrossRef]
27. Xing, Z.; Jiang, A.L.; Xie, J.J.; Feng, Y.C. Benefit analysis and Surface Engineering Application of Automobile Engine Remanufacturing. *China Surf. Eng.* **2004**, *9*, 1–5.
28. Chang, X.; Zhong, Y.; Wang, Y.; Chen, Z. Research of low-carbon policy to promote automotive parts remanufacturing in China: A case study of auto engine remanufacturing. *Syst. Eng. Theory Pract.* **2013**, *33*, 2811–2821.
29. Zhang, Y.F.; Li, M.; Liu, S.C. IoT-based production process optimization method for automobile engine remanufacturing. In Proceedings of the ICRA 2014—IEEE International Conference on Robotics and Automation, Hong Kong, China, 31 May–5 June 2014.
30. Shi, J.; Hu, J.; Jin, Z.; Ma, Y.; Ma, M.; Wang, H.; Ma, Q. Recycling Mode and Remanufacturing Cost Analysis of Used Automobile Engine Based on System Dynamics. *J. Phys. Conf. Ser.* **2021**, *1986*, 012041. [CrossRef]
31. Zhang, Y.; Liu, S.; Liu, Y.; Yang, H.; Li, M.; Huisingh, D.; Wang, L. The 'Internet of Things' enabled real-time scheduling for remanufacturing of automobile engines. *J. Clean. Prod.* **2018**, *185*, 562–575. [CrossRef]
32. Yu, X.; Jiang, N.; Wang, L.P. Characteristics and Approaches to the Information System of Automobile Engine Remanufacturing. *J. Cent. South Univ. For. Technol.* **2009**, *29*, 176–179.
33. Ju, J.; Li, W.; Wang, Y.; Fan, M.; Liu, Y. Optimization Design of Vibration Control Parameters of Flexible Manipulators Based on Multi-Population Genetic Algorithm. *Mach. Tool Hydraul.* **2016**, *44*, 94–97.
34. Liu, L. Research on assembling method of remanufacturing engine. Ph.D. Thesis, Tianjin University of Science and Technology, Tianjin, China, 2017.
35. Wen, H.; Meng, X.; Zeng, A.; Guo, X.; Xu, X. Fatigue life prediction of neural network remanufactured based on second-order particle swarm optimization. *Sci. Technol. Eng.* **2019**, *19*, 21–26.
36. Yuan, S.; Li, T.; Wang, B.; Liu, Q. Co-evolutionary iterated greedy algorithm for the two-stage flow shop group scheduling problem. *Syst. Eng. Theory Pract.* **2020**, *40*, 2707–2716.

Article

Identification of Key Brittleness Factors for the Lean–Green Manufacturing System in a Manufacturing Company in the Context of Industry 4.0, Based on the DEMATEL-ISM-MICMAC Method

Xiaoyong Zhu [1], Yu Liang [2], Yongmao Xiao [3,4,5,*], Gongwei Xiao [1] and Xiaojuan Deng [1]

1 School of Economics & Management, Shaoyang University, Shaoyang 422000, China
2 Shaoyang University Library, Shaoyang University, Shaoyang 422000, China
3 School of Computer and Information, Qiannan Normal University for Nationalities, Duyun 558000, China
4 Key Laboratory of Complex Systems and Intelligent Optimization of Guizhou Province, Duyun 558000, China
5 Key Laboratory of Complex Systems and Intelligent Optimization of Qiannan, Duyun 558000, China
* Correspondence: xym198302@163.com

Abstract: In the context of Industry 4.0, the lean–green manufacturing system has brought many advantages and challenges to industrial participants. Security is one of the main challenges encountered in the new industrial environment, because smart factory applications can easily expose the vulnerability of manufacturing and threaten the operational security of the whole system. It is difficult to address the problem of the brittleness factor in manufacturing systems. Therefore, building on vulnerability theory, this study proposes a vulnerability index system for lean–green manufacturing systems in a manufacturing company in the context of Industry 4.0. The index has four dimensions: human factors, equipment factors, environmental factors, and other factors. The Decision-Making Trial and Evaluation Laboratory (DEMATEL) approach was used to calculate the degree of influence, the degree of being influenced, and the centrality and causes of the factors. The causal relationships and key influences between the factors were identified. Then, the dependence and hierarchy of each of the key influencing factors were analyzed using the Matrix-Based Cross-Impact Multiplication Applied to Classification (MICMAC) and Interpretative Structural Model (ISM) methods, and a hierarchical structural model of the factors was constructed. Finally, an intelligent manufacturing system that produces a micro-acoustic material and device was used as an example to verify the accuracy of the proposed method. The results show that the method not only identifies the key brittleness factors in a lean–green manufacturing system but can also provide a guarantee for the safe operation of a manufacturing system. This study provides theoretical guidance for the effective management of intelligent manufacturing systems; moreover, it lays a foundation and provides a new methodology for assessing the vulnerability of manufacturing systems.

Keywords: brittleness factor; lean–green manufacturing; Industry 4.0; DEMATEL-ISM-MICMAC; sustainable development

Citation: Zhu, X.; Liang, Y.; Xiao, Y.; Xiao, G.; Deng, X. Identification of Key Brittleness Factors for the Lean–Green Manufacturing System in a Manufacturing Company in the Context of Industry 4.0, Based on the DEMATEL-ISM-MICMAC Method. *Processes* **2023**, *11*, 499. https://doi.org/10.3390/pr11020499

Academic Editor: Sergey Y. Yurish

Received: 11 January 2023
Revised: 3 February 2023
Accepted: 6 February 2023
Published: 7 February 2023

1. Introduction

Manufacturing has become a key industry for creating wealth and is the basis of the material and social development of human society. With the rapid development of science and technology, the manufacturing industry in China has grown rapidly. In 2007, the total global manufacturing output was USD 9.324 trillion, and China's was USD 1.15 trillion. By 2021, China's total manufacturing output will be USD 4.864 trillion, accounting for 29.75% of the world's manufacturing added value. A strong manufacturing industry is the surest way to enhance comprehensive national power and defend national security. China's manufacturing industry has achieved remarkable results in the past 10 years through the in-depth implementation of the manufacturing power strategy; the accelerated transformation and upgrading of the manufacturing industry has led manufacturing in an intelligent, green, service-based direction, amid other transformations and upgrades. In May 2015, the State Council issued "Made in China 2025", a comprehensive overview of the implementation of the manufacturing power strategy, which aims to promote industrial and technological change and optimization. The main aspect of the strategy concerns upgrading to intelligent manufacturing, promoting a new model for the manufacturing industry model and allowing enterprise to undergo a fundamental change. In the context of Industry 4.0, smart manufacturing enterprises use modern information and communication technologies and integrate them with virtual technologies in cyberspace; this new model makes full use of the IOT information system to implement data-driven and intelligent improvement methods for manufacturing, supply and sales in traditional industries [1] Today, Industry 4.0 has made it possible to eliminate the intermediate links between consumers and manufacturers. In addition, the high degree of digitalization, automation and informatization has made the customization of the periphery of goods shorter and has significantly improved production efficiency, reducing labor and production costs and resulting in a significant increase in the overall production efficiency of the manufacturing industry. It is estimated that the connections and collaboration between machines and people in the Industry 4.0 production model will increase the speed of operation of the entire production system by 30% and the efficiency by 25% [2].

At the outset, Industry 4.0 was a high-tech strategic plan proposed by Germany [3] In the era of Industry 4.0, industrial development is not the only goal; other aims include creating a modern enterprise development model based on intelligence, digitalization and personalization, continuously optimizing product quality, transforming production and service models, moving away from traditional business solutions, and improving the over all level of development of enterprise [4]. Industry 4.0 changes the relationship between the various elements of the production process; meanwhile, changes in management philoso phy and management technology work together to drive changes in the organization o labor, such as in the integration of digital manufacturing technologies with advanced lean management technology, i.e., the integration of lean management with Industry 4.0. The result of the convergence of lean management and Industry 4.0 is an adaptive process o advanced artificial intelligence working in collaboration with people [5]. Lean management facilitates the exploitation of the potential of Industry 4.0, while avoiding "automation waste" (unnecessary waste in the automation process, such as repeatedly moving machines and adjusting layouts) [6]. On the other hand, new technologies are necessary to realize the concept of lean management, to reduce the pressure on shop floor workers, and to overcome the effects and impacts of lean management [7]. This integration is known as "Lean Industry 4.0" or "Lean Automation", and its benefits are mainly clustered in the five areas of flexibility, high performance, efficiency, quality, and safety [8]. Surveys have shown that companies that have adopted the Lean Industry 4.0 production model reduce production costs by nearly 40% over a 5–10-year period [9]. However, on the other hand, manufacturing consumes a great deal of the limited resources available to human society and causes serious environmental damage through the process of transforming manufacturing resources into products, as well as in the use and disposal of products Due to the large volume and scale of the manufacturing industry, its overall impact on the environment is significant. The question of how the manufacturing industry can reduce re source consumption and produce as little environmental pollution as possible is a pressing issue. Green manufacturing is key to solving the problem of environmental pollution in the manufacturing industry, and it is crucial to controlling the sources of environmental pollution. Green manufacturing is the essence of the sustainable development strategy of modern manufacturing. Green manufacturing is committed to developing a harmonious relationship between human technological innovation and productivity enhancement on the one hand, and the natural environment on the other, in line with the current need for sustainable development. Industry 4.0 was created to address this global challenge. Thus Industry 4.0 is, at its core, smart, green, and humanized [10]. Industry 4.0 manufacturing companies should first replace traditional energy sources with alternative, non-traditional sources of clean energy in the production process, to alleviate the problems of energy depletion and environmental pollution. Additionally, they should produce less pollution in the production and consumption of products that can be recovered and recycled to achieve sustainable development. Industry 4.0 has many benefits and creates many opportunities for industrial players. Many organizations are developing strategies to shift toward digitalization and intelligence, and manufacturing companies are responding quickly to diverse and uncertain market demands through lean–green manufacturing systems, with the rapid manufacturing of multiple varieties and small batches of products to provide environmentally friendly goods that meet customer demands.

In the context of Industry 4.0, the lean–green system for manufacturing companies is a complex system: an organic whole that encompasses the production processes of the manufacturing industry. Industry 4.0 systems aim to realize the transformation of the manufacturing industry from informatization to wisdom through cyber-physical systems, manufacturing driven by big data, cloud platforms, the Internet of Things, and other technologies. Through the deep integration of information and physical systems, and using mobile terminal and wireless communication, a virtual network world can be realized, facilitating barrier-free communication and intelligent human–computer interactions in large complex systems [11]. The lean–green manufacturing system consists of the superposition of different participant systems, with interaction effects between these systems, and the existence of nonlinear characteristics. Brittleness is a fundamental characteristic of complex systems. At the same time, a system is in an uncertain environment, where rapid changes in the external environment can result in dramatic changes in the complexity and scale of production management. The dynamic nature of the manufacturing environment is enhanced during periods of environmental change, which are more likely to stimulate the vulnerability of the manufacturing system, threatening the operational security of the entire system. Such shifts can cause the collapse of one or several subsystems (units) in the system, with the transmission and expansion of collapse behavior leading the entire system to collapse. The brittleness of manufacturing systems can change from implicit to explicit as the complexity and scale of the system increases and as the system evolves. Once triggered, brittleness can threaten the safety of the entire system operation.

Industry 4.0 has brought many advantages and also many challenges to industrial participants. Many organizations are developing strategies to move in a digital and intelligent direction. Although intelligent manufacturing enterprises can quickly respond to the di-versified and uncertain demands of the market through lean–green manufacturing systems, and provide environmentally friendly products to meet customer needs through fast manufacturing, rapid changes in the external environment have led to significant changes in the complexity and scale of production management. The dynamic characteristics of the manufacturing environment are enhanced, which can more easily stimulate the vulnerability of manufacturing systems and threaten the operational security of the whole system. With the application and development of digitalization and intelligence, manufacturing systems are becoming larger and more complex. Although highly complex systems are robust, they are more likely to experience system fragility, which can threaten the safety of the whole system. In order to ensure the safe operation of manufacturing systems, a deeper study of system brittleness and its key causative factors is required. In complex manufacturing processes, many factors affect system brittleness; these factors are often interrelated, so it is difficult to conduct a quantitative analysis and evaluation. An objective, comprehensive, and accurate method of analysis is needed to effectively identify the key brittleness factors in the system and to ensure the essential safety of system operations. This paper constructs an improved DEMATEL-ISM-MICMAC integration method and validates it with reference to a case study of a company.

This article is structured as follows. Section 2 discusses relevant studies of manufacturing system brittleness and describes the research questions. Section 3 presents an analysis of complex system brittleness factors for lean–green manufacturing in the context of Industry 4.0. Section 4 briefly describes the methods and materials. Section 5 presents an intelligent manufacturing system for the production of a micro-acoustic material and device; this is taken as an example to verify the accuracy of the proposed method of identifying the key brittleness factors of a manufacturing system. The paper ends by summarizing the research outcomes and indicating directions for future work.

2. Related Works and Research Questions

The realization of leanness and greenness, the core concepts of Industry 4.0, requires a combination of advanced technologies. The lean–green manufacturing system, in the context of Industry 4.0, is a smart manufacturing system with vertical integration, made possible by the Internet of Things. The system is complex and changeable, and the subsystem (software) evolves quickly. In order to guarantee the safe operation of the manufacturing system, we studied system brittleness and explored the key factors related to it. The vulnerability of a manufacturing system comes from both its software and its hardware. Coding errors, process defects, system software with poorly designed interactive features, and design defects and failures in the system hardware are the root causes of a system's vulnerability. At present, research on the combination of manufacturing systems and fragility theory is still in its initial stages, and the relevant literature focuses on the following areas. Some scholars have studied the fragility of just-in-time production systems under lean manufacturing systems [12], while others have studied the opportunities and challenges of manufacturing logistics systems under dynamic uncertainty [13]. Other studies focus on the fragility of industrial networks and use software tools to build robust systems. The brittleness of industrial networks has been studied using software tools to establish robust network systems, [14] and the brittleness of the Internet of Things has been studied to establish a brittleness model for different application scenarios of manufacturing IOT [15]. Elsewhere, the brittleness of manufacturing system equipment has been studied to evaluate the performance parameter index system for the state of manufacturing system equipment [16,17]. Of course, some scholars have also investigated the effect of combination of spatial modeling and fragility theory on the fragility excitation mechanisms of manufacturing systems to assess the reliability of these systems [18]. Gac Guibing et al. proposed three different methods using a generic generating function based on state entropy; by considering the variation in performance parameters, they undertook a structural fragility assessment of mixed-flow manufacturing systems, providing a reference for the safe operation and monitoring of manufacturing systems [19–21]. The main methods used for the assessment of system fragility are empirical analysis [22], agent-based methods [23], network-based methods [24], and methods based on the dynamic properties of the system [25]. The assessment methods vary according to the researcher's field and interdisciplinarity. It can be seen that in previous studies of complex networks of manufacturing systems, the problem of identifying the key vulnerability factors of manufacturing systems and the coupling relationships between the vulnerability factors have rarely been studied. Currently, as manufacturing systems are moving towards the era of Industry 4.0/smart manufacturing, the safety of manufacturing systems is more susceptible to interference from various factors, and research on the vulnerability aspects of lean and green manufacturing systems in the context of Industry 4.0 is paying more attention. Constructing a vulnerability indicator system is a prerequisite for vulnerability evaluation, while analyzing vulnerability indicator factors is an effective way to find ways to reduce vulnerability. In this study, the vulnerability factors of the manufacturing system will be identified and the evaluation system will be constructed, while not only analyzing the correlation between the factors causing the vulnerability of the manufacturing system and the degree of influence but also identifying the logical structure and influence mechanism between the factors.

Smart manufacturing is becoming an increasingly important trend and a core element of manufacturing development. Due to the deep integration of information technology and industrialization in contemporary society, control networks, production networks, management networks, and networked interconnections have become the norm. Production networks are increasingly integrated, and common protocols, common hardware and common software are increasingly used. As such, information security in production control systems is becoming increasingly prominent, and information security threats are becoming correspondingly complex. The security of systems is one of the main challenges faced in the new industrial environment. The security of Industry 4.0 or the Internet of Things has already been studied and discussed in a number of works [25,26]. In smart manufacturing, a new form of manufacturing, security is mainly concerned with the following four areas: (1) Network security: the use of deep integration with the Internet, the network IP, wireless networks, and flexible networking for smart manufacturing brings greater security risks. (2) Data security: open, mobile, and shared data, as well as privacy protection, are facing unprecedented threats. The diversification of business applications, such as network collaboration and personalization, has placed higher demands on application security. (3) Control security: the openness of the control environment has allowed external internet threats to penetrate the production control environment. (4) Device security: the intelligence of devices leaves production equipment and products more vulnerable to attack, which, in turn, affects normal production. The vulnerability of manufacturing systems lies in these security gaps. To ensure the normal operation of the manufacturing system, we must first ensure the security of the system; the fragility factors in a manufacturing system must be identified and considered, with a focus on the security impact factors of intelligent manufacturing systems. In summary, scholars have studied the vulnerability of manufacturing systems and provided methods and tools for preventing fragility in traditional manufacturing systems. In complex manufacturing processes, it is crucial to effectively identify, quantitatively analyze, and evaluate the key fragility factors, and the interactions and interconnections among these factors, to ensure the operational safety of manufacturing systems. However, Industry 4.0 manufacturing systems have increased in complexity and scale. The goal is to realize the transformation of the manufacturing industry from informatization to wisdom through information–physical systems at the work site; the main characteristics of these systems are interconnectivity, innovation, integration, and big data. It is difficult for traditional fragility assessment methods and tools to cope with big data, the random diversification of production information, and dynamic fluctuations in the manufacturing environment. The development of vulnerability theory has, until now, encompassed environmental, resource, social, economic, and management aspects of vulnerability, which means that the vulnerability indicator system of Industry 4.0 manufacturing system is necessarily complex. In order to build up a clearer understanding of complex systems, it is first necessary to clarify the relationship and hierarchy between the many intricate factors involved, to analyze the fragility of manufacturing systems, and to ensure the normal and safe production of the system. Therefore, it is necessary to identify the factors involved in the brittleness of lean–green manufacturing systems in a smart manufacturing/Industry 4.0 scenario, as well as elucidating the relationships between the factors. This framework addresses the following research questions:

Question 1: What are the brittleness factors of lean–green manufacturing systems in Chinese manufacturing companies in the context of Industry 4.0?

Question 2: What is the causal relationship between these factors?

Question 3: What are the key brittleness factors in the lean–green manufacturing system for Chinese manufacturing companies in the context of Industry 4.0?

Question 4: What measures need to be taken to improve the functioning of lean–green manufacturing system in Chinese manufacturing companies in the context of Industry 4.0?

In this paper, combined with vulnerability theory, a vulnerability index of lean–green manufacturing systems in manufacturing companies in the context of Industry 4.0 is established. This index has four dimensions: human factors, equipment factors, environmental factors, and other factors. The interpretative structural model (ISM) method is used to analyze the correlation relationships and influence mechanisms between these factors. The Decision-Making trial and Evaluation Laboratory (DEMATEL) is used to simplify the operation of ISM and to analyze the importance and mutual influence relationships of system factors. The Matrix-Based Cross-Impact Multiplication Applied to Classification (MIC MAC) approach is used to analyze the dependence-driving relationships of system factors. Considering the subjectivity and fuzziness inherent in the process of system analysis, triangular fuzzy numbers are introduced into the DEMATEL method to construct an improved DEMATEL-ISM-MICMAC integration method, which can be effectively used to research the above proposed problem. Finally, an intelligent manufacturing system used to produce a micro-acoustic material and device is taken as a case study to verify the accuracy of the proposed method. The key brittleness factors in the lean–green manufacturing system are identified to provide a guarantee for its safe operation. This study identifies the key factors that affect the vulnerability of lean–green intelligent manufacturing systems and provides theoretical guidance for the effective management of intelligent manufacturing systems; moreover, it lays a foundation for assessing the vulnerability of manufacturing systems.

3. Analysis of Complex System Brittleness Factors for Lean–Green Manufacturing with Industry 4.0

As an inherent property of complex systems, brittleness does not disappear with the evolution of the system or due to changes in the external environment. During system operations, system brittleness, once triggered, can cause the collapse of a subsystem or unit of the system, which can lead to the collapse of other subsystems or units associated with it and, eventually, the collapse of the whole system. As a fully automated manufacturing system, a lean–green manufacturing system in the context of Industry 4.0 is susceptible to the interference of various internal and external random factors during its operation and processing. This stimulates the brittleness of the manufacturing system and produces a collapse, resulting in the stagnation of the production system and delays in fulfilling customer orders or the generation of product quality defects. The process whereby manufacturing system fragility is triggered by fragility factors, leading to system collapse, is shown in Figure 1. Figure 1 shows that the process by which system fragility is triggered and the system collapses can be divided into two parts: the implicit layer and the explicit layer. The recessive layer consists of the interrelated fragility factors that affect each other and the system fragility events that result from each fragility factor. The upper layer is the dominant layer, and it contains the structure of the system and the fragility risk resulting from the fragility events acting on the manufacturing system. In order to ensure the normal operation of the system, without collapse, it is necessary to comprehensively analyze the factors affecting the system fragility, as well as determining the internal relationship between the influencing factors and the key factors for management and monitoring; at the same time, relevant reasonable measures must be put in place for continuous improvement

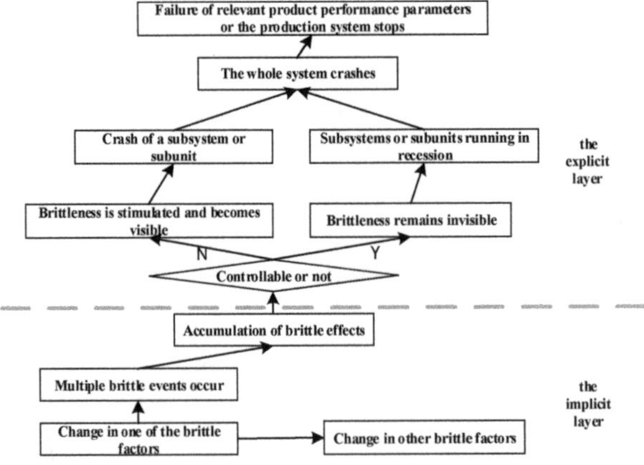

Figure 1. Analysis of the process whereby manufacturing system brittleness is stimulated until collapse occurs.

The brittleness effect in manufacturing systems is not resolved by the improvement of these systems or changes in the external environment, and may be stimulated at any time. As such, a manufacturing system with high reliability can also be brittle; the reduction in the reliability of the manufacturing system is a manifestation of the brittleness effect [27]. Brittleness is also an inherent property of manufacturing equipment, and even highly reliable equipment is brittle [28]. When the brittleness effect accumulates to a certain degree, it will be stimulated and become visible, and the working state of the equipment unit will continuously decline, leading to the failure of the relevant performance parameters of the product output of the system. Under the cumulative effects of brittleness, the equipment may collapse, eventually performing at a lower level than the specified performance level, and stop the operation of the manufacturing system.

Lean–green manufacturing systems under Industry 4.0 are typically large complex systems. With increasing digitalization, networking, and the use of other intelligent devices, the complexity of the manufacturing environment is becoming more and more difficult to predict; meanwhile, with diverse manufacturing tasks and dynamic and uncertain external environmental perturbations, the brittleness factors that affect manufacturing systems are increasingly obscure, diverse, hazardous, and interlocking. A lean–green manufacturing system in the context of Industry 4.0 has many internal elements that are closely interconnected, and each subsystem or subunit coordinates and interconnects to accomplish the system's multitasking goals in its daily operations. Therefore, from the perspective of systems theory, we identify and analyze the factors affecting the brittleness of lean–green manufacturing systems in smart manufacturing environments. Our analysis is based on four areas: "human–machine, environment, and management". In today's manufacturing system, 5M1E (Man/Manpower, Machine, Material, Method, Measurement, Environment) analysis of manufacturing systems' operational processes, such as processing, personnel, and environmental factors, proves instructive. Brittleness in equipment units can occur due to the randomness of each manufacturing task event in the manufacturing system service process; one factor can have an impact on other factors, prompting the acceleration of brittleness in equipment units, which is reflected in the reduction of the reliability of multi-state manufacturing systems. This is the result of a combination of brittle excitation factors, which show a coupling relationship. For example, a manager's poor decision making can lead to the inefficient scheduling of processing tasks, which increases the workload of the manufacturing system's processing equipment and causes excessive wear and tear on its key functional components, thus reducing its working performance status until failure occurs [29]. To ensure the reliability and objectivity of the analyzed factors, a total of 15 experts and scholars in the field were invited to determine the causal factors of brittleness in lean–green manufacturing systems in the context of Industry 4.0. Responses were collected using a questionnaire based on the relevant literature and on the actual production context. After repeated discussions, four major categories of human factors, equipment factors, environmental factors, and other factors were identified, and 16 specific causal factors were analyzed, as shown in Table 1.

Table 1. Causal factors of brittleness in lean–green manufacturing systems in the context of Industry 4.0.

Factor Classification	No.	Brittleness Factor	Factor Classification	No.	Brittleness Factor
	S_1	Mismanagement		S_9	Software Device resilience
	S_2	Personnel intrusion		S_{10}	Line failure and repair
Human Factors	S_3	Personnel operation and handling capabilities	Equipment factors	S_{11}	Amount and status of equipment
	S_4	Personnel skills		S_{12}	Equipment processing capacity
	S_5	Personnel experience		S_{13}	Production equipment breakdown and repair
	S_6	Number of personnel		S_{14}	Foreign body intrusion
Other factors	S_7	Sudden emergency orders	Environmental factors	S_{15}	Temperature and humidity
	S_8	Inadequate emergency management system		S_{16}	Laws and regulations, etc.

(1) Human factors. The human factors that affect the brittleness of lean–green manufacturing systems in smart factories arise from human actions inside and outside the manufacturing system. The internal personnel factors of the system mainly concern the front-line personnel involved in the product manufacturing and assembly process, etc.

Compared with traditional production lines, the production line in the context of Industry 4.0 is highly automated and requires different ratios and comprehensive qualities for various types of personnel in the manufacturing system. During the operation of the automated production line, the improper management of internal personnel or the improper operation of staff are factors involved in brittleness. External factors include personnel intrusion and other related human factors. These internal and external objective conditions are the basic requirement to ensure the smooth operation of a production line, and any human factors that disturb the normal operation of the production line and production conditions will lead to the excitation of brittleness in the manufacturing system, thus producing a system collapse or a production stoppage.

(2) Equipment factors. The lean–green manufacturing system in the context of Industry 4.0 is equipment intensive and sophisticated, with a high degree of information technology, a complex production system structure, and demanding equipment operation and maintenance conditions. Automated production systems require a large number of tooling fixtures for automated rapid positioning and clamping and tooling gauges for product processing quality inspection. As the production system becomes increasingly automated, the degree of complexity of its structure affects the accuracy of product assembly and the efficiency of the production line. In the vast majority of cases, when equipment failure occurs during the operation of a manufacturing system, it leads to a brittle collapse or a forced shutdown of the production line.

(3) Environmental factors. Environmental factors are all external causes of manufacturing systems' fragility. Temperature and humidity inside smart factories, industry quality standards, and foreign object intrusion in automation systems can all lead to the initiation of manufacturing system brittleness, triggering a collapse or stoppage of the manufacturing system.

(4) Other factors. Other factors include unscheduled surge production orders, the lack of effective emergency management in the face of unforeseen events, etc. These factors can interfere with the normal operation of a manufacturing system and lead to the excitation of system brittleness, and so on.

The factors outlined in Table 1 are the specific brittle influencing factors of lean–green manufacturing systems in the context of Industry 4.0 that were ultimately identified through an extensive data review, based on the principle that certain factors have been studied more than twice by different scholars. The proposed set of influencing factors were sent to relevant experts for assessment [30–40]. These influencing factors affect each other and are coupled with each other. It is impossible to form a wholly scientific and objective understanding of the causes of the brittleness of manufacturing systems through a simple qualitative analysis; it is therefore necessary to use the corresponding mathematical models for in-depth research.

4. Research Methodology

The DEMATEL method mainly comprises graph theory and uses the matrix algorithm for constructing graphs. It is a methodology for analyzing the factors of uncertain relationships in a system on the basis of expert cognition; it is mainly used to evaluate the relationships between factors and the magnitude of their influence. In other words, DEMATEL is a method based on graph theory and matrix tools, which makes full use of experts' knowledge and experience to construct a relationship matrix for analyzing the influence relationships between system elements; it also represents these relationships with specific values. The DEMATEL method reflects not only the relationships between factors, but also the degree of action. The main purpose of the method is to study the logical relationships between the factors of influence in a complex system, so as to construct a direct influence matrix, and to accurately analyze the importance of these factors by calculating the degree of influence, the degree of cause, and the degree of centrality of each factor of influence in the whole complex system, and to determine the cause-and-effect factors.

The ISM technique is a qualitative and interpretive approach to solving complex problems by identifying the main study variables based on the complex interconnected structural mapping of the system's constituent elements. It can also help to systematically identify the solutions to complex problems with causal feedback relationships between the variables involved in the system [41]. ISM is a suitable tool for identifying the contextual relationships between these identified barriers, relying on a relationship describing the interconnection between elements, supporting the identification and ranking of complex relationships between elements in a system, and thus analyzing the influence between elements, using a systematic approach to transform unclear models into well-structured and structured models [42,43].

The MICMAC method allows for the analysis of the position and role of factors in a system, and it facilitates the assessment of the dependence and drive of factors [44,45]. The MICMAC method classifies indicators into correlated, adjusting, driving, dependent, and autonomous factors according to their roles, which can be expressed visually through the directed connecting lines of the skeleton diagram. The quadrant diagram is obtained based on the ISM reachability matrix by dividing the system into a clear hierarchy [46]. The reachable matrix with zeros and ones, ignoring the relationship between weak influencing factors in the system, offers insight into the dependency-driver role in terms of the range of influence expressed, according to strengths and weaknesses, up to a certain level of influence, and according to the value of the cumulative absoluteness of the mode of action. The ISM model was constructed to divide the factors into levels. The MICMAC method is used to analyze the influencing factors, and the position and role of the influencing factors are deeply divided. The corresponding dependencies and driving forces are determined, and targeted countermeasures are proposed. The current domestic and international literature mainly uses DEMATEL, ISM, and MICMAC alone or with two methods combined; fewer studies have combined all three methods, especially in the study of manufacturing systems [47–50]. In this paper, for the first time, three methods (DEMATEL, ISM, and MICMAC) are organically combined to form the DEMATEL-ISM-MICMAC method to study the structure of the vulnerability index system and the association between the index factors of lean–green manufacturing systems in the context of Industry 4.0. Additionally, the specific differences and mutual benefits of the organic combination of these three methods are analyzed in depth. The specific technology roadmap is shown in Figure 2. Considering that the lean–green manufacturing system in the context of Industry 4.0 has many fragile influencing factors, with strong coupling among them, the combination of DEMATEL and ISM not only helps to identify the key elements of the index system and the degree of influence, but also constructs the hierarchical structure of the index system.

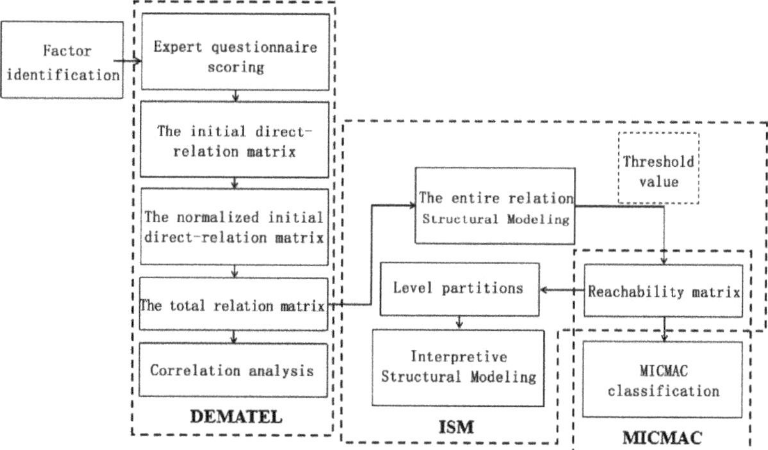

Figure 2. Method flowchart.

Since the DEMATEL method is based on expert experience for scoring, its results are influenced by individual differences and expert subjectivity, so combining fuzzy theory and the DEMATEL method can eliminate problems such as the semanticization and fuzzification of expert evaluative information. The direct influence matrix is obtained by converting the expert scores into the corresponding Triangular Fuzzy Numbers (TFNs); when fuzzified the Triangular Fuzzy Number is converted into an accurate value using the conversion method (converting fuzzy numbers into crisp scores, CFCS) and then integrated using the ISM and MICMAC methods. The specific steps for the construction of the corresponding method-specific model are as follows.

4.1. Improved Integrated DEMATEL-ISM Method

(1) Step 1: Determine the correspondence between the linguistic variables and TFN. The results of experts' ratings of the relationships between the evaluation indicators constitute the evaluation set. The mapping relationship between the linguistic variables and fuzziness is established, as shown in Table 2.

Table 2. Semantic transformation table.

Language Variables	Triangular Fuzzy Number
No effect (NO)	$(0, 0, 1)$
Very low impact (VL)	$(0, 1, 2)$
Low impact (L)	$(1, 2, 3)$
High impact (H)	$(2, 3, 4)$
Very high impact (VH)	$(3, 4, 4)$

(2) Step 2: Construct the TFN direct influence matrix $Z^{(k)} = \left[\tilde{x}_{ij}^{(k)} \right]_{n \times n}$ between the factors related to the vulnerability indicators of manufacturing systems, where $\tilde{x}_{ij}^{(k)} = \left(a_{ij}^k, b_{ij}^k, c_{ij}^k \right)$ is the TFN of the k-th ($k = 1, 2...., q$) TFN of the degree of influence of fragility factor a with factor b, according to the expert.

(3) Step 3: the TFN of the degree of influence between the fragility factors is first standardized and its calculation formula is expressed in Equations (1)–(3).

$$l_{ij}^k = \frac{a_{ij}^k - \min\limits_{1 \le k \le q} a_{ij}^k}{\max\limits_{1 \le k \le q} c_{ij}^k - \min\limits_{1 \le k \le q} a_{ij}^k} \tag{1}$$

$$m_{ij}^k = \frac{b_{ij}^k - \min\limits_{1 \le k \le q} b_{ij}^k}{\max\limits_{1 \le k \le q} c_{ij}^k - \min\limits_{1 \le k \le q} a_{ij}^k} \tag{2}$$

$$r_{ij}^k = \frac{c_{ij}^k - \min\limits_{1 \le k \le q} c_{ij}^k}{\max\limits_{1 \le k \le q} c_{ij}^k - \min\limits_{1 \le k \le q} a_{ij}^k} \tag{3}$$

(4) Step 4: Calculate the standardized clear value of the upper and lower boundaries of the triangular fuzzy set. Its calculation formula is expressed in (4) and (5).

$$u_{ij}^k = \frac{m_{ij}^k}{1 + m_{ij}^k - l_{ij}^k} \tag{4}$$

$$v_{ij}^k = \frac{r_{ij}^k}{1 + m_{ij}^k - l_{ij}^k} \tag{5}$$

(5) Step 5: Calculate the clear value of TNF $z^{(k)}$; its calculation formula is (6):

$$z^{(k)} = \min_{1 \leq k \leq q} a_{ij}^k + \frac{\left(\min\limits_{1 \leq k \leq q} c_{ij}^k - \min\limits_{1 \leq k \leq q} b_{ij}^k \right) \left(u_{ij}^k \left(1 - u_{ij}^k \right) + v_{ij}^k v_{ij}^k \right)}{1 - u_{ij}^k - v_{ij}^k} \tag{6}$$

(6) Step 6: Calculate the average value of $z^{(k)}$ to obtain the direct impact matrix $M = \left[z_{ij} \right]_{m \times n}$; its calculation formula is (7):

$$z_{ij} = \left(z_{ij}^1 + z_{ij}^2 + \cdots + z_{ij}^k \right) / k \tag{7}$$

(7) Step 7: The direct impact matrix is normalized to obtain the matrix $M' = \left[\chi_{ij} \right]_{m \times n}$; its calculation formula is (8):

$$M' = \frac{z_{ij}}{\max(\sum_{j=1}^n z_{ij})} \tag{8}$$

(8) Step 8: In order to analyze the indirect influence relationship between the factors, it is necessary to solve the integrated influence matrix M'', where I is the unit matrix. This can be found using Equation (9):

$$M'' = M' + M'^2 + \cdots + M'^n = \frac{M'(I - M''^n)}{(I - M')} = M'(I - M')^{-1} \tag{9}$$

(9) Step 9: Calculate the cause degree ($R_i - C_i$) and the center degree ($R_i + C_i$) of the driving strength between the factors that influence the fragility of the manufacturing system. From the comprehensive influence matrix $M'' = \left[t_{ij} \right]_{n \times n}$, the influence degree and the influenced degree of each factor index can be calculated as R_i and C_i, respectively (see Equation (10)); then, w deduce the centrality degree $mn = R + C$, which is used to indicate the role size (importance) of each factor in all evaluation indexes, and the cause degree $rn = R - C$, which is used to indicate the internal structure.

$$R_i = \sum_{j=1}^n t_{ij} \qquad C_j = \sum_{i=1}^n t_{ij} \tag{10}$$

Here, the influence degree is the sum of the elements in each row, which is the combined influence value of the corresponding element in that row on all other elements; it is referred to as the influence degree. (2) The influence degree is the sum of each element in each column, which is the combined influence value of the corresponding element in that column by all other elements; this is called the influence degree. ③ Centrality is the sum of the influence degree of each element and the influence degree is called the centrality of the element, which indicates the position of the element in the system, and the role of the size. ④ The difference between the degree of influence and the degree of being influenced of each element is the cause degree of the element. ⑤ The cause element is the cause degree > 0, which indicates that the element has a great influence on other elements; this is called the cause element. ⑥ The result element is the cause degree < 0, which indicates that the element is influenced by other elements; this is called the result element. Through the above calculations, we can judge the degree of influence of each factor on the magnitude of the manufacturing system's brittle force, according to the factors' degree of influence and the degree of being influenced. Then, according to the central degree, we can determine the importance of each indicator in the manufacturing system's brittleness.

(10) Step 10: For the 16 influencing factors in Table 1, the inter-influence relationships among the factors were evaluated by means of questionnaires and expert scoring to initially determine the correlations between the factors. The mapped inter-influence relationships among the brittle factors can be expressed using a box plot, such as that shown in Figure 3. A binary relationship box plot is a graphical representation of the intrinsic connections between the elements within a complex system, with the corresponding letters indicate the

interrelationships between the elements. It is abbreviated as box plot diagram, which can also be referred to as a block diagram. In the box plot, A indicates that the column factors have an influence on the row factors, V indicates that the row factors have an influence on the column factors, and X indicates that the row and column factors have an influence on each other.

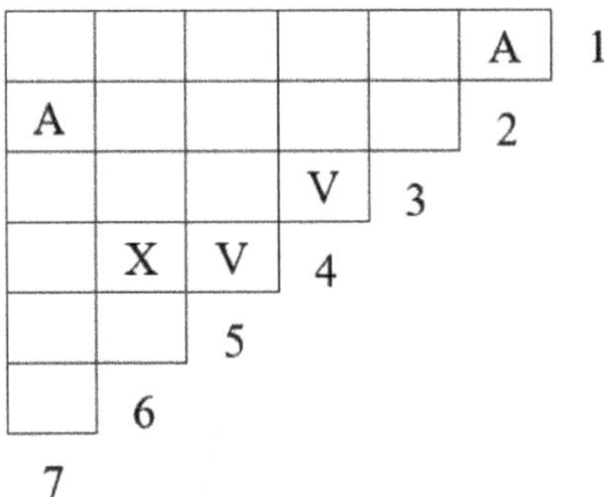

Figure 3. Box plot of the interactions between the factors.

(11) Step 11: Construct the adjacency matrix based on the interrelationship equation between the fragility factors, and, on the basis of the adjacency matrix, find the reachable matrix $H = \left[h_{ij}\right]_{n \times n}$ through matrix operations.

(12) Step 12: Based on the reachable matrix, region division, level division, and skeleton matrix extraction are carried out to establish the interpretative structural model of manufacturing systems' brittle factors.

4.2. MICMAC Validation Analysis of Key Brittleness-Influencing Factors

Using the MICMAC method, quadrant diagrams are drawn according to the calculated dependency and driving force results. This enables further analysis of the status and the role played by the brittleness-influencing factors of the manufacturing system; additionally, it allows for the elucidation of different characteristics of each brittleness-influencing factor and provides a basis for making suggestions or taking measures. The method applies the principle of matrix multiplication to analyze the degree of influence and correlation between factors by calculating their drivers and dependencies; it is often used in combination with ISM to identify factors with high dynamics and high dependencies in the system.

Stratification among indicators is performed by calculating the driving force (Q) and dependency power (Y) of each indicator, where the driving force is the sum of the elements of each row of the reachability matrix and the dependency is the sum of the elements of each column of the reachability matrix. Accordingly, the Q–Y classification diagram of influencing factors was drawn, and the mean values of drive and dependence were used as the dividing line; the diagram was finally divided into four quadrants, and quadrants I, II, III, and IV were the autonomy, dependence, association, and drive quadrants, respectively, as shown in Figure 4 [51]. Greater dependence indicates greater reliance on other factors when addressing a given factor, and greater drive indicates the extent to which this factor can help to address other factors [52]. Thus, factors in the independent quadrant are more driven and less dependent; their factors are weakly influenced by the remaining elements but they have a greater impact on other elements. Therefore, the identification of such factors is important for assessing the brittleness of a manufacturing system and is the basis for determining whether an accidental manufacturing system collapse occurs [53].

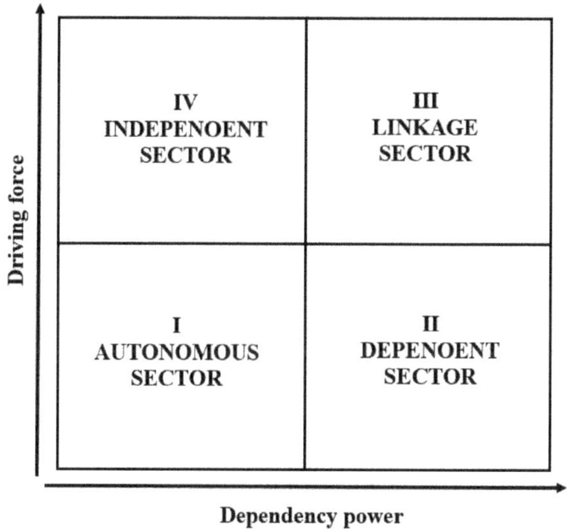

IV
INDEPENOENT
SECTOR

III
LINKAGE
SECTOR

I
AUTONOMOUS
SECTOR

II
DEPENOENT
SECTOR

Driving force — Dependency power

Figure 4. Matrices Impact Croises-Multiplication Appliance Classement (MICMAC).

The results of MICMAC analysis can be visualized using a two-dimensional coordinate diagram, with the vertical axis representing the driving forces and the horizontal axis representing the dependencies. For each factor in the whole complex system, the numerical magnitudes of the driving force $DF(X_i)$ and the dependency $DP(X_i)$ can be calculated based on the reachability matrix H. This can be found using Equations (11) and (12):

$$DF(X_i) = \sum_{j=1}^{n} h_{ij} (i = 1, 2, \cdots, n) \tag{11}$$

$$DP(X_i) = \sum_{i=1}^{n} h_{ij} (i = 1, 2, \cdots, n) \tag{12}$$

Based on the MICMAC analysis of the causes of brittleness in lean–green manufacturing systems in the context of Industry 4.0, a dependency matrix of indicator drivers can be derived. A dependency and driver diagram of brittleness in manufacturing systems can also be calculated, allowing for the analysis of the position and role of the influencing factors in order to suggest targeted improvements.

5. Case Study

With the goal of building a world-leading lean–green factory for the intelligent manufacturing of a micro-acoustic material and device, a smart factory proposes the following manufacturing system modules: (1) technological innovation; (2) information systems; (3) core equipment; (4) a resource strategy; and (5) basic security [54]. The technology innovation module is used to break through the key short-board equipment and to apply artificial intelligence technology, such as data mining and machine vision, to enhance the processing level, improve operation efficiency, and reduce energy consumption. The information system module integrates PDM/ERP/MES/APSWMS/SCADA and other information systems to achieve core equipment networking and monitoring, as well as the collaborative management and control of the entire production process. The core equipment module introduces and integrates dozens of core pieces of intelligent equipment, upgrades chip lines, expands packaging lines, and builds new test lines to significantly increase production capacity and automation levels. The resource strategy module upgrades the ERP system and introduces the OA system to realize the unified and collaborative management of human resources, social resources, information resources, and financial resources. The basic guarantee module, which strengthens organizational, technical, personal, mechanical and financial factors, provides a strong guarantee for project implementation. Under the synergistic operation of each module, through carefully sorting out the process layout and material flow process, an efficient operation flow and material pulling mechanism can be established; this completely eliminates the phenomenon of material stagnation and stoppage, improves the logistic speed and production beat, and realizes lean management and the green transformation of production operations and material logistics. Through management innovations and technological innovations, and by facilitating the deep integration of information technology and industrialization technology, we will gradually generate intelligent work stations, processes, and workshops, and build a modern, intelligent factory that integrates information and industry from point to point [55].

This intelligent manufacturing system is a large and highly complex system. In order to analyze the causal factors of brittleness in this lean–green manufacturing system in the context of Industry 4.0, the pool of survey interviewers was further expanded beyond the previous 15 experts in order to assess the results more accurately and scientifically. Each of the interviewees has more than ten years of work experience; in total, 200 people were surveyed. Including enterprise staff, teachers in research institutes, and industry consultants. The job titles, work units, and education levels of the respondents are shown in Table 3.

Table 3. Distribution of respondents' basic information.

Basic Information	Category	Number of People (pcs)	Percentage
	Research Institutes	40	20.0%
Work Unit	Professional consulting company	22	11.0%
	Manufacturing Company	138	69.0%
	University professors	35	17.5%
Position Information	Business leaders, department managers, supervisors	30	15.0%
	General front-line employees	135	67.5%
	College and below	78	39.0%
Education level	Bachelor's degree	67	33.5%
	Masters and above	55	27.5%

The surveyed interviewers increased their basic knowledge of the company and their own expertise on the 16 brittle influencing factors of the manufacturing system described in the case study; this was achieved by applying the linguistic variables in Table 2 to the field TNF assessment. Then, according to steps 1 to 6, after defuzzification by CFCS, the direct influence matrix of the brittle influence factors of lean–green manufacturing system in the context of Industry 4.0 was obtained, as shown in Table 4.

Table 4. Direct impact matrix of the causal factors of brittleness in lean–green manufacturing system in the context of Industry 4.0.

	S_1	S_2	S_3	S_4	S_5	S_6	S_7	S_8	S_9	S_{10}	S_{11}	S_{12}	S_{13}	S_{14}	S_{15}	S_{16}
S_1	0	0.36	0.24	0	0	0.1	0.17	0.55	0.11	0.15	0.14	0.11	0.13	0.35	0	0
S_2	0.22	0	0	0	0	0	0.41	0.01	0	0.35	0.38	0.1	0.09	0	0	0
S_3	0	0	0	0.33	0.34	0.28	0.41	0.02	0.52	0.5	0.43	0.48	0.55	0	0	0
S_4	0	0	0	0	0.1	0	0	0	0	0	0	0.42	0	0	0	0
S_5	0	0	0	0.02	0	0	0	0	0	0	0	0	0	0	0	0
S_6	0	0	0	0	0	0	0.23	0	0.28	0.11	0.02	0.21	0	0.45	0	0
S_7	0.01	0.03	0	0	0	0	0	0	0	0	0	0	0	0	0	0
S_8	0.33	0.35	0.18	0.2	0.01	0.29	0.29	0	0.15	0.22	0.28	0.27	0.09	0.31	0	0
S_9	0.02	0.24	0	0.03	0.01	0	0.12	0	0	0.21	0.22	0.47	0.03	0	0	0
S_{10}	0	0.09	0	0.02	0.01	0	0.1	0	0.1	0	0.31	0.13	0	0	0	0
S_{11}	0.05	0.32	0	0.31	0.13	0	0.25	0	0	0	0	0	0	0	0	0
S_{12}	0.52	0.36	0	0	0	0	0.32	0	0	0	0	0	0	0	0	0
S_{13}	0	0.19	0.23	0	0	0	0.27	0	0	0	0	0	0	0	0	0
S_{14}	0.2	0.15	0	0	0	0	0.21	0	0.21	0.57	0.18	0.32	0.01	0	0	0
S_{15}	0.07	0.11	0	0	0.17	0	0.42	0	0.45	0.52	0.38	0.59	0.29	0.31	0	0
S_{16}	0.25	0.18	0	0.06	0	0	0.41	0	0.29	0.31	0.33	0.28	0.16	0.01	0.08	0

Next, the influence, affectedness, centrality, and causality of the causal factors of brittleness in the lean–green manufacturing system are calculated according to steps 7 to 9; the results of the influence degree, influenced degree, and centrality–cause degree of each causal factor are shown in Table 5.

Table 5. Influence degree, influenced degree, centrality degree, and cause degree of each causative factor.

No.	Influence Degree	Influenced Degree	Centrality Degree	Cause Degree	No.	Influence Degree	Influenced Degree	Centrality Degree	Cause Degree
S_1	0.903	0.716	1.619	0.186	S_9	0.485	0.678	1.164	−0.193
S_2	0.538	1.020	1.558	−0.482	S_{10}	0.268	1.036	1.304	−0.769
S_3	1.299	0.243	1.542	1.057	S_{11}	0.347	1.008	1.356	−0.661
S_4	0.188	0.378	0.566	−0.191	S_{12}	0.485	1.180	1.665	−0.695
S_5	0.006	0.271	0.277	−0.264	S_{13}	0.284	0.446	0.730	−0.162
S_6	0.487	0.229	0.716	0.258	S_{14}	0.671	0.421	1.092	0.250
S_7	0.017	1.401	1.418	−1.384	S_{15}	1.007	0.000	1.188	1.147
S_8	1.152	0.256	1.409	0.896	S_{16}	0.873	0.257	1.007	1.007

According to the centrality and cause degrees, the relationships between the causal factors of the manufacturing system were plotted, as shown in Figure 5. It can be seen that the cause degrees of S_2, S_4, S_5, S_7, S_9, S_{10}, S_{11}, S_{12}, and S_{13} are negative, which means that these factors will be influenced by other factors, thus causing the manufacturing system to crash or stop. The high centrality of S_1, S_2, S_3, S_7, S_8, and S_{12} indicates that these factors are key factors in the occurrence of brittleness in manufacturing systems and need to be taken seriously by managers.

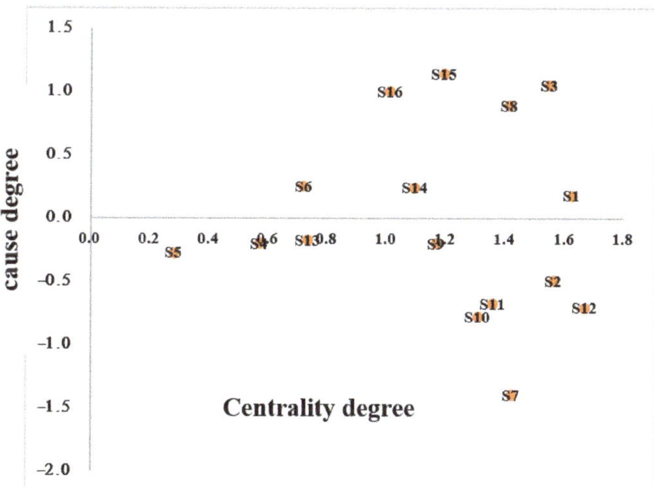

Figure 5. Centrality-Cause degree distribution of brittleness index factors in lean–green manufacturing systems in the context of Industry 4.0.

According to step 10, the 16 influencing factors detailed in Table 1 were evaluated by means of questionnaires and expert scoring, in order to initially determine the correlations between the factors and draw a variogram of the interactions among the brittle factors, as shown in Figure 6. According to step 11, the adjacency matrix was constructed based on the interrelationship equation between the brittle factors; on the basis of the adjacency matrix, the reachability matrix is obtained by matrix operations and is shown in Table 6.

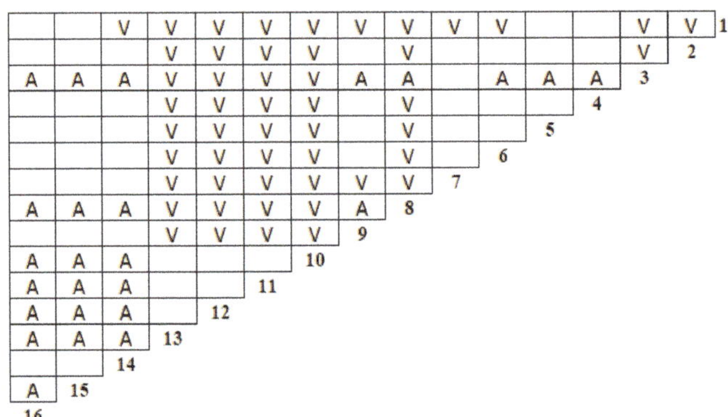

Figure 6. Box plot of the interactions between the brittleness factors of lean–green manufacturing systems in the context of Industry 4.0.

Table 6. The reachable matrix of factors influencing the brittleness of lean–green manufacturing systems in the context of Industry 4.0.

	S_1	S_2	S_3	S_4	S_5	S_6	S_7	S_8	S_9	S_{10}	S_{11}	S_{12}	S_{13}	S_{14}	S_{15}	S_{16}
S_1	1	1	1	0	0	1	0	1	1	1	1	1	1	1	0	0
S_2	0	1	1	0	0	0	0	1	0	1	1	1	1	0	0	0
S_3	0	0	1	0	0	0	0	0	0	1	1	1	1	0	0	0
S_4	0	0	1	1	0	0	0	1	0	1	1	1	1	0	0	0
S_5	0	0	1	0	1	0	0	1	0	1	1	1	1	0	0	0
S_6	0	0	1	0	0	1	0	1	0	1	1	1	1	0	0	0
S_7	0	0	1	0	0	0	1	1	1	1	1	1	1	0	0	0
S_8	0	0	1	0	0	0	0	1	0	1	1	1	1	0	0	0
S_9	0	0	1	0	0	0	0	1	1	1	1	1	1	0	0	0
S_{10}	0	0	0	0	0	0	0	0	0	1	0	0	0	0	0	0
S_{11}	0	0	0	0	0	0	0	0	0	0	1	0	0	0	0	0
S_{12}	0	0	0	0	0	0	0	0	0	0	0	1	0	0	0	0
S_{13}	0	0	0	0	0	0	0	0	0	0	0	0	1	0	0	0
S_{14}	0	0	1	0	0	0	0	1	0	1	1	1	1	1	0	0
S_{15}	0	0	1	0	0	0	0	1	0	1	1	1	1	0	1	0
S_{16}	0	0	1	0	0	0	0	1	0	1	1	1	1	0	0	1

Based on the reachable matrix in Table 6 and Step 12, a model diagram of the ISM explanatory structure of the causal factors of brittleness is obtained, as shown in Figure 7.

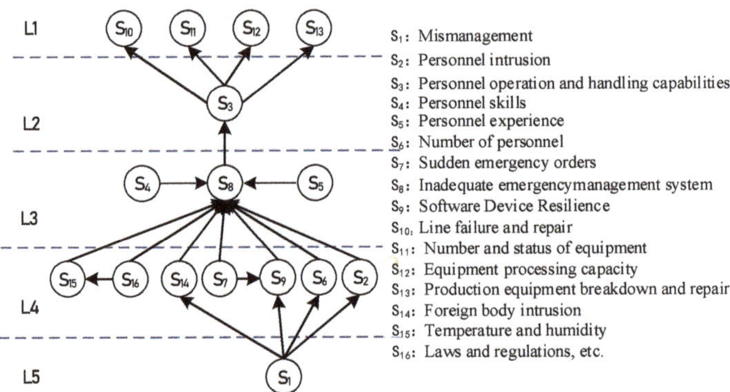

Figure 7. Multi-level recursive interpretative structural model of the brittleness of lean–green manufacturing systems in the context of Industry 4.0.

Based on the MICMAC principle and the reachability matrix H, the driving forces and dependencies of each factor that influences the vulnerability of the lean–green manufacturing system in the context of Industry 4.0 can be calculated, as shown in Table 7. Based on this, a driving-force-dependency power diagram of the factors influencing the fragility of the example manufacturing system is drawn, as shown in Figure 8.

Table 7. Driving-force-dependency power of factors influencing the brittleness of lean–green manufacturing systems in the context of Industry 4.0.

Factor	Dependency Power	Driving Force	Factor	Dependency Power	Driving Force
S_1	1	11	S_9	3	7
S_2	2	7	S_{10}	13	1
S_3	12	5	S_{11}	13	1
S_4	1	7	S_{12}	13	1
S_5	1	7	S_{13}	13	1
S_6	2	7	S_{14}	2	7
S_7	1	8	S_{15}	1	7
S_8	11	6	S_{16}	1	7

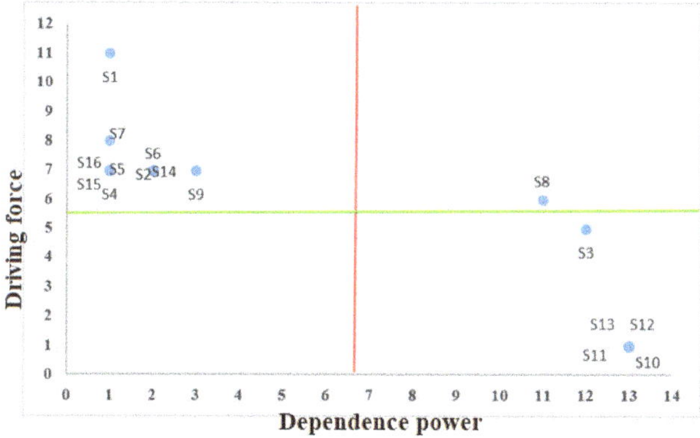

Figure 8. Driving-force-dependency power of factors influencing the brittleness of lean–green manufacturing systems in the context of Industry 4.0.

6. Discussion

6.1. Analysis of the Correlation and Importance of the Influencing Factors

The direct influence matrix M can be obtained according to Equations (1)–(7). According to the direct influence matrix M of factors that influence manufacturing systems' fragility and Equations (8)–(10), Table 5 can be obtained. Table 5 shows the influence degree, influenced degree, centrality degree, and cause degree of each causative factor of brittleness in lean–green manufacturing systems, and can be used to analyze the correlation and importance between factors. The role and importance of each factor are shown in Figure 5. The magnitude of centrality indicates the degree of association of a given factor with other factors and its importance in the system; the higher the magnitude of the value in Table 5, the more closely the factor is associated with the vulnerability indicators and the stronger the relevance; the same is true for the values situated towards the right of Figure 5. The ranking of the numerical sizes of the centrality degrees is as follows: equipment handling capacity (S_{12}), improper management (S_1), personnel intrusion (S_2), personnel operation and handling capacity (S_3), sudden emergency orders (S_7), and inadequate emergency management system (S_8). Meanwhile, there is a significant difference between the degree of influence and the degree of being influenced for temperature and humidity (S15), the improper operation of personnel (S_3), laws and regulations (S_{16}), sudden emergency orders

(S$_7$), equipment failure and maintenance (S$_9$), line failure and maintenance (S$_{10}$), equipment quantity and status (S$_{11}$), equipment handling capacity (S$_{12}$), and production equipment failure and maintenance (S$_{13}$). The first three are less strongly influenced and mainly influence the other factors, while the latter five factors are more likely to be influenced for the rest of the factors, the degree of influence and the degree of being influenced are comparable. Among them, personnel operation and handling ability (S$_3$) has the greatest influence degree of 1.299, and it is the main causal factor of the excitation of manufacturing systems' brittleness leading to system collapse or stoppage, followed by temperature and humidity (S$_{15}$) and laws and regulations (S$_{16}$). Attention must also be paid to these last two factors. In terms of the roles played by different factors, the centrality degree values of mismanagement (S$_1$), laws and regulations (S16), and temperature and humidity (S$_{15}$) are relatively large and are all greater than 0. Moreover, the values for the degree of affectedness are also relatively small, indicating that these factors influence to a greater degree than they are influenced, and play a largely causative role in the system. The other factors are essentially outcome factors. From Figure 5, it can be seen that personnel operation and disposal ability (S$_3$), mismanagement (S$_1$), laws and regulations (S$_{16}$), and other factors are situated in the upper right part of Figure 5, meaning that their comprehensive influence on other factors is relatively large. Meanwhile, the five factors in the lower right corner are dependent factors, or passive factors; these factors are more deeply influenced by other factors and are also the direct cause of brittleness excitation.

6.2. System Hierarchy Analysis

The interpretative structure model is capable of grading factors and revealing the internal structure of a system [56]. As can be seen from the interpretative structure model the ISM divides all factor indicators into five levels, from top to bottom, representing proximate causation (L1), transitional causation (L2, L3), deep causation (L4), and essential causation (L5). The connection between the causal factors of brittleness in lean–green manufacturing systems in the context of Industry 4.0 is highly complex, and this complexity is reflected in the following four aspects.

(1) Line failure and maintenance, equipment quantity and status, equipment handling capacity and production equipment failure, and maintenance are direct causes of system breakdown or stoppage. At the same time, the improper operation of personnel or inadequate emergency management systems are also important causes of system breakdowns. Personnel skills and experience are the factors that most directly determine the quality of a manufacturing system; they can also directly affect the productivity of production equipment and non-production equipment (software system) and the chance of equipment failure and maintenance taking place, thus affecting the scheduling of production plans and the timely completion of orders, and even lead to the collapse of the system and stop production. Most automated equipment is highly dependent on fixtures and jigs, which affects the efficiency of the production line and increases the risk of brittleness. Therefore reducing equipment failures (including those of software systems and information systems) is one of the most important means of preventing system crashes or stoppages.

(2) Flaws in relevant personnel operations and the disposal abilities of emergency management systems are also critical. For the causal factors of the L2 and L3 layers in the ISM model, these two factors are of great importance. In particular, when an unexpected event occurs, efficient emergency response capabilities and various emergency management systems are crucial to preventing the occurrence or further spread of adverse events. Therefore, it is important to improve the business handling abilities of relevant personnel and to develop relevant contingency plans for the occurrence of an exceptional event. By utilizing the skills of relevant personnel and implementing functional emergency management systems, the probability of a system crashing or stopping can be reduced; if a crash or stop does occur, it can be controlled such that the local minimum scope of the system is maintained, or else can be resolved after a short period of time. Therefore, these

factors not only affect the efficiency of the system, but can even rectify situations where the production line goes offline.

(3) Temperature and environmental factors, such as humidity, and laws and regulations have a serious impact on the excitation of brittleness in manufacturing systems. These factors essentially belong to the L4 layer of causative factors. When these two types of factors do not meet production conditions, a manufacturing system is at high risk of brittleness. For example, if temperature and humidity conditions are not up to standard, and equipment or software information systems cannot function properly, a chain collapse phenomenon could occur in each system. The manufacturing system is certain to cease operations if one part of the product or service in the manufacturing process violates the relevant laws and regulations related to the quality standard. For instance, as a microelectronics company, the company used as a case study in this research has higher requirements for temperature and humidity; moreover, most of the connections are made in a clean and quiet workshop, which has higher requirements for humidity and temperature. Therefore, it is important to monitor such causative factors in real time and to prepare for emergencies in advance, to minimize the probability that they will violate the relevant parameters.

(4) Mismanagement is the fundamental cause of the excitation of systemic brittleness. The above-mentioned "human–machine-environment", along with the other three types of factors, can directly lead to the excitation of a system's brittleness and cause the manufacturing system to collapse or stop; however, management factors are the root cause of accidents caused by the above three types of factors. It is therefore necessary to improve the quality of managers and management systems, and to use advanced management models to establish a scientific management system, clear responsibilities, and smooth communication and feedback channels. Additionally, to achieve continuous improvement it is necessary to undertake continuous management innovation, improve the management level, and update the relevant management systems and processes.

6.3. Driving Force and Dependency Power Relationship Analysis

The dependency $DF(X_i)$ and driver $DP(X_i)$ obtained from MICMAC can be used as the x-axis and y-axis, respectively, to obtain Figure 8. On this basis, the driving force and dependency power relationship between the factors can be analyzed and classified accordingly; the relevant results can be seen from Figure 8.

Quadrant area I contains autonomous factors, which are either relatively independent and have little correlation with other factors, or are not influential enough to trigger a chain reaction; there were no factors in this category in our case study. Quadrant area II contains the dependent factors, which are generally more strongly linked to and easily controlled by other factors; however, the driving force is not strong. In this category are personnel operation and disposal ability, line failure and maintenance, equipment quantity and status, equipment handling ability, and production equipment failure and maintenance; these issues can generally be solved by first resolving other factors. Quadrant III contains the associated factors, including imperfect emergency management systems, which have high dependence and a strong driving force. Finally, quadrant IV contains the driving factors; all of the remaining 10 influencing factors fall into this category, meaning that they have a greater impact on the other factors in the system and are generally at the lower level of the ISM progression structure.

6.4. Conclusions

Based on the selection and establishment of the vulnerability indicators of lean–green manufacturing systems in the context of Industry 4.0 and the development of the implementation system, the vulnerability indicator system can be constructed from sixteen specific indicators in four major areas, namely, "human–machine, environment and management". The most important factors are generally those factors with high relevance to the other factors, and factors that play a driving role in the system; controlling these two major categories of indicators are the key to reducing a system's vulnerability. One of the most important indicators of effective manufacturing operations is the ability of manufacturing systems to meet customer needs on time, efficiently, and with high quality. However, manufacturing systems are susceptible to brittle factors that can cause them to crash or stop functioning. In order to deepen the scientific understanding of factors that cause brittleness in manufacturing systems, an improved DEMATEL-ISM-MICMAC analysis method was constructed to explore the internal connections between factors that generate fragility in lean–green manufacturing systems in the context of Industry 4.0. This study aimed to clearly determine the hierarchy of key factors and influence pathways, and to distinguish the dependency power and driving force of each fragility-influencing factor.

From the perspective of actual system in operation, the factors corresponding to the top five critical nodes in the network in terms of node importance can all be classified as related to production equipment, information transmission, and information systems, etc. As the core of system operations, the failure frequency, maintenance guarantee strategy, operation status, and recovery and processing capabilities of the equipment are the key factors affecting a system's fragility. The data indicate that these factors are also key to ensuring the safe and normal operation of the system, as is consistent with the actual operations of the system. The nodes ranked sixth and seventh in node importance correspond to the fragility factors related to work skills and work experience, respectively, indicating that the business ability of employees is also a key fragility factor that affects the normal operation of the system. As such, the comprehensive abilities of employees should be strengthened to enhance their effectiveness, and an effective job-posting assessment system should be developed to focus on the cultivation of high-quality talents and to highlight the importance of talents in the system. Additional objective and specific quantitative factors involved in the brittleness and collapse of manufacturing systems include mismanagement, industry quality standards, laws and regulations, the working system of the plant, temperature and humidity, and the operation methods and quantity of existing equipment and information systems. Although these factors have a relatively small impact on system brittleness, they are still essential to ensuring the normal operation of manufacturing systems and are important safeguards to achieving the overall function of a system. Therefore, even as managers pay increased attention to key brittleness factors, the other relevant factors should not be ignored, so that safe operation measures can be formulated more efficiently and accurately to ensure the safe and stable operation of the system.

6.5. Managerial Implications

The focus of this paper was to construct a vulnerability indicator system for lean–green manufacturing systems in the context of Industry 4.0 in Chinese manufacturing enterprises. We also aimed to analyze the relationship between various vulnerability indicator factors, establish a hierarchy of factors and classify them according to driving forces and dependencies, and finally use the proposed method to identify the key vulnerability factors that affect system fragility. For this last step, rather than considering all factors, we used a manufacturing system from an example company to identify key indicator factors and specific factors that play a role in the monitoring and management of lean–green manufacturing systems in the manufacturing industry and its enterprises. According to the relationships between the factors and the reachable pathways, an index system was built into a five-level hierarchy from bottom to top, L5 referred to the essential causes, L4 to the deep causes, L2–3 to the transitional causes, and L1 to proximate causes. The different levels of indicators have their own status and characteristics in the system, according to which different stages of management planning can be implemented. Additionally, according to the roles of the different factors, the indicators can be divided into five categories: correlation factors, adjustment factors, driving factors, dependency factors, and autonomous factors. The roles of the factors in these categories can be visually expressed as connecting paths in the hierarchy. Based on how many paths and pointers can be managed for indicators on a primary or secondary level, managers can focus their attentions on correlation factors, adjustment factors, and driving factors, followed by dependency factors; autonomous factors do not require much attention. Based on the research presented in this paper, managers can implement different adaptive management strategies based on indicator relationships, roles, and hierarchies.

However, this study has certain limitations: the data are derived from experts' experience and scoring, and although they are authoritative and representative, they inevitably contains some degree of subjectivity and uncertainty, and may deviate from the real situation. Further refinement of the validation models can be attempted for verification. The 0 and 1 values of the ISM reachable matrix indicate that the obscure relationship is ignored, which can be corrected in the future by combining this model with fuzzy theory. At the same time, further in-depth research can be conducted based on the key fragility factors combined with the actual production situation, to enable a deeper analysis of the operation status of the equipment in the system and to develop a more reasonable maintenance guarantee strategy to guide real-world production. Additionally, to better understand the importance of each brittleness factor at a later stage of the research process, it can be measured by calculating the relevant weight classes of the factors, either by using the ANP method or the AHP method.

In addition, in order to classify the brittleness factor index system, the classification can be managed using the cluster analysis method. Manufacturing systems' brittle factor indicators can be classified as input class indicators and output class indicators, which can be analyzed using the Data Envelopment Analysis (DEA) method to determine the relevant validity of a unit. In addition, system fragility factor indicators can be ranked using the TOPSIS or VIKOR methods [57]. These multi-criteria decision-making tools can be used individually or in combination. The combined use of these methods can produce more scientific, verifiable, and robust findings.

Author Contributions: Conceptualization, X.Z., Y.L., Y.X. and G.X.; methodology, X.Z. and Y.X.; formal analysis, Y.X., X.D. and G.X.; investigation, X.Z., Y.L.; writing—original draft preparation, X.Z., Y.L., Y.X., X.D. and G.X.; writing—review and editing, X.Z., Y.L. and Y.X. All authors have read and agreed to the published version of the manuscript.

Funding: This research was funded by the Natural Science Foundation of Hunan Province, China (Project number: 2022JJ50244); the Education Department of Hunan Province (Project number: 21B0695; 21A0475); the Project of Hunan social science achievement evaluation committee in 2022 (Project number: XSP22YBC081); the Project of Shaoyang social science achievement evaluation committee in 2022 (Project number: 22YBB10).

Institutional Review Board Statement: Not applicable.

Informed Consent Statement: Informed consent was obtained from all subjects involved in the study.

Data Availability Statement: Not applicable.

Conflicts of Interest: The authors declare no conflict of interest.

References

People's Daily. 2021 Manufacturing Value Added Reached 31.4 Trillion-Yuan, Accounting for Nearly 30% of the Global Proportion of China's Comprehensive Manufacturing Strength Continues to Improve [EB/OL]. 2022. Available online: http://www.zjsjw.gov.cn/shizhengzhaibao/202208/t20220803_6639840.shtml (accessed on 18 December 2022).

Pfeiffer, S. *Effects of Industry 4.0 on Vocational Education and Training*; Austrian Academy of Science: Vienna, Austria, 2015; Volume 6, p. 20.

Li, C. Analysis of the impact of Industry 4.0 on modern equipment management. *China Equip. Eng.* **2021**, *3*, 20–21. (In Chinese)

Tang, L.; Huang, S. From "machine for man" to "man-machine dance"—The role of engineering talents and the shape of education in the process of Industry 4.0. *Res. High. Eng. Educ.* **2020**, *4*, 75–82. (In Chinese)

Dougherty, P.; Wilson, J. *Machine and Man: Accenture on the New Artificial Intelligence*; Zhao, Y., Translator; CITIC Press: Beijing, China, 2018; pp. 31–97. (In Chinese)

When Lean Meets Industry 4.0: The Next Level of Operational Excellence; The Boston Consulting Group: Boston, MA, USA, 2017; Volume 1.

Zuehlke, D. Smart Factory-Towards a Factory-of-things. *Annu. Rev. Control* **2010**, *34*, 129–138. [CrossRef]

8. *Industry 4.0: The Future of Productivity and Growth in Manufacturing Industries of Productivity and Growth in Manufacturing Industries*; The Boston Consulting Group: Boston, MA, USA, 2015; Volume 2, p. 4.
9. *Five Lessons from the Frontlines of Industry 4.0*; The Boston Consulting Group: Boston, MA, USA, 2017; Volume 7.
10. Zhang, S. How Chinese manufacturing enterprises are moving towards Industry 4.0. *Mech. Des. Manuf. Eng.* **2014**, *43*, 1–5. (In Chinese)
11. Liao, G.; Yi, S.; Zhou, J.; Wen, P.; Xiong, S. Human factors and ergonomics in industry 4.0 environment. *Technol. Innov. Manag.* **2016**, *37*, 270–275. (In Chinese)
12. Albino, V.; Garavelli, A.C. A methodology for the vulnerability analysis of just-in-time production systems. *Int. J. Prod. Econ.* **1995**, *41*, 71–80. [CrossRef]
13. Nof, S.Y.; Morel, G.; Monostori, L.; Molina, A.; Filip, F. From plant and logistics control to multi-enterprise collaboration. *Annu. Rev. Control* **2006**, *30*, 55–68. [CrossRef]
14. Cheminod, M.; Bertolotti, I.C.; Durante, L.; Valenzano, A. On the analysis of vulnerability chains in industrial networks. In Proceedings of the IEEE International Workshop on Factory Communication Systems, Dresden, Germany, 21–23 May 2008; pp. 215–224.
15. DeSmit, Z.; Elhabashy, A.E.; Wells, L.J.; Camelio, J.A. An approach to cyber-physical vulnerability assessment for intelligent to cyber-physical vulnerability assessment for intelligent manufacturing systems. *J. Manuf. Syst.* **2017**, *43*, 339–351. [CrossRef]
16. Liu, W.; Xu, L.; Chen, Y. Brittle measurement and evaluation analysis of manufacturing equipment based on brittle risk entropy. *Comput. Integr. Manuf. Syst.* **2019**, *25*, 2820–2830. (In Chinese)
17. Liu, J.; Zhang, G.B.; Li, D.Y.; Li, Y.; Qian, B.M. Reliability analysis for multi-state manufacturing system based on brittleness theory. *Comput. Integr. Manuf. Syst.* **2014**, *20*, 155–164. (In Chinese)
18. Qin, Y.; Zhao, L.; Yao, Y. Quality dynamic characteristics analysis based on brittleness theory in complex manufacturing processes. *Comput. Integr. Manuf. Syst.* **2010**, *16*, 2240–2249. (In Chinese)
19. Gao, G.; Yue, W.; Zhang, R. Structural vulnerability assessment method of manufacturing systems vulnerability assessment method of manufacturing systems based on the state entropy. *Comput. Integr. Manuf. Syst.* **2017**, *23*, 134–143. (In Chinese)
20. Gao, G.; Yue, W.; Ou, W. Vulnerability assessment method for mixed-model manufacturing systems method for mixed-model manufacturing systems based on UGF. *China Mech. Eng.* **2018**, *29*, 2087–2093. (In Chinese)
21. Gao, G.; Yue, W.; Wang, F. Intelligent diagnosis on health status of manufacturing systems based on embedded CPS method and vulnerability assessment. *China Mech. Eng.* **2019**, *30*, 212–219. (In Chinese)
22. Yin, H.; Li, B.; Zhu, J.; Guo, T. Measurement method and empirical research on systemic vulnerability of environmental sustainable development capability. *Pol. J. Environ. Stud.* **2014**, *23*, 243–253.
23. Kizhakkedath, A.; Tai, K.; Sim, M.S.; Tiong, R.L.; Lin, J. *An Agent-Based Modeling and Evolutionary Optimization Approach for Vulnerability Analysis of Critical Infrastructure Networks*; Springer: Berlin/Heidelberg, Germany, 2013.
24. Cavdaroglu, B.; Hammel, E.; Mitchell, J.E.; Sharkey, T.C.; Wallace, W.A. Integrating restoration and scheduling decisions for disrupted interdependent infrastructure systems. *Ann. Oper. Res.* **2013**, *203*, 279–294. [CrossRef]
25. Krichen, M.; Lahami, M.; Cheikhrouhou, O.; Alroobaea, R.; Maâlej, A.J. *Security Testing of Internet of Things for Smart City Applications: A Formal Approach*; EAI/Springer Innovations in Communication and Computing; Springer: Cham, Switzerland, 2019; pp. 629–653. [CrossRef]
26. Jamai, I.; Ben Azzouz, L.; Saidane, L.A. Security issues in Industry 4.0. In Proceedings of the 2020 International Wireless Communications and Mobile Computing (IWCMC), Limassol, Cyprus, 15–19 June 2020; pp. 481–488. [CrossRef]
27. Ruiz-Agudelo, C.A.; Bonilla-Uribe, O.D.; Páez, C.A. The vulnerability of agricultural and livestock systems to climate variability. Using dynamic system models in the Rancheria upper basin (Sierra Nevada de Santa Marta). *J. Prot. Mt. Areas Res.* **2015**, *7*, 50–60. [CrossRef]
28. Rong, P.; Jin, H.; Wei, Q. Brittleness research on complex system based on brittle link entropy. *J. Mar. Sci. Appl.* **2006**, *5*, 51–54.
29. Xiang, R.P.; Zhang, J.H.; Qi, W.; Mei, Y.L. Research on the characteristic of complex systems based on brittle linkage. *Electr. Mach. Control.* **2005**, *9*, 111–115. (In Chinese)
30. Li, M.; Wang, W.; Zhang, Z. Study on construction risk factors based on ISM and MICMAC. *J. Saf. Environ.* **2022**, *22*, 22–28. (In Chinese) [CrossRef]
31. Wang, W.; Liu, X.; Qin, Y.; Huang, J.; Liu, Y. Assessing contributory factors in potential systemic accidents using AcciMap and integrated fuzzy ISM—MICMAC approach. *Int. J. Ind. Ergon.* **2018**, *68*, 311–326. [CrossRef]
32. Becker, T.; Meyer, M.; Windt, K. A manufacturing systems network model for the evaluation of complex manufacturing systems. *Int. J. Prod. Perform. Manag.* **2014**, *63*, 324–340. [CrossRef]
33. Vrabic, R.; Skulj, G.; Butala, P. Anomaly detection in shop floor material flow: A network theory approach. *CIRP Ann.-Manuf. Technol.* **2013**, *62*, 487–490. [CrossRef]
34. Gao, G.; Rong, T.; Yue, W. Vulnerability assessment method for the manufacturing system based on complex method for the manufacturing system based on complex network. *Comput. Integr. Manuf. Syst.* **2018**, *24*, 160–168. (In Chinese)
35. Jiang, H.; Gao, J.; Chen, F.; Gao, Z. Vulnerability analysis to distributed and complex electromechanical analysis to distributed and complex electromechanical system based on network property. *Comput. Integr. Comput. Integr. Manuf. Syst.* **2009**, *15*, 791–79. (In Chinese)

36. Li, X.; Yuan, Y.; Sun, W.; Feng, H. Bottleneck identification in job-shop based on network structure characteristic. *Comput. Integr. Manuf. Syst.* **2016**, *22*, 1088–1096. (In Chinese)
37. Li, H.; Lu, S.; Yang, T. Process route optimization of production line of complex products based on vulnerability of production line of complex products based on vulnerability analysis. *China Mech. Eng.* **2014**, *25*, 168–3173. (In Chinese)
38. Liu, W.; Xu, L. Identification of key brittleness factor for manufacturing system based on ISM and complex network. *Comput. Integr. Manuf. Syst.* **2021**, *27*, 3076–3092. (In Chinese)
39. Sun, Y.; Zhang, T.; Liu, N.; Zhao, Y.; Li, A. Analysis of the factors contributing to the late train service of high-speed railroads based on the improved ISM-DEMATEL. *Railw. Transp. Econ.* **2022**, *44*, 1–7. (In Chinese)
40. Zhang, Y.; Chen, X. Selection and analysis of vulnerability indicators for China's pelagic squid fishing based on DEMATEL-ISM-MICMAC method. *J. Shanghai Ocean. Univ.* **2022**, *2*, 479–490. (In Chinese)
41. Lee, K.H. Why and how to adopt green management into business organizations? The case study of Korean SMEs in manufacturing industry. *Manag. Decis.* **2009**, *47*, 1101–1121. [CrossRef]
42. Herron, C.; Hicks, C. The transfer of selected lean manufacturing techniques from Japanese automotive manufacturing into general manufacturing (UK) through change agents. *Robot. Comput. Manuf.* **2008**, *24*, 524–531. [CrossRef]
43. Kumar, S.; Luthra, S.; Haleem, A. Critical success factors of customer involvement in greening the supply chain: An empirical study. *Int. J. Logist. Syst. Manag.* **2014**, *19*, 283. [CrossRef]
44. Govindan, K.; Palaniappan, M.; Zhu, Q.; Kannan, D. Analysis of third party reverse logistics provider using interpretive structural modeling. *Int. J. Prod. Econ.* **2012**, *140*, 204–211. [CrossRef]
45. Kadam, S.; Bandyopadhyay, P.K. Modelling passenger interaction process (PIP) framework using ISM and MICMAC approach. *J. Rail Transp. Plan. Manag.* **2020**, *14*, 100171. [CrossRef]
46. Agarwal, R.; Shirke, A.; Panackal, N. Enablers of the collective bargaining in industrial relations: A study of India' s industrial policies through ISM and MIC.s industrial policies through ISM and MICMAC Analysis. *Indian J. Labour Econ.* **2020**, *63*, 781–798. [CrossRef]
47. Soner, O. Application of fuzzy dematel method for analyzing of accidents in enclosed spaces onboard ships. *Ocea Eng.* **2021**, *220*, 108507. [CrossRef]
48. Quiñones, R.S.; Caladcad, J.A.; Himang, C.M.; Quiñones, H.G.; Castro, C.J.; Caballes, S.A.; Abellana, D.P.; Jabilles, E.M.; Ocampo, L.A. Using Delphi and fuzzy DEMATEL for analyzing the intertwined relationships of the barriers of university technology transfer: Evidence from a developing economy. *Int. J. Innov. Stud.* **2020**, *4*, 85–104. [CrossRef]
49. Akhavan, A.; Ghatromi, A.R.; Azar, A. Mapping sustainable production model using ISM and fuzzy DEMATEl. *Ind. Manag. Stud.* **2017**, *15*, 1–26.
50. Singh, K.; Misra, M. Developing an agricultural entrepreneur inclination model for sustainable agriculture by integrating expert mining and ISM–MICMAC. *Environ. Dev. Sustain.* **2020**, *23*, 5122–5150. [CrossRef]
51. Wu, N.Z.; Xu, L. Analysis of factors influencing construction quality based on the improved explanatory structural model and cross-influence matrix multiplication. *Sci. Technol. Eng.* **2020**, *20*, 3222–3230. (In Chinese)
52. Xue, X.; Xin, C.; Xu, D.; Liu, T. Cause Analysis of Falling Accidents in Construction Engineering Based on Improved ISM-MICMAC. *J. Saf. Environment.* **2023**, 1–8. (In Chinese) [CrossRef]
53. Zhao, J.; Wang, D.; Xu, Y. Research on the Development of New Building Industrial Development Barrier Based on Fuzzy ISM-MICMAC Model. *J. Beijing Univ. J.* **2021**, *37*, 103–112. (In Chinese)
54. Ma, J.; Yao, B.; Li, Y. Smart manufacturing factory model. *Manuf. Autom.* **2019**, *41*, 24–36. (In Chinese)
55. Li, M.; Gan, N.F.; Yu, Y. Analysis of the construction of intelligent manufacturing demonstration factory. *Robot. Ind.* **2022**, *5*, 75–82. (In Chinese)
56. Virmani, N.; Sharma, V. Prioritisation and assessment of leagile manufacturing enablers using interpretive structural modelling approach. *Eur. J. Ind. Eng.* **2019**, *13*, 701. [CrossRef]
57. Opricovic, S.; Tzeng, G.-H. Compromise solution by MCDM methods: A comparative analysis of VIKOR and TOPSIS. *Eur. J. Oper. Res.* **2004**, *156*, 445–455. [CrossRef]

Article

Application of PLC-Based Spectrophotometric System Nitrogen Protection Device to Automated Direct Measurement of Target Substances in Zinc Hydrometallurgy

Xuefei Zhang [1], Ning Duan [1,2,3,*], Linhua Jiang [1,2,3,*], Fuyuan Xu [2,3], Zhaosheng Yu [4], Wen Cheng [2], Wenbao Lv [1] and Yibing Qiu [1]

[1] School of Materials Science and Engineering, Anhui University of Science and Technology, Huainan 232001, China
[2] State Key Laboratory of Pollution Control and Resources Reuse, College of Environmental Science and Engineering, Tongji University, Shanghai 200092, China
[3] Shanghai Institute of Pollution Control and Ecological Security, Shanghai 200092, China
[4] Tianjin Xinke Environmental Protection Technology Co., Ltd., Tianjin 300457, China
* Correspondence: ningduan2020@163.com (N.D.); jianglinhuann@163.com (L.J.); Tel.: +86-138-010-149-10 (N.D.); +86-132-693-671-32 (L.J.)

Citation: Zhang, X.; Duan, N.; Jiang, L.; Xu, F.; Yu, Z.; Cheng, W.; Lv, W.; Qiu, Y. Application of PLC-Based Spectrophotometric System Nitrogen Protection Device to Automated Direct Measurement of Target Substances in Zinc Hydrometallurgy. *Processes* **2023**, *11*, 672. https://doi.org/10.3390/pr11030672

Academic Editor: Sergey Y. Yurish

Received: 24 January 2023
Revised: 17 February 2023
Accepted: 20 February 2023
Published: 22 February 2023

Abstract: Due to the fast material reaction in zinc hydrometallurgy, the traditional national standard photometric method cannot capture the characteristic information of target substances in real time. Herein, a nitrogen protection device is built based on ultraviolet spectrophotometry, supplemented by a programmable logic controller (PLC), to form an automatic control system for the direct detection of target substances (SO_4^{2-}, Pb^{2+} and S^{2-}) in zinc hydrometallurgy. The baseline straightness comparison results show that the nitrogen atmosphere can effectively improve the stability of the instrument. Furthermore, the detection sensitivity of SO_4^{2-}, Pb^{2+} and S^{2-} under the nitrogen atmosphere is higher than that of the air atmosphere, manifesting in sensitivity increases of 16.23%, 18.05% and 17.91%, respectively. Additionally, devices based on PLC systems show advantages over manual control both in states feedback and information backtrack. Moreover, the regulation time and nitrogen consumption during the regulation process are reduced by 80% and 75%, respectively, which effectively reduces the test cost and improves the equipment utilization rate (from four cycles per day to six cycles per day). The device can meet the requirements of different target substances and different process conditions by changing the electronic control parts and air source, so it has great application potential in the automatic direct measurement of target substances in zinc hydrometallurgy.

Keywords: zinc hydrometallurgy; PLC; spectrophotometric system; nitrogen protection device; automated control

1. Introduction

Zinc is an important, indispensable economic pillar in the process of national production and construction [1]. It is also one of the most important metal elements in the world [2], and its consumption ranks fourth in global metal consumption [3–5]. According to incomplete statistics, more than 85% of the world's annual zinc production is produced by hydrometallurgy [6–8]. However, zinc hydrometallurgy will result in much pollution such as industrial wastewater, anode sludge and acid mist, which is caused by extensive management of substance concentration [9]. Based on the process management in the process industry, real-time monitoring of target substances provides a good idea for industrial cleaner production because this idea reduces the source pollution and also reduces the burden of end-of-pipe treatment [10]. Therefore, the real-time monitoring of the target substances is very important for the cleaner production of zinc hydrometallurgy.

Due to the general characteristics of the fast reaction speed of the liquid phase system in zinc hydrometallurgy, the current analysis of target substances in zinc hydrometallurgy mainly adopts the national standard spectrophotometric method. This method requires pretreatment of test samples in the early stage of detection, such as dilution, digestion [11,12], complexation and color development [13–15], which cannot obtain accurate photometric information of the target substance in real time within the reaction time of the material and leads to a serious lag in the detection results relative to the real-time state of the liquid phase system in zinc hydrometallurgy, making it difficult to realize real-time feedback and regulation of industrial process [16]. However, the key to realizing the rapid determination of liquid phase flow systems in the process industry is to accurately obtain the luminosity information of target substances at a specific wavelength. The preliminary study shows that UV spectrophotometry based on light absorption is an alternative technology for the direct measurement of liquid in zinc hydrometallurgy without pollution because of its simplicity, rapidity and environmental friendliness [17]. However, the characteristic absorption wavelength of many substances in zinc hydrometallurgy is located in the ultraviolet region. The detection of target substances in the ultraviolet region (SO_4^{2-} [18], S^{2-} [19], etc.) also faces interference caused by ultraviolet attenuation, which has been confirmed to be caused by ultraviolet absorption by oxygen [20,21]. Specifically, oxygen has strong absorption characteristics in the ultraviolet band and readily absorbs the wavelength <240 nm UV light. When oxygen absorbs ultraviolet light, the chemical bonds between oxygen molecules are broken and recombined to produce ozone molecules (Equations (1)–(3)). However, ozone molecules can also be destroyed by absorbing ultraviolet light in the 200~350 nm band (Equations (4)–(6)) [22]. Therefore, the reversible conversion between oxygen and ozone will not only consume UV energy (Equation (7)) [21] but also cause poor accuracy of the detection results of substances in the ultraviolet region [23]. One of the effective means to reduce the absorption of ultraviolet by oxygen is to adopt inert gases (such as nitrogen, argon, etc.) to isolate oxygen from ultraviolet.

$$O_2 + hv \rightarrow O(^3P) + O(^1D) \tag{1}$$

$$O(^3P) + O(^3P) + M \rightarrow O_2 + M \tag{2}$$

$$O(^3P) + O_2 + M \rightarrow O_3 + M \tag{3}$$

$$O_3 + hv\,(200 \sim 300\ nm) \rightarrow O_2 + O(^1D) \tag{4}$$

$$O_3 + hv\,(300 \sim 350\ nm) \rightarrow O_2 + O(^1D) \tag{5}$$

$$O_3 + hv\,(300 \sim 350\ nm) \rightarrow O_2 + O(^3P) \tag{6}$$

$$3O_2 \xleftrightarrow{UV\ light} 2O_3 \tag{7}$$

where O_2 is the oxygen. O_3 is the ozone. h is the Planck constant, $h = 6.626 \cdot 10^{-34}$ J·s. v is the frequency, Hz, s^{-1}. $O\,(^3P)$ is the ground -state oxygen atom. $O\,(^1D)$ is the electrically excited oxygen atom. M is the neutral third body, generally composed of nitrogen molecules or oxygen molecules, that are used for energy transfer but do not participate in chemical reactions.

Although UV spectrophotometry has great application potential in the field of direct measurement, at present, nitrogen replacement of oxygen is still done manually, resulting in a low degree of automation. A programmable logic controller (PLC) is a ruggedized computer used for industrial automation with strong programmability, high reliability, strong anti-interference ability, perfect functions and strong practicability, specifically designed for practical application in a complex industrial environment. Therefore, PLC is widely applied in electric power [24], energy [25], coal [26], chemical industry [27], machinery manufacturing [28], transportation [29], environmental protection [30], medicine [31] and other practical industrial environments. It is found that for the ultraviolet spectrophotometer, which is a medium and small scientific research instrument and equipment, there is no clear report on the relevant research of using PLC technology to carry out the automatic control of the atmosphere displacement of the instrument for the direct measurement of target substances.

Therefore, based on the ultraviolet spectrophotometer equipped with a nitrogen protection device, this work designs a control system for the direct-detection of the target substances in the zinc hydrometallurgy process industry. The system adopts PLC technology to avoid the interference of manual control with the nitrogen protection device. The stability of the instrument under two atmospheres is compared by baseline flatness. The applicability of manual control and PLC control in the direct measurement of zinc hydrometallurgy is evaluated by means of the test method, parameter setting, test condition, test result, state change and historical traceability. The economic benefits of manual control and PLC control are evaluated comprehensively regarding regulation time and nitrogen amount during the regulation period, test cycle and equipment utilization rate. In the end, the application potential of the whole apparatus in the direct measurement of different substances in multiple scenarios of the zinc hydrometallurgy industry is explored. In short, this study has realized the functions of the overall setting of control parameters, diverse detection methods, timely response to state changes, and historical information backtracking. It has the advantages of good stability, high control accuracy, low test cost, and wide application scenarios, and it can provide reliable case references for the application of PLC in the automatic control of small and medium-sized scientific research instruments and equipment.

2. Relevant Research

2.1. The Setting of Nitrogen Protection Device

The nitrogen protection device is built based on an ultraviolet spectrophotometer, as shown in Figure 1 [32]. In order to prevent the interference of dust and temperature and humidity changes in the environment to the operation of the instrument, and to stabilize the nitrogen atmosphere, a closed organic glass cover body (1200 mm × 850 mm × 437 mm) is installed outside the instrument to ensure the recycling of nitrogen in the instrument. The whole device is equipped with five mass flow control meters (M.F.C.) with different ranges: four hood air intakes (ø = 5 mm) are arranged at four corners of the hood bottom plate, and one mass flow controller controls the nitrogen flow into the hood (range 0~30 L/min). There is an optical system area, sample room and data -receiving area in the ultraviolet spectrophotometer. However, the distances ultraviolet rays travel in each region and the volume of each region are different. Therefore, the bottom plate of the hood body is also equipped with three -chamber air intakes, and the nitrogen flow into the optical system area, sample room and data receiving area is accurately controlled by three M.F.C. (ranges are: 0~10 L/min, 0~5 L/min and 0~5 L/min, respectively). The outlet flow is monitored by an M.F.C. (range 0~30 L/min) (All the M.F.C. models are S48-32/HMT, HORIBA Precision Instruments (Beijing) Co., Ltd., Beijing, China). Monitoring instruments such as the oxygen concentration detector and differential pressure meter and gas outlet of the device are located above the hood body.

In order to realize the automatic sample injection function under the sealed condition of the device, the sample injection tray and the flow cell of different specifications and size are designed, and the two ends of the flow cell are connected with inlet/outlet pipelines. During the testing process, the peristaltic pump (BT100-2J, Baoding Longer Precision Pump Co., Ltd., Baoding, China) drives the inlet pipeline forward to make the samples to be tested enter and fill the flow cell. In order to avoid bubbles in the flow cell caused by too fast injection speed, 30 r/min is selected as the injection speed in this experiment. After the detection, the peristaltic pump drives the liquid outlet pipe backward to discharge the liquid to be measured from the flow cell.

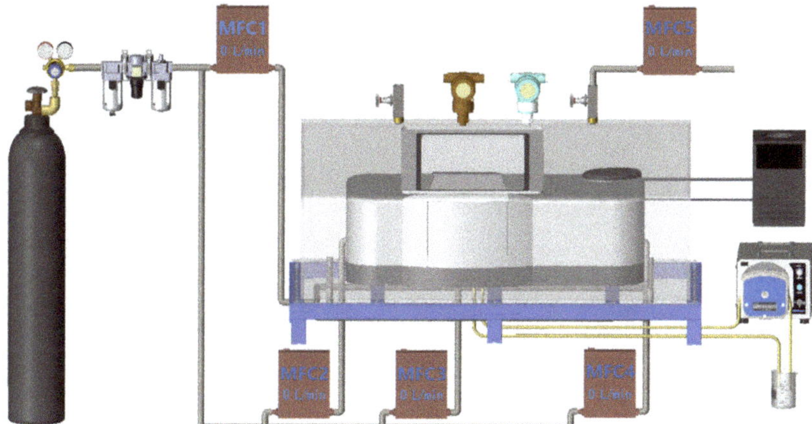

Figure 1. Manual operation diagram of the nitrogen protection device of the spectrophotometric system.

The whole device is also equipped with a chiller and a heat sink. The heat sink is installed on the back of the cover body. When the internal temperature of the cover body is too high, the cooling function can be achieved through the heat sink and the chiller.

2.2. PLC Logic Control

There are still technical limitations to nitrogen flow rate control and nitrogen amount control in the unit. Specifically, the existing nitrogen flow control is completed manually, resulting in low operating efficiency and frequent misoperation of the equipment. Therefore, the equipment is supplemented by a PLC system to achieve automatic control.

Figure 2 shows the block diagram of the control system with S7-1200 PLC as the core, which controls all the controls based on PLC signal transmission to realize the automatic control of the nitrogen protection device of the spectrophotometric system.

As can be seen from Figure 2, the whole system adopts 220 V AC (purple line) and 24 V DC (red line) power supply modes. The peristaltic pump, ultraviolet spectrophotometer and chiller use 220 V AC, while PLC, upper computer, intermediate relay and gas solenoid valve use 24V DC. Each intermediate relay controls a gas solenoid valve because the whole system contains multiple intermediate relays and gas solenoid valves, shown by ellipses. At the same time, the air switch and fuse can realize power failure protection when the internal current of the system is too large.

This system takes the S7-1200PLC as the core and is equipped with an extended 485 module for reading, receiving and transmitting communication signals (shown by the yellow line in Figure 2). Among them, 485-1 is used to monitor and synchronously read the instrument data of oxygen concentration, temperature and humidity in the current oxygen concentration detector. 485-2 transmits the control signal from PLC to the peristaltic pump to realize the control of start, positive and negative rotation, stop and rotation speed. In addition, the system is equipped with an extended analog input module SM1231 and an analog output module SM1232, which are used to feedback on the current flow of the M.F.C. and control the opening and closing of its valve (shown by green and blue lines in Figure 2).

2.3. Instrumentation

According to the contents of the design scheme and the specific requirements of the actual setting, the relevant components are selected and assembled. The names, models, quantities and manufacturers of the required components are shown in Table 1. Each of the control is arranged and combined according to its installation size and space area size, and the final overall structure of the electric control cabinet is shown in Figure S1.

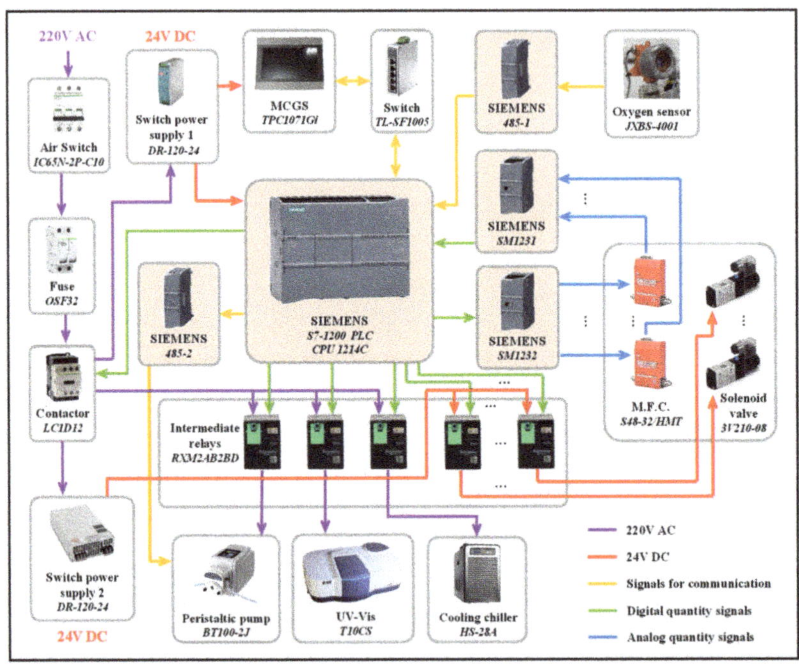

Figure 2. Block diagram of PLC control system.

Table 1. A detailed list of the internal control components in the electronic control cabinet.

Name	Type	Quantity	Unit	Manufacturer
PLC	S7-1200	1	Piece	SIEMENS AG of Germany, Berlin, Germany
485 Communication module	6ES7-241-1CH32-0XB0	2	Piece	SIEMENS AG of Germany, Berlin, Germany
Analog input module	SM1231	1	Piece	SIEMENS AG of Germany, Berlin, Germany
Analog output module	SM1232	1	Piece	SIEMENS AG of Germany, Berlin, Germany
Switchboard	TL-SF1005	1	Set	TP-LINK Technology Co., Ltd., Shenzhen, China
Air switch	IC65N-2P-C10	1	Set	Schneider Electric, Rue, France
Fuse	OSF32-2P-10A	1	Set	Schneider Electric, Rue, France
Contactor	LC1D12	1	Set	Schneider Electric, Rue, France
Intermediate relay	RXM2AB2BD	Some	Set	Schneider Electric, Rue, France
Wiring terminal	-	Some	Set	Phoenix Contact Electric Group, Bloomberg, Germany
Gas solenoid valve	3V210-08	5	Set	Airtac International Group, Taipei, Taiwan
Gas M.F.C.	S48-32/HMT	5	Set	HORIBA Precision Instruments (Beijing) Co., Ltd., Beijing, China
Switching power supply	DR-100-12	1	Set	MEAN WELL (Guangzhou) Electronics Co., Ltd., Guangzhou, China
Switching power supply	DR-120-24	1	Set	MEAN WELL (Guangzhou) Electronics Co., Ltd., Guangzhou, China

The whole PLC control program adopts a modular structure design, and the main program contains subroutine modules which include instrument communication, analog input, flow setting, optical path selection, peristaltic pump control, automatic sampling time control, alarm prompt, output control, etc. Figure S2 shows some program segments in the flow setting sub-module, which can complete the setting of the optimal flow value in the optical system area, sample room and data receiving area by the upper computer and transmit the setting value to the corresponding mass flow controller through the extension module of PLC. After the oxygen concentration, temperature, humidity and other parameters inside the hood meet the preset conditions, the PLC will execute relevant operations.

The MCGSPro TPC1071 Gi produced by Shenzhen Kunlun Tongtai Automation Software Technology Co., Ltd. (Shenzhen, China) is configured as the upper computer (The definition is: the computer issuing the operation command. Cortex A8 600 MHz processor, 128 MB memory). Five HMI (human–machine interface) interfaces are designed, including operation monitoring, optical path selection, parameter setting, manual control and alarm prompt. The whole system can be controlled remotely during the test, and the interface is easy to operate and readable. Supplementary Figures S3–S7 show the specific HMI interface.

Taking into account the requirements of logic design, hardware assembly and system construction, the three-layer framework is designed for placing the ultraviolet spectrophotometer, the organic glass hood, the upper computer, the peristaltic pump, the water chiller, the PLC electric control cabinet, the power supply, the test control terminal and other controls. The system renderings and physical drawings of the nitrogen protection device of the spectrophotometric system are shown in Figure S8 and Figure 3. The size of the entire system is 1500 mm × 1200 mm × 1350 mm. Meanwhile, the operation flow chart of the upper computer program of the whole device is shown in Figure S9.

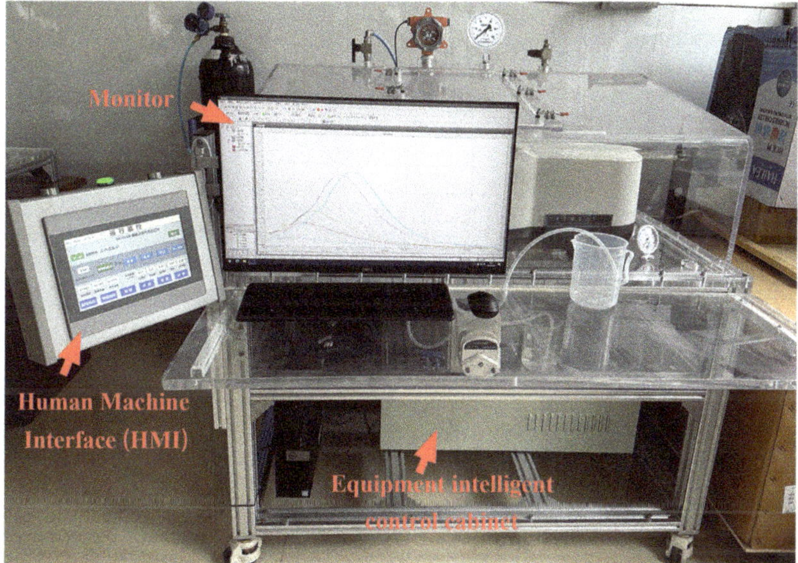

Figure 3. Spectrophotometric system nitrogen protection automation device physical picture.

3. Results and Discussion

3.1. Nitrogen Effect Verification

This section may be divided into subheadings. It should provide a concise and precise description of the experimental results, their interpretation, as well as the experimental conclusions that can be drawn.

Baseline flatness is a measure of the fluctuation noise of the environment itself when no samples are placed inside the instrument of an ultraviolet spectrophotometer [33]. The closed nitrogen atmosphere in this device is realized by controlling the numerical size of the M.F.C. in different ventilation stages by PLC. Therefore, the difference in the baseline flatness index under the two different atmospheres of air and nitrogen can be compared to verify the stability of the instrument before and after PLC control. Figure 4 shows the baseline flatness spectral curve under two atmospheres of air and nitrogen. The instrument parameters are set as follows: wavelength range of 180~360 nm, wavelength scanning interval of 1 nm, slow scanning speed, and continuous scanning six times without delay. As can be seen from Figure 4, the average baseline flatness value of the six scanning results under nitrogen atmosphere is 0.009 Abs, which attenuates by 92.11% compared with the average baseline flatness under air atmosphere (0.114 Abs). The significant difference in baseline straightness shown in Figure 4 is obviously attributed to the nitrogen atmosphere inhibiting the absorption of oxygen to UV and significantly reducing the luminosity noise of the spectral curve, which proves that the nitrogen atmosphere can effectively improve the stability of the detection system.

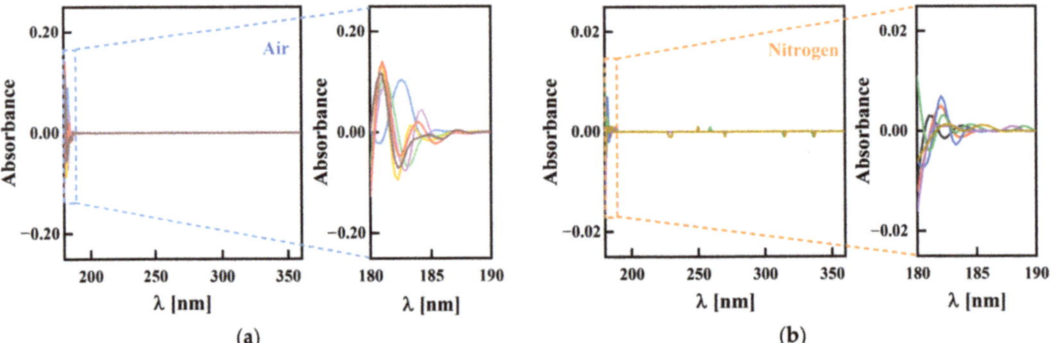

Figure 4. Results of 6 baseline straightness tests under (**a**) air and nitrogen (**b**) atmospheres.

The commonly used optical path of 10 mm is selected to obtain the spectral absorption curves of acid groups (SO_4^{2-}), metal lead ions (Pb^{2+}) and nonmetal sulfur ions (S^{2-}) in the ultraviolet region, respectively, under two atmospheres of air and nitrogen, as shown in Figure 5a,b,d,e,g,h. In the spectral curve, the absorbance at the spectral peak changes most significantly with the concentration. Therefore, the concentration-absorption peak intensity curve (C-A curve) is established according to the spectral curves of different concentrations of SO_4^{2-}, Pb^{2+} and S^{2-}, as shown in Figure 5c,f,i. Among them, SO_4^{2-} the solution is diluted with concentrated sulfuric acid (98%, AR) produced by Sinopharm Chemical Reagent Co., LTD. The Pb^{2+} solution and S^{2-} solution are both obtained by dissolving the $PbSO_4$ sample (AR) and $Na_2S \cdot 9H_2O$ sample (AR) produced by Shanghai Maclin Biochemical Technology Co., LTD. (Shanghai, China) in ultra-pure water.

It can be seen from Figure 5a,b,d,e,g,h that within a certain concentration range of SO_4^{2-}, Pb^{2+} and S^{2-}, the absorption intensity of SO_4^{2-}, Pb^{2+} and S^{2-} solutions under air and nitrogen atmosphere shows an increasing trend along with the increasing concentration of SO_4^{2-}, Pb^{2+} and S^{2-}, respectively, which is in line with Beer-Lambert law (Equations (8)–(10)) [33].

$$I_t = I_0 \cdot 10^{-A} = 10^{-\varepsilon(\lambda) \cdot b \cdot C} \tag{8}$$

$$A = lg\left(\frac{I_0}{I_t}\right) = lg\left(\frac{1}{T}\right) = -lg(T) = \varepsilon(\lambda) \cdot b \cdot C \tag{9}$$

$$k = \frac{dA}{dC} = \frac{\varepsilon(\lambda)b\Delta C}{\Delta C} = \varepsilon(\lambda) \cdot b \tag{10}$$

where I_t is the Intensity of transmitted light. I_0 is the Intensity of incident light. A is the response value of the instrument. T is the transmittance. $\varepsilon(\lambda)$ represents the molar absorption coefficient at a specific wavelength and is a characteristic constant. b is the thickness of the cuvette, that is, the optical path length. C is the concentration of the substance to be measured. k is the method sensitivity, i.e., the slope of the fitting curve.

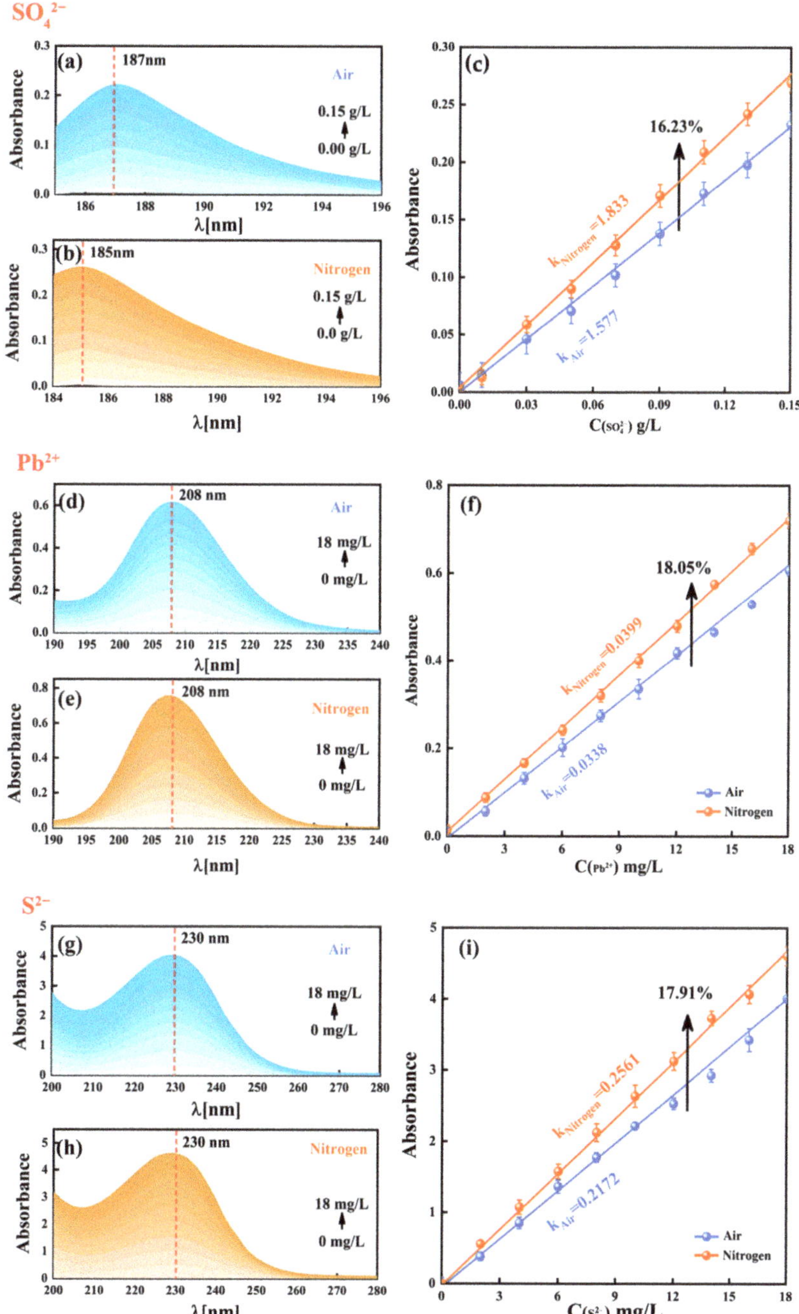

Figure 5. Spectral curves and C−A fitting curves of SO_4^{2-}, Pb^{2+} and S^{2-} in air and nitrogen atmosphere at b = 10 mm, where (**a**,**b**) are spectral curves of SO_4^{2-} in air and nitrogen atmosphere, (**c**) C−A fitting curves of SO_4^{2-}, (**d**,**e**) are spectral curves of Pb^{2+} in air and nitrogen atmosphere, (**f**) C−A fitting curves of Pb^{2+}, (**g**,**h**) are spectral curves of S^{2-} in air and nitrogen atmosphere, (**i**) C−A fitting curves of S^{2-}.

By comparison with Figure 5c, it can be found that when the concentration of SO_4^{2-} is the same, the absorption intensity value detected in the nitrogen atmosphere is always significantly higher than that in the air atmosphere, with the increase of the C-A curve slope of SO_4^{2-} under air and nitrogen atmospheres from 1.577 to 1.833, an increase of 16.23%. Similarly, the C-A curve slope obtained by Pb^{2+} and S^{2-} in the nitrogen atmosphere increases by 18.05% and 17.91%, respectively, compared with that in the air atmosphere (Figure 5f,i), which is attributed to the fact that the additional light attenuation caused by the absorption of ultraviolet light by oxygen is significantly inhibited by the nitrogen atmosphere. Therefore, under a given optical path and sample concentration, the detection process under nitrogen atmosphere shows higher sensitivity due to lower baseline noise and higher absorbance detection value compared with air atmosphere. The difference in fitted linear slope is attributed to the difference in absorbance of samples of a certain concentration under different atmospheres.

The above phenomenon should be attributed to the fact that the nitrogen atmosphere inhibits the absorption of ultraviolet by oxygen, and the reduced attenuation degree of ultraviolet leads to the absorption of ultraviolet with more light intensity by SO_4^{2-}, Pb^{2+} and S^{2-} solutions, which indicates that nitrogen atmosphere can provide a higher signal-to-noise ratio for substance detection and obtain more accurate detection results, and it is necessary to regulate nitrogen atmosphere environment [34,35] accurately.

3.2. Comparison of PLC Control and Manual Control

PLC plays an important role in strengthening the function of the nitrogen protection device. In order to demonstrate the superiority of PLC automatic control, the differences between PLC and manual control in detection mode, parameter setting, stability, accuracy and other aspects are compared. The detailed results are shown in Table 2.

Table 2. Comparison of manual control and PLC control.

Comparison Items	Manual Control	PLC Control
Detection methods	Single-step operation	Single-step operation/Continuous operation
Parameter setting flexibility	Multiple repetitive settings during detection	One-time set-up
Stability of detection conditions	Vulnerable to environmental changes	Stable
Accuracy of detection results	Large influence of human factors	Precise
Timeliness of status changes	Relying on human judgment	Timely feedback
Historical tracing possibilities	Not traceable	Traceable

Under manual control, the sample import, sample testing, sample recovery/waste discharge, cleaning and other steps need to be manually judged and completed, while under PLC control, it's possible to choose either completing in steps or completing all operations continuously, which greatly shortens the sample testing cycle and also reduces the misoperation. The operation monitoring interface of the upper computer in Figure S3 distinguishes the operation buttons controlled by manual control and PLC control. At the same time, Figure 6 also shows in detail the logic flow behind the automatic sample testing under PLC control. In the early stage of the test, the detection times are set according to the sample test cycle, and all samples can be completed periodically (8 min for a single sample test and 24 min for three repeated tests) through the control of the timing/delay module inside the PLC program (sample recovery and cleaning operation after each detection are set inside the program), which demonstrates the advantages of the PLC automation control in reducing manual intervention and significantly reducing system errors introduced by human operation.

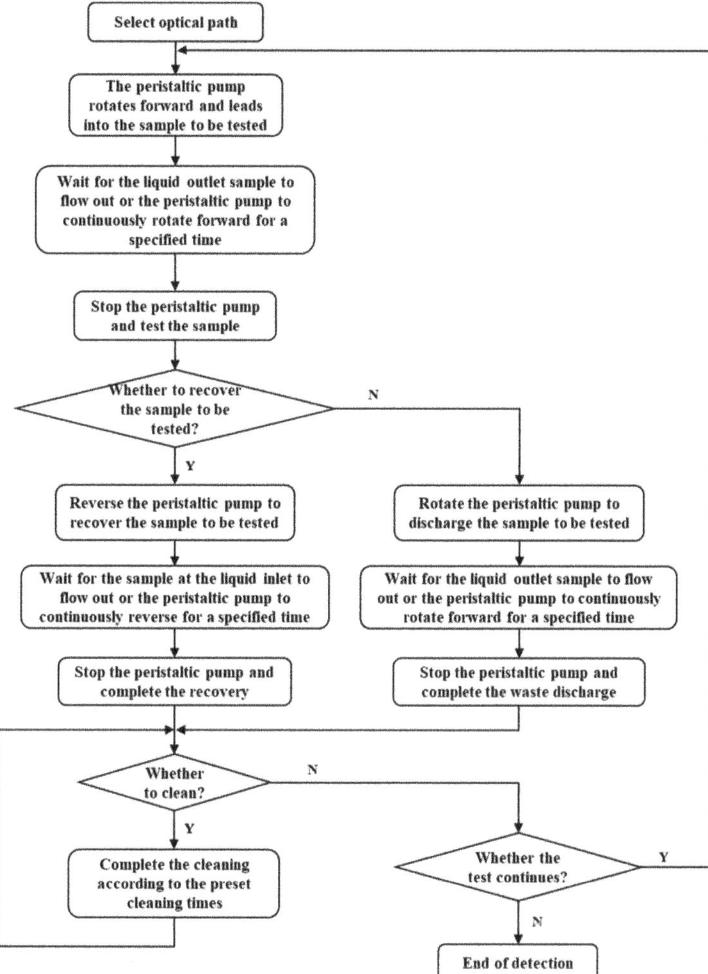

Figure 6. PLC logic control flow.

Compared to the multiple manual controls of the peristaltic pump speed, direction and M.F.C. to regulate the opening and closing of its valves, the PLC control of the parameters just needs to be set for one time in the early stages of the test and can be automatically called during the use. The parameter setting interface in Figures S4 and S5 can flexibly set parameters according to the actual needs, thus replacing the multiple repeated operations under manual control, which is the prerequisite for automation. At the same time, the parameters of each detection under PLC control remain constant, and the benchmark of multiple test parameters is consistent, which not only improves the stability of test conditions during the test but also greatly improves the accuracy compared with manual control.

Under manual control, the adjustment of the nitrogen flow rate is realized by rotating the adjusting knob of the M.F.C. and observing whether the real-time flow value on the M.F.C. flow display reaches the predetermined value. When the real-time flow value is consistent with the predetermined value, the valve of the M.F.C. reaches the specified opening and closing degree. However, the control of PLC simplifies the steps of regulating the flow display. The flow value set by the upper computer is output to the mass flow control meter through the analog output module SM1232 to directly regulates the opening and closing of the valve through the analog output module SM1232, as shown in Figure

S5. At the same time, the precision of PLC control has been improved to a certain extent compared with manual control. Take the full range of 30 L/min as an example, the flow indicator MT52 (display digit: $3\frac{1}{2}$, HORIBA Precision Instruments (Beijing) Co. Ltd., Beijing, China) is used for manual control to display the output, and its accuracy is 0.1 L/min. However, the accuracy of reading the mass flow control meter through SM1231 (conversion accuracy: 13 bits) can reach 0.00366 L/min. Under manual control, it is necessary to observe and monitor the current state of instruments, M.F.C. and peristaltic pump controls, while PLC is to send instructions to the controlled device and receive feedback signals, and can timely collect the current state information of the controlled device and the environmental changes. Therefore, when the detection environment or the state of the controlled components meets the specified conditions, PLC can start the next operation in time, which is more timely and significantly more accurate compared to human observation.

According to the previous research results, in order to make nitrogen quickly replace the internal air in the system in a short time, the replacement is divided into two stages. The air inside the hood is quickly replaced with a large flow rate of 30 L/min, and the air inside the optical system area, sample room and data receiving area of the ultraviolet spectrophotometer is quickly replaced with an optimal flow rate of 6, 2 and 3 L/min respectively. At the same time, nitrogen in the detection stage is injected into the optical system area, sample room and data-receiving area at a micro-flow rate of 0.6 L/min. The operating status of each control inside the device can be clearly and accurately understood through the human-computer interaction interface. Compared with artificial monitoring under manual control, the automatic state switching of PLC has been improved to a certain extent. Figure S6 shows the operating status of the M.F.C. under different states.

In addition, the upper computer under PLC control has the function of recording, storing and backtracking the current status, alarm and other information, and can provide an inspection basis for the identification of causes of abnormal test results, such as the abnormal test results caused by high temperature without alarm, high humidity and insufficient nitrogen gas source, as shown in Figure S7, which is a major innovation and improvement compared with manual control.

To sum up, PLC control has good advantages over manual control in detection mode, parameter setting, stability, accuracy, timeliness and history traceability, which can effectively improve the stability of detection conditions and improve the accuracy of detection results.

3.3. Economic Benefits

In order to better demonstrate the application effect of PLC technology in the nitrogen protection device of the spectrophotometric system, three indicators such as the required time of regulation, the amount of nitrogen in the regulation process and the single test cycle are selected to compare the differences between manual control and PLC control. The comparison results are shown in Table 3.

Table 3. Manual regulation vs. Integrated PLC control.

Comparison Items	Manual Regulation	Integrated PLC Control
Large flow adjustment/s	6	1
Optimal flow adjustment/s	9	1
Micro flow adjustment/s	12	1
Stability time of M.F.C./s	3	3
Sum of regulation time/s	30	6
Sum of nitrogen amount/L·min^{-1}	5.72	1.43

According to the above analysis, there are three adjustment times in total for the 1~4 M.F.C. under manual control. In addition, either manual regulation or PLC integrated control will take 1 s to reach a stable output after each regulation of the M.F.C. Therefore, the regulation time calculated in Table 3 is the sum of the three regulation times plus the time from signal reception to the stability of the M.F.C.

In which the sum of the stability time and nitrogen amount of the M.F.C. is calculated as follows:

$$\text{Stability time of M.F.C. (3 s)} = \text{time required to reach stable output} \\ ((1\ \text{s}) \cdot \text{regulation times 3 times}) \tag{11}$$

$$\text{Total nitrogen consumption} = \text{nitrogen flow rate at each stage} \cdot \\ (\text{regulation time at each stage} + \text{stability time of M.F.C.}) \tag{12}$$

Take manual control as an example:

$$\frac{(6+1)\cdot 30}{60} + \frac{(9+1)\cdot(6+2+3)}{60} + \frac{(12+1)\cdot(0.6+0.6+0.6)}{60} = 5.72\ \text{L/min} \tag{13}$$

Similarly, PLC integrated control is calculated in the same way.

As can be seen from Table 3, compared with manual regulation, the regulation time of PLC automatic control is reduced from 30 s to 6 s, reducing the control time by 80%. On the one hand, the sum of nitrogen consumption during the regulation period is also reduced from 5.72 L/min to 1.43 L/min, saving 75% nitrogen consumption, with obvious effects of nitrogen saving. On the other hand, nitrogen consumption caused by inaccurate operation during manual control is reduced. The automatic control of PLC not only reduces the degree of fatigue in the repetitive manual operation, decreases the test error introduced due to human reasons during the regulation period, but also reduces the nitrogen consumption during the regulation period and saves the detection cost.

A single test cycle represents the time it takes for the operator to turn on the instrument until the entire sample test is completed. The test counts the single test cycle that measures the same batch of samples under manual control and PLC control. According to the daily working time of the instrument of 12 h, the single test cycle under manual control takes about 2.8~3.2 h, and four cycles of sample detection can be completed every day. Regardless of the interval time between operators, the single test cycle can be shortened to 1.8~2.0 h after PLC control, and the instrument allows the completion of 6 cycles of sample detection every day. As a result, the efficiency of the instrument is improved.

Combined with the above analysis, the comparison results of the four indexes, including time required for regulation under manual control and PLC control, nitrogen amount in the regulation process, single test cycle and instrument use efficiency, are recorded in Figure 7. As can be seen from Figure 7, compared with manual control, PLC control shows good advantages and better reflects the PLC control of fast response time, good stability, and high -reliability advantages, which can provide a good detection environment for the accurate determination of target substances, reduce testing costs, and improve economic benefits. It can be predicted that with the continuous development of automation degree in modern society, automatic control based on PLC will play a greater role in more fields in the future social development.

The advantages of the PLC-based nitrogen protection device of the spectrophotometric system compared with manual control in stability, accuracy and economic benefits are analyzed above. In addition, the device can also be expanded from the following points:

(1) Because S7-1200 PLC itself has a certain expansion function, it can reasonably add other controls according to the requirements of the zinc hydrometallurgy industrial system, such as an automatic sampling device, long-distance transmission and distribution system, acid mist sensor and others, in order to meet the actual working conditions of the detection; namely, the device has high flexibility, strong scalability.

(2) The light source applied in this device is not limited to the combination of a deuterium lamp and tungsten lamp inside the instrument and can be replaced in accordance with the band range of the target substance to realize the qualitative and quantitative analysis of different target substances in different band regions. Taking a zinc hy-

drometallurgy electrolysis system containing a certain amount of Mn^{2+}, for example, the characteristic wavelength of Mn^{2+} is located at 401 nm in the visible region. In order to quantitatively study the influence degree of Mn^{2+} concentration on the reaction process, the quantitative relationship between the absorbance and concentration of Mn^{2+} can be directly established after the replacement of 401 nm single point light source, which simplifies the operation procedure and shortens the measurement time, timely feeding back the change rule of manganese element in the electrolyte, and provides a basis for accurately regulating the mass balance in the electrolyte.

(3) The automatic control terminal of zinc hydrometallurgy is established based on PLC to feed back the real-time dynamic monitoring results of the target substances. According to the results, the dosing sequence and amount of the materials are reasonably controlled, thus truly controlling the ion network of the whole production process.

(4) Currently, the nitrogen protection device of the spectrophotometric system established in this paper is being used in the laboratory. It can accurately determine target substances such as SO_4^{2-}, S^{2-}, Pb^{2+}, F^- and Cl^- whose characteristic absorption wavelength is located in the ultraviolet region. Since the characteristic absorption wavelengths of many target substances in the electrolysis industry are located in the ultraviolet region, they are susceptible to oxygen interference in the air. Therefore, the establishment of a nitrogen protection device in the spectrophotometric system can provide research and development ideas for online monitors, promote the application of nitrogen protection devices in online monitors, provide reliable technical support for the application of online monitors in more process industries, and facilitate the sustainable development and cleaner production of process industries.

(5) Due to various abnormal conditions that may occur during actual use, such as the peristaltic pump, water chiller and other actuators failing to respond to the instructions issued by PLC, the system expansion of the whole device can be carried out to increase the monitoring of the operating status of the actuator, and timely respond to the abnormal conditions, so as to ensure the normal operation of the whole device.

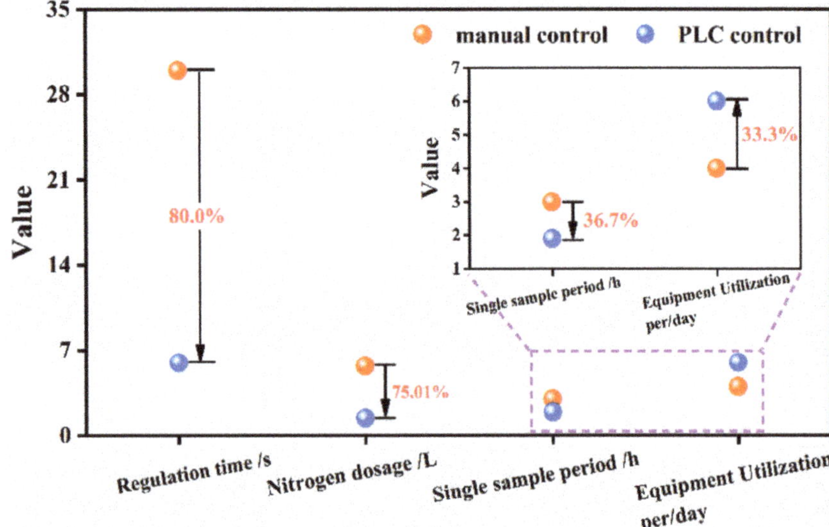

Figure 7. Effect of PLC control vs. manual control.4 Exploitable Potential.

To sum up, the research in this paper is not the end point. The multi-disciplinary crossing is the trend of rapid development of modern society, which is also the key and difficult point of continuous breakthrough in the future scientific research field.

4. Conclusions

In order to quickly obtain the characteristic spectral information of the target substance in real time in the process of zinc hydrometallurgy, the paper establishes an ultraviolet spectrophotometer equipped with a nitrogen protection device and PLC system. The PLC system includes PLC, upper computer, electric control cabinet, peristaltic pump, oxygen concentration detector, M.F.C., solenoid valve and other controls. Compared with air, the baseline flatness under the nitrogen atmosphere is reduced from 0.114 Abs to 0.009 Abs (attenuation of 92.11%), indicating that the nitrogen atmosphere can effectively improve the stability of instrument operation. The sensitivity of $SO_4{}^{2-}$, Pb^{2+} and S^{2-} detection results is better than that of air atmosphere, with the sensitivity increasing by 16.23%, 18.05% and 17.91%, respectively. However, the ultraviolet spectrophotometer based on manual control still faces the issues of operating accuracy and economic benefit in the application of real-time substance monitoring. Compared with manual control, the PLC system realizes continuous sample testing through logic programming control of electric control accessories, improves the flexibility of parameter setting, stability of detection conditions, the accuracy of test results and timeliness of status feedback, and especially provides the function of historical information backtracking, demonstrating significant advantages of automatic control. At the same time, the regulation time and nitrogen consumption in the regulation process are reduced by 80% and 75%, respectively, which improves the efficiency of the instrument (from 4 cycles/day to 6 cycles/day) and can effectively reduce the test cost and reflect the advantages of fast response time, good stability and high reliability of automatic control. In addition, some development potential and application scenarios of the device are prospected. Therefore, this study provides theoretical guidance and a case reference for the application of a UV spectrophotometer in the automatic direct measurement of substances in zinc hydrometallurgy and promotes the sustainable development and cleaner production of the process industry.

Supplementary Materials: The following supporting information can be downloaded at: https://www.mdpi.com/article/10.3390/pr11030672/s1, Figure S1: Internal physical picture of electric control cabinet; Figure S2: Current test environment detection procedure. Figure S3: Operation monitoring interface. Figure S4: Optical path selection interface. Figure S5: Parameter setting interface. Figure S6: Manual control interface under different conditions, including (a) nitrogen, replaces the air inside the hood at a high flow rate of 30 L/min; (b) Nitrogen displaces the air inside the light source system area, sample room and data receiving area of the instrument at the optimal flow rate of 6, 2 and 3 L/min; (c) At the detection stage, the nitrogen flow is 0.6 L/min. Figure S7: Alarm prompt interface. Figure S8: The system rendering of the device. Figure S9: Upper computer program operation flow chart.

Author Contributions: Conceptualization, X.Z.; methodology, X.Z. and N.D.; writing–original draft, X.Z.; data curation, X.Z.; visualization, X.Z.; supervision, N.D., L.J. and F.X.; funding acquisition, N.D.; project administration, N.D. and L.J.; writing–review and editing, N.D.; resources, L.J. and F.X.; investigation, X.Z., Z.Y., W.C., W.L. and Y.Q. All authors have read and agreed to the published version of the manuscript.

Funding: This work is supported by the Fundamental Research Funds for the Central Universities, Tongji University (No. 22120220166) and National Major Independent Project of Water Pollution Control and Treatment of the "13th Five-Year Plan": Whole Process Control Technology Integration and Engineering Demonstration of Water Pollution in Key Industries (No. 2017ZX07402004).

Data Availability Statement: The data presented in this study are available on request from the corresponding authors. The data are not publicly available because of continuous research.

Acknowledgments: This article will be supported by Anhui University of Science and Technology and Tongji University.

Conflicts of Interest: The authors declare no conflict of interest.

References

1. Xue, Y.; Hao, X.; Liu, X.; Zhang, N. Recovery of Zinc and Iron from Steel Mill Dust—An Overview of Available Technologies. *Materials* **2022**, *15*, 4127. [CrossRef]
2. Terrones-Saeta, J.M.; Suárez-Macías, J.; Moreno-López, E.R.; Corpas-Iglesias, F.A. Leaching of Zinc for Subsequent Recovery by Hydrometallurgical Techniques from Electric Arc Furnace Dusts and Utilisation of the Leaching Process Residues for Ceramic Materials for Construction Purposes. *Metals* **2021**, *11*, 1603. [CrossRef]
3. Ye, W.; Xu, F.; Jiang, L.; Duan, N.; Li, J.; Zhang, F.; Zhang, G.; Chen, L. A novel functional lead-based anode for efficient lead dissolution inhibition and slime generation reduction in zinc electrowinning. *J. Clean. Prod.* **2021**, *284*, 124767. [CrossRef]
4. Luo, J.; Duan, N.; Xu, F.; Jiang, L.; Zhang, C.; Ye, W. System-level analysis of the generation and distribution for Pb, Cu, and Ag in the process network of zinc hydrometallurgy: Implications for sustainability. *J. Clean. Prod.* **2019**, *234*, 755–766. [CrossRef]
5. Qin, S.; Jiang, K.; Wang, H.; Zhang, B.; Wang, Y.; Zhang, X. Research on Behavior of Iron in the Zinc Sulfide Pressure Leaching Process. *Minerals* **2020**, *10*, 224. [CrossRef]
6. Rybarczyk, P.; Kawalec-Pietrenko, B. Simultaneous Removal of Al, Cu and Zn Ions from Aqueous Solutions Using Ion and Precipitate Flotation Methods. *Processes* **2021**, *9*, 1769. [CrossRef]
7. Ye, W.; Xu, F.; Jiang, L.; Duan, N.; Li, J.; Ma, Z.; Zhang, F.; Chen, L. Lead release kinetics and film transformation of Pb-MnO$_2$ pre-coated anode in long-term zinc electrowinning. *J. Hazard. Mater.* **2021**, *408*, 124931. [CrossRef]
8. Priyadarshini, J.; Elangovan, M.; Mahdal, M.; Jayasudha, M. Machine-Learning-Assisted Prediction of Maximum Metal Recovery from Spent Zinc—Manganese Batteries. *Processes* **2022**, *10*, 1034. [CrossRef]
9. Ma, Z.; Duan, L.; Jiang, J.; Deng, J.; Xu, F.; Jiang, L.; Li, J.; Wang, G.; Huang, X.; Ye, W.; et al. Characteristics and threats of particulate matter from zinc electrolysis manufacturing facilities. *J. Clean. Prod.* **2020**, *259*, 120874. [CrossRef]
10. Duan, N.; Jiang, L.; Xu, F.; Zhang, G. A non-contact original-state online real-time monitoring method for complex liquids in industrial processes. *Engineering* **2018**, *4*, 392–397. [CrossRef]
11. Lee, J.; Park, Y.-S.; Lee, H.-J.; Koo, Y.E. Microwave-assisted digestion method using diluted nitric acid and hydrogen peroxide for the determination of major and minor elements in milk samples by ICP-OES and ICP-MS. *Food Chem.* **2022**, *373*, 131483. [CrossRef]
12. Ju, T.; Han, S.; Meng, Y.; Song, M.; Jiang, J. Occurrences and patterns of major elements in coal fly ash under multi-acid system during microwave digestion processes. *J. Clean. Prod.* **2022**, *359*, 131950. [CrossRef]
13. Yang, R.; Tian, J.; Liu, Y.; Zhu, L.; Sun, J.; Meng, D.; Wang, Z.; Wang, C.; Zhou, Z.; Chen, L. Interaction mechanism of ferritin protein with chlorogenic acid and iron ion: The structure, iron redox, and polymerization evaluation. *Food Chem.* **2021**, *349*, 129144. [CrossRef]
14. Singh, K.; Kumar, A. Kinetics of complex formation of Fe(III) with caffeic acid: Experimental and theoretical study. *Spectrochim. Acta Part A.* **2019**, *211*, 148–153. [CrossRef]
15. Singh, K.; Kumar, A. Kinetics of complex formation of Fe(III) with syringic acid: Experimental and theoretical study. *Food Chem.* **2018**, *265*, 96–100. [CrossRef]
16. Du, J.; Zhu, H.; Li, Y.; Zhang, T.; Yang, C. Simultaneous determination of trace Cu^{2+}, Cd^{2+}, Ni^{2+} and Co^{2+} in zinc electrolytes by oscillopolarographic second derivative waves. *Trans. Nonferrous Met. Soc. China* **2018**, *28*, 2592–2598. [CrossRef]
17. Cheng, W.; Zhang, X.; Duan, N.; Jiang, L.; Xu, Y.; Chen, Y.; Liu, Y.; Fan, P. Direct-determination of high-concentration sulfate by serial differential spectrophotometry with multiple optical pathlengths. *Sci. Total Environ.* **2022**, *811*, 152121. [CrossRef]
18. Ai, X.; Zhang, Z. Quantitative measurement model for sulfate in the temperature range of 298.15–343.15K based on vacuum ultraviolet absorption spectroscopy. *Vacuum* **2022**, *200*, 111035. [CrossRef]
19. Szczuka, A.; Berglund-Brown, J.P.; Macdonald, J.A.; Mitch, W.A. Control of sulfides and coliphage MS$_2$ using hydrogen peroxide and UV disinfection for non-potable reuse of pilot-scale anaerobic membrane bioreactor effluent. *Water Res. X.* **2021**, *11*, 100097. [CrossRef]
20. Krish, A.; Streicher, J.W.; Hanson, R.K. Ultraviolet absorption cross-section measurements of shock-heated O$_2$ from 2000~8400 K using a tunable laser. *J. Quant. Spectrosc. Radiat. Transf.* **2020**, *247*, 106959. [CrossRef]
21. Wang, L.; Zhang, Y.; Zhou, X.; Zhang, Z. Sensitive dual sensing system for oxygen and pressure based on deep ultraviolet absorption spectroscopy. *Sens. Actuators B.* **2019**, *281*, 514–519. [CrossRef]
22. Wang, M.; Zheng, X. *Introduction to Atmospheric Chemistry*; China Meteorological Press: Beijing, China, 2005.
23. Ai, X. *Ultraviolet Absorption Spectroscopy Properties of Inorganic Anions Such as Sulphate and Nitrate*; Harbin Institute of Technology: Harbin, China, 2019.
24. Pongswatd, S.; Smerpitak, K.; Asadi, F.; Thepmanee, T. Design of PLC-based system for linearity output voltage of AC–DC converter. *Energy Rep.* **2022**, *8*, 972–978. [CrossRef]
25. Ting, C.; Lai, C.; Huang, C. Developing the dual system of wind chiller integrated with wind generator. *Appl. Energy* **2011**, *88*, 741–747. [CrossRef]
26. Kazagic, A.; Smajevic, I. Experimental investigation of ash behavior and emissions during combustion of Bosnian coal and biomass. *Energy* **2007**, *32*, 2006–2016. [CrossRef]
27. Aller, F.; Blázquez, L.F.; Miguel, L.J. Online monitoring of an industrial semi-batch vinyl acetate polymerization reaction by programmable logic controllers. *IFAC Proc. Vol.* **2014**, *47*, 1290–1295. [CrossRef]

28. Anusha, R.; Chandrashekar Murthy, B.N. Automatic trimming machine for valve stem seal. *Mater. Today Proc.* **2021**, *46*, 4993–5000. [CrossRef]
29. Aziz, N.; Tanoli, S.A.K.; Nawaz, F. A programmable logic controller based remote pipeline monitoring system. *Process Saf. Environ. Prot.* **2021**, *149*, 894–904. [CrossRef]
30. Manesis, S.A.; Sapidis, D.J.; King, R.E. Intelligent control of wastewater treatment plants. *Artif. Intell. Eng.* **1998**, *12*, 275–281. [CrossRef]
31. Güler, H.; Ata, F. Design and Implementation of Training Mechanical Ventilator Set for Clinicians and Students. *Proc.—Soc. Behav. Sci.* **2013**, *83*, 493–496. [CrossRef]
32. Zhang, X.; Duan, N.; Jiang, L.; Cheng, W.; Yu, Z.; Li, W.; Zhu, G.; Xu, Y. Study on stability and sensitivity of deep ultraviolet spectrophotometry detection system. *Spectrosc. Spectral Anal.* **2022**, *42*, 3802–3810.
33. University Wuhan. *Analytical Chemistry Volume II*, 5th ed.; Higher Education Press: Beijing, China, 2007.
34. Chen, G.; Huang, X.; Liu, W.; Zheng, Z.; Wang, Z. *Ultraviolet-Visible Spectrophotometry*; Atomic Energy Press: Beijing, China, 1983; Volume I.
35. Zhou, X.; Yu, J.; Wang, L.; Gao, Q.; Zhang, Z. Sensitive detection of oxygen using a diffused integrating cavity as a gas absorption cell. *Sens. Actuators B* **2017**, *241*, 1076–1081. [CrossRef]

processes

Article

Structural Optimization of High-Pressure Polyethylene Cyclone Separator Based on Energy Efficiency Parameters

Baisong Hu [1], Shuo Liu [1], Chuanzhi Wang [2] and Bingjun Gao [1,*]

1   School of Chemical Engineering and Technology, Hebei University of Technology, Tianjin 300130, China
2   Beijing Yanhua Engineering Construction Company, Beijing 102502, China
*   Correspondence: bjgao@hebut.edu.cn

Abstract: The high-pressure polyethylene process uses cyclone separators to separate ethylene gas, polyethylene, and its oligomers. The oligomers larger than 10 microns that cannot be separated must be filtered through a filter to prevent them from entering the compressor and affecting its normal operation. When the separation efficiency of the cyclone separator is low, the filter must be cleaned more frequently, which will reduce production efficiency. Research shows that improving the separation efficiency of the separator is beneficial for the separation of small-particle oligomers and reduces the frequency of filter cleaning. For this reason, Computational Fluid Dynamics simulations were performed for 27 sets of cyclone separators to determine the effects of eight structural factors (cylinder diameter, cylinder height, cone diameter, cone height, guide vane height, guide vane angle, exhaust pipe extension length, and umbrella structure height) on separation efficiency and pressure drop. The equations for separation efficiency and pressure drop using these eight factors and the equations based on energy-efficiency parameters were determined. The optimization analysis showed that separation efficiency can be improved by 98.7% under the premise that the pressure drop is only increased by 8.2%. By applying the improved structure to the high-pressure polyethylene process, separation efficiency is increased by 17.7%, which could effectively reduce the frequency of filter cleaning for this process, and thereby greatly improve production efficiency.

Keywords: high-pressure polyethylene; cyclone separator; computational fluid dynamics; design of experiment; separation efficiency; pressure drop

Citation: Hu, B.; Liu, S.; Wang, C.; Gao, B. Structural Optimization of High-Pressure Polyethylene Cyclone Separator Based on Energy Efficiency Parameters. *Processes* **2023**, *11*, 691. https://doi.org/10.3390/pr11030691

Academic Editor: Sergey Y. Yurish

Received: 7 February 2023
Revised: 18 February 2023
Accepted: 20 February 2023
Published: 24 February 2023

1. Introduction

In the production of high-pressure polyethylene, the main processes are divided into ethylene compression, polymerization, separation, granulation, mixing and air delivery processing, packaging and palletizing, and other processes. The corresponding process flow is shown in Figure 1. The conversion rate of ethylene is about 20%. After the ethylene is compressed and polymerized, the polyethylene/ethylene needs to be separated. The mixture enters the high-pressure and low-pressure separation systems, and the separated polyethylene is made into products after granulation, mixing, air delivery, processing and packaging and palletizing. The ethylene obtained from the high-pressure separation process is mixed and filtered with fresh ethylene and then enters the secondary compressor of the compression system for recycling.

The process of the high-pressure separation system is shown in Figure 1. The polyethylene/ethylene mixture from the reactor is first separated from the polyethylene product via separator A, the oligomer enters the high-pressure primary separators B and C, and then it enters the secondary separation coolers $E_{1\sim4}$ and separators $D_{1\sim4}$ in turn with the ethylene. The separation process is divided into three stages. First, the mixture enters separator A through the a at 29 MPa and 230 °C to separate the polyethylene from the unreacted ethylene. By this means, 91 to 94% of the polyethylene can be separated. The separated polyethylene enters the low-pressure separation system through the n. The unseparated oligomers are entrained in the form of mist entrainment or a homogeneous phase through b and c in series to the high-pressure primary separators B and C, which separate the larger oligomers. Subsequently, the fine oligomers that cannot be separated enter the coolers $E_{1\sim4}$ and separators $D_{1\sim4}$ in series in the order of $d{\rightarrow}E_1{\rightarrow}e{\rightarrow}D_1{\rightarrow}f{\rightarrow}E_2{\rightarrow}g{\rightarrow}D_2{\rightarrow}h{\rightarrow}E_3{\rightarrow}i{\rightarrow}D_3{\rightarrow}j{\rightarrow}E_4{\rightarrow}k{\rightarrow}D_4$ for cooling and separation, in which some homogeneous substances will change into two-phase states of gas and liquid phases in the coolers when the temperature decreases, which will be separated in the separators. The oligomers separated in separators B, C, and $D_{1\sim4}$ are stored in the discharge tank after o, p, q, r, s and t. The final oligomers that fail to be separated will be filtered through l in the mixing filter F to remove oligomers of 10 microns and larger to prevent larger oligomer particles from entering the secondary compressor and causing damage. The filter needs to be cleaned regularly, and when the separation efficiency of the separator is low, the frequency of filter cleaning increases, which is very inconvenient for the ultra-high-pressure production device. For this reason, it is necessary to improve the separation efficiency of the separation system. Research shows that droplets with a large particle size are easy to separate in gas-liquid separation, while droplets with a small particle size are difficult to separate [1–3]. Therefore, separators $D_{1\sim4}$ are the key to the separation of small-particle oligomers.

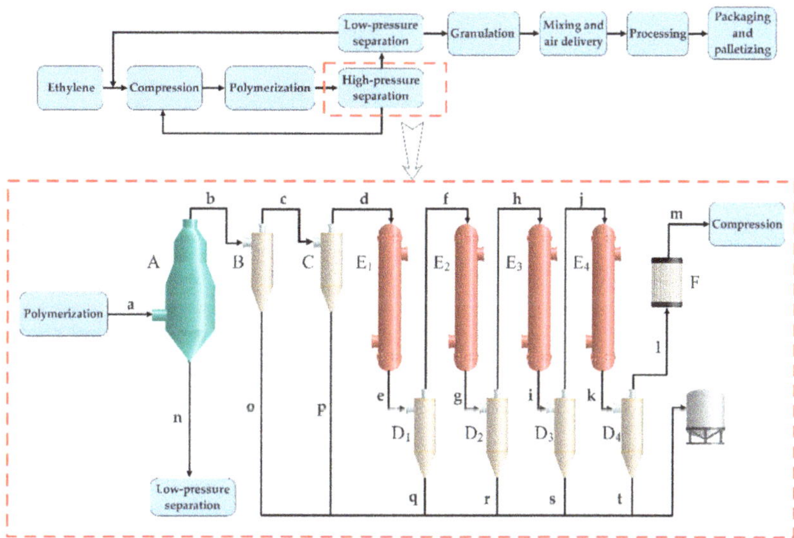

Figure 1. High-pressure polyethylene process flow and high-pressure separation system.

The $D_{1\sim4}$ separators each use cyclone separators, a kind of centrifugal separator, which use the centrifugal force field generated by strong cyclones to achieve relative movement of mixed materials with different densities to achieve the purpose of gas-liquid separation. This system is advantageous due to its compact structure, low volume capacity, low cost, and ability to cope with high temperature and high pressure [4–6].

The cyclone separator has no moving parts and its overall form is simple, but the internal flow field is very complex, and the dimensions between the various structures are very closely matched and mutually constrained. In recent decades, a great deal of experimental and theoretical research has been carried out in order to make cyclones smaller and more compact in size, more efficient in separating materials, and more energy-efficient [7–10]. However, it is very difficult to test the efficiency of a high-pressure cyclone separator with a real cyclone separator model. Many scholars have struggled to derive a suitable mathematical model to study and predict the flow behavior and to find a suitable cyclone geometry. Therefore, due to the complexity of cyclone separators, no accurate mathematical model of the cyclone separator has been proposed yet. Computational Fluid Dynamics (CFD), as an effective numerical method for calculating complex flow, has been widely used in the study of cyclone separators [11]. In recent years, scholars have conducted a large number of studies on various cyclonic separators using CFD [12–18]. In CFD, the selection of the turbulence model is very important. For the calculation of swirling flow, the existing turbulence models used are mainly the standard k-ε model, RNG k-ε model, Reynolds Stress Model (RSM), and Large Eddy Simulation (LES)—and only the RSM and LES can simulate the main characteristics of highly complex vortex flow in the swirling flow field. Many scholars have studied which turbulence model is most suitable for cyclone separators [14,19]. Among various turbulence models, RSM is the best model for predicting airflow turbulence and simulating the flow pattern inside the cyclone separator [20–22], because it ignores the assumption of isotropic flow. The results from this turbulence model were in good agreement with the experimental data [23].

The main evaluation indices of cyclone separator performance are separation efficiency and pressure drop [24]. Separation efficiency directly determines productivity and is the first indicator used to evaluate performance; pressure drop is an indicator used to evaluate the energy dissipation rate of the cyclone separator. When the pressure of the gas flowing from the inlet to the outlet of the cyclone separator decreases, pressure drop occurs in the cyclone separator. The reason for this pressure drop is that when the mixture hits or makes contact with the guide vanes, walls, and other components of the cyclone separator, the pressure decreases due to friction and resistance. The geometry of the cyclone separator has a significant effect on its performance. Various scholars have studied the effect of geometric parameters on the performance of cyclone separators [5,25,26]. The results show that there are many secondary flows inside the cyclone separator, such as short-circuit flow, circulating flow, and back-mixing flow at the bottom of the straight pipe section [27,28]. In CFD simulations, Elsayed, K. and Lacor, C. found that changing the extension length of the exhaust pipe has a significant effect on the pressure drop and separation efficiency of the cyclone separator [29]. Hamdy, O. and Shastri, R. et al. discovered that the cone length and cone angle had a significant effect on the flow pattern inside the cyclone separator [30,31]. Misiulia, D. and Zhou, F. et al. found that spiral guide vanes also have a great influence on the velocity distribution, turbulence intensity, pressure drop, and collection efficiency of cyclone separators [32,33]. In the research of many scholars [34,35], new designs have been thoroughly studied by adjusting one parameter each time to optimize the cyclone separator's geometric parameters, keeping all other parameters unchanged until the optimum working conditions are found. The industrial application of cyclone separators generally requires the simultaneous consideration of several factors as well as the complexity of the flow field inside the cyclone separator and the interplay between various structural factors. Therefore, it is necessary to establish the restriction and coordination relationship between the structural dimensions in order to optimize the cyclone separator. Ficici, F. and Ari, V. have used the Taguchi method to optimize the preheater cyclone separator [36]. They optimized the swirl performance by changing four parameters: diameter, vortex probe length, velocity inlet, and particle concentration. Significant energy savings could be achieved through the modification of the preheater cyclone separator design. Safikhani, H. obtained the pressure drop and medium particle size of the cyclone separator via CFD calculation and then optimized the performance of the cyclone separator using a multi-objective optimization method with an artificial neural network to determine the objective function [37]. Sankar, P.S. and Prasad, K. used response surface methodology to optimize various geometric parameters of the Stairmand cyclone separator [38]. Mariani, F. et al. optimized the length and angle of the overflow pipe to improve separation efficiency [39]. Venkatesh, S. analyzed the particle size and pressure drop performance of the square cyclone separator by changing five important geometric parameters via the CFD method [40]. Many studies on cyclones have been conducted by Elsayed, K. and Lacor, S. [41–44], who performed a multi-objective optimization of cyclone performance with the adjoint method, using an artificial neural network and genetic algorithms to analyze the pressure drop, separation efficiency, and cut-off diameter. The results showed that the inlet velocity and various geometric dimensions have a significant effect on the flow pattern, collection efficiency, pressure drop, and acoustic noise of axial cyclones. The main body of a cyclone, the vortex finder diameter and its insertion length, the height of the conical segment, the cone tip diameter, the dipleg length, the dustbin height, and the dustbin diameter had significant effects on cyclone efficiency. Some experts have also studied other types of cyclone separators. Yao, X. et al. designed a gas-liquid cyclone separator with a simple structure, low pressure drop, and high separation efficiency to achieve efficient long-term separation of gas and liquid droplets [45]. He found that the vortex can be strengthened by changing the operating conditions and annular zone height, but inlet velocity does not impact vortex strength. The screw pitch is not affected by either inlet velocity or annular zone height. The total pressure drop is only weakly affected by the annular zone height but is obviously affected by the inlet velocity. Baltrėnas, P. and Chlebnikovas, A. designed a multi-channel cyclone separator to avoid the adhesion of sticky and moist solid particles on the inner surfaces of the cyclone by improving three structures (the secondary inlet, inner slit, and convex bottom) [46], which was effectively used in heat production equipment in private households. Zhou, W. et al. studied the effect of operating parameters such as inlet gas velocity and inlet liquid concentration on a gas-liquid cyclone separator in WGS [47], developed a model to predict the pressure drop, and also proposed an improved weighting method to calculate the droplet separation efficiency.

There are many studies on the optimization of cyclone separator geometries in the existing literature [48–50], but they are all related to the traditional cyclone separator. When using different fluid parameters and working spaces, the separator has different separation conditions and application scopes. In the context of the production of high-pressure polyethylene, the optimization of the high-pressure environment and high-density ethylene gas has not yet been studied. In high-pressure polyethylene production, the separation efficiency and pressure drop in the cyclone separator are important objective functions to be optimized simultaneously. For this, CFD simulations were carried out for separators $D_{1\sim4}$ to solve the problem of small-particle droplets being difficult to separate, to find the main structural factors that affect separation efficiency and pressure drop, and to investigate the influence of the structural parameters of the cyclone separator on the separation efficiency in combination with the relevant process conditions, which can also reduce the frequency of cleaning required for the oligomer filter. In order to optimize the results more comprehensively, a dual-objective optimization study of performance parameters (separation efficiency and pressure drop) was carried out. The eight main size parameters of cyclone separators, cylinder section diameter, cylinder section length, cone section diameter, cone section height, guide vane height, guide vane angle, exhaust pipe extension length, and umbrella structure height, were analyzed using the Taguchi method. Fluent software was used to simulate the experimental combination model. The optimal combination of cyclone separator structures for the high-pressure polyethylene process, with respect to the eight dimensional parameters described, was determined based on energy-efficiency parameters using the unified objective function method. This will provide a feasible optimization method for the optimal design of cyclone separators. At the same time, it will have important research value for cyclone separators in high-pressure environments involving high-density gas phase media, and have important engineering application value for cyclone separators in the high-pressure polyethylene process.

2. Numerical Simulation

2.1. Cyclone Geometry

In the high-pressure polyethylene process, the cyclone separators B, C, and $D_{1\sim4}$ have the same structural model. Their models are shown in Figure 2. The total height (H) of the cyclone separator is 1750 mm and the inner diameter is 250 mm. The main size parameters for the cyclone separator are listed in Table 1, which include the cylinder height (H_{cy}), cylinder diameter (D), cone height (H_c), cone diameter (D_c), inlet diameter (d), exhaust pipe diameter (D_e), bottom flow pipe diameter (D_b) and height (H_b), guide vane height (h_v) and angle (α_v), exhaust pipe extension length (h_e), and umbrella structure angle (β_{um}) and height (h_{um}). The inlet pipe, guide vane, cylinder, cone, exhaust pipe, underflow pipe and umbrella structure are the main components of the cyclone separator.

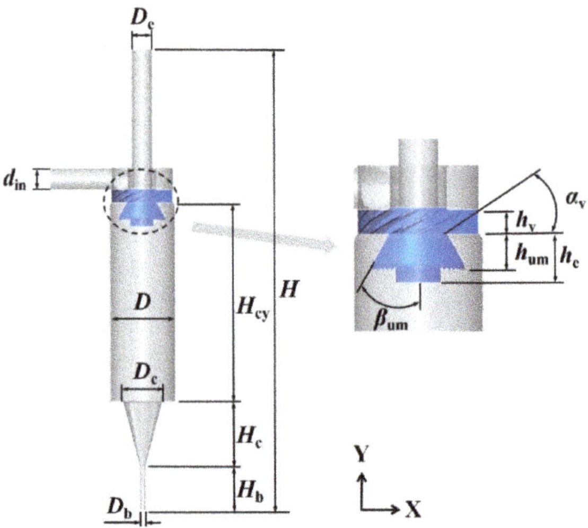

Figure 2. Cyclone separator model.

Table 1. Main structural parameters of cyclone separators.

H (mm)	H_{cy} (mm)	H_c (mm)	H_b (mm)	D (mm)	D_c (mm)	d (mm)	D_e (mm)	D_b (mm)	h_v (mm)	h_e (mm)	h_{um} (mm)	α_v (°)	β_{um} (°)
1750	741	242	169	250	150	77	77	20	45	91	68	30	30

2.2. The Governing Equations

It is assumed that the gas flow inside the cyclone is an incompressible isothermal flow. Therefore, according to the Reynolds-averaged Navier–Stokes (RANS) equations, the continuity equation of the averaged flow is as follows [51]:

$$\frac{\partial u_i}{\partial x_i} = 0 \tag{1}$$

The time-averaged Navier–Stokes equation is as follows [51]:

$$\frac{\partial}{\partial x_j}(\rho u_i u_j) = -\frac{\partial p}{\partial x_i} + \frac{\partial}{\partial x_j}\left[\mu\left(\frac{\partial u_i}{\partial x_j} + \frac{\partial u_j}{\partial x_i} - \frac{2}{3}\delta_{ij}\frac{\partial u_k}{\partial x_k}\right)\right] + \frac{\partial}{\partial x_j}\left(-\overline{\rho u'_j u'_k}\right) \tag{2}$$

The last term is defined as the Reynolds stress tensor, which reflects the effect of turbulence intensity in the fluid flow. In the RSM, the transport equation is written as follows [48]:

$$\frac{\partial}{\partial x_k}\left(\rho u_k \overline{u'_i u'_j}\right) = D_{ij} + P_{ij} + \Phi_{ij} + \varepsilon_{ij} + S \tag{3}$$

The expression on the left side of the equation represents the convective transport term. The five terms on the right side of the equation are the diffusion term, yield term pressure strain term, dissipation term, and source term. The final transport equation for the RSM can be written as follows [51]:

$$\frac{\partial}{\partial x_k}\left(\rho u_k \overline{u_i' u_j'}\right) = \frac{\partial}{\partial x_k}\left[\frac{\mu_t}{\sigma_k}\left(\frac{\partial}{\partial x_k}\overline{u_i' u_j'}\right)\right] - \rho\left[\overline{u_i' u_k'}\frac{\partial u_j}{\partial x_k} + \overline{u_j' u_k'}\frac{\partial u_i}{\partial x_k}\right]$$
$$+ p\left(\frac{\partial u_i'}{\partial x_j} + \frac{\partial u_j'}{\partial x_i}\right) - 2\mu\frac{\partial u_i'}{\partial x_k}\frac{\partial u_j'}{\partial x_k} \tag{4}$$

The eddy viscosity, μ_t, was calculated using the following equation:

$$\mu_t = \rho C_\mu \frac{k^2}{\varepsilon} \tag{5}$$

where C_μ is 0.09.

2.3. Boundary Conditions

The inlet boundary condition was set as the inlet velocity. It is known that the volume flow rate of the gas-liquid mixture from separator B was 7.5 kg/s, the inlet velocity was 9.053 m/s, the mass concentration of the oligomers was 0.0033%, and the hydraulic diameter was 77 mm. The outlet boundary condition was set as the outlet pressure; the working pressure was 25 MPa and the atmospheric pressure was the standard atmospheric pressure. The inlet temperature was 200 °C. Since the fluid was gradually cooled, the density and viscosity of the gas phase gradually changed as the temperature decreased, but the oligomer was less affected, irrespective of changes in density and viscosity. In different cyclone separators, the corresponding calculation boundary conditions changed. The specific boundary conditions for each model are shown in Table 2.

Table 2. Boundary conditions of the cyclone separator.

Models	B	C	D_1	D_2	D_3	D_4
Inlet velocity (m/s)	9.05	9.05	7.32	6.86	6.52	6.37
Working temperature (°C)	200	200	110	80.0	55.0	45.0
Gaseous density (g/cm^3)	0.178	0.178	0.220	0.235	0.247	0.253
Gaseous viscosity (cP)	0.0150	0.0150	0.0130	0.0130	0.0110	0.0110
Liquid density (g/cm^3)	0.740	0.740	0.740	0.740	0.740	0.740
Liquid viscosity (cP)	290	290	290	290	290	290

Since the liquid phase only accounted for a very low volume fraction, the Euler–Lagrange method, based on the discrete phase model (DPM), was used to solve the time-averaged N–S equation with the fluid phase as the continuous phase and the liquid phase as the discrete phase. The wall flow boundary adopted the no-slip solid wall condition, and the standard wall function method was used to determine the flow near the solid wall [52]. This also assumed that there was no particle–particle interaction and that the flow field was bidirectionally coupled between the particles and the flow field [51,53]. To consider the effect of turbulent fluctuations on particles, the discrete random walk (DRW) model was used [20]. The inlet injection source for the droplet phase was a surface injection source with velocity consistent with that of the gas phase. Neglecting the aggregation and fragmentation caused by the collision between droplets, the outlet of the exhaust pipe was set as the escape condition, and the bottom outlet of the cyclone separator was set as the trap condition. Using the Rosin–Rammler distribution for the dispersed liquid phase, the inlet droplet size distribution from the filed feedback was used as the inlet boundary condition for the dispersed droplets, and the inlet of separator B obeyed the Rosin–Rammler droplet size distribution, as shown in Figure 3a. The inlet of separator D_1 obeyed the Rosin–Rammler droplet size distribution, as shown in Figure 3b.

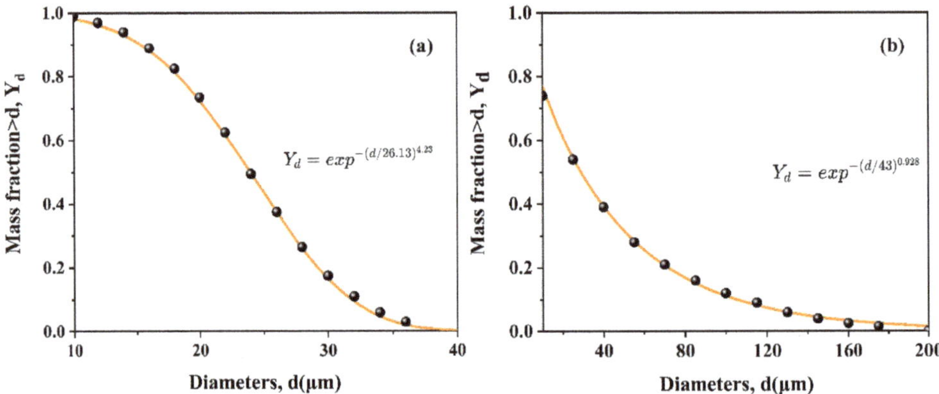

Figure 3. Rosin–Rammler droplet size distribution at the inlet of cyclone separator (**a**) B and (**b**) D_1

Using the commercial CFD software Fluent, the variables on the surface of the contro
body were interpolated using the second-order upwind format. The governing equation
were solved numerically using the finite volume method. For pressure-velocity coupling
the Semi-Implicit Method for Pressure-Linked Equations Consistent (SIMPLEC) algorithm
was used. Due to the highly rotating flow in the cyclone separator, the pressures wer
interpolated with the PREssure STaggered Option (PRESTO). In order to measure the flow
in the cyclone separator, the Quadratic Upstream Interpolation for Convective Kinetic
(QUICK) format of the momentum equation was used. The second-order upwind schem
was used to discretize the Reynolds stress equation. When the residual dropped below
10^{-4}, the calculation was considered to be convergent. In this study, all of the simulation
were performed using these discretization schemes.

Grid independence analysis is necessary in CFD simulation, which starts with a coars
grid and gradually refines the grid until the variation observed in the results is less than
a predefined acceptable error. The cyclone separator results were calculated five time
using the Fluent meshing method with different mesh sizes. In order to ensure that the
continuously smaller grid size was used to enhance the results in the calculation, the gri
independence analysis was carried out. Each calculation was analyzed with differen
grid sizes under the same boundary and operating conditions. The grid sizes range
from 583,461 (i.e., coarse grid) to 3,016,754 (i.e., fine grid). The total pressure drop an
tangential velocity distributions calculated at different grid sizes are shown in Figure
It should be noted that the tangential velocity profiles given here were evaluated on th
plane y = 950 mm from the bottom of the cyclone separator. It can be seen from the figur
that the difference in pressure drop between the finest grids and the coarsest grids was les
than 3.1%. The 2,362,457 and 3,016,754 grids have a pressure drop difference of 0.2% an
the gas tangential velocity curves almost overlap, indicating that the simulation result
within the grid range are less affected by the grid. In summary, 2,362,457 polyhedra
meshes were chosen to reduce the computational effort in the simulation, as shown i
Figure 5. For the other cyclonic separators discussed using different structures, the sam
CFD mesh generation method was also used to ensure acceptable computational time an
grid independence.

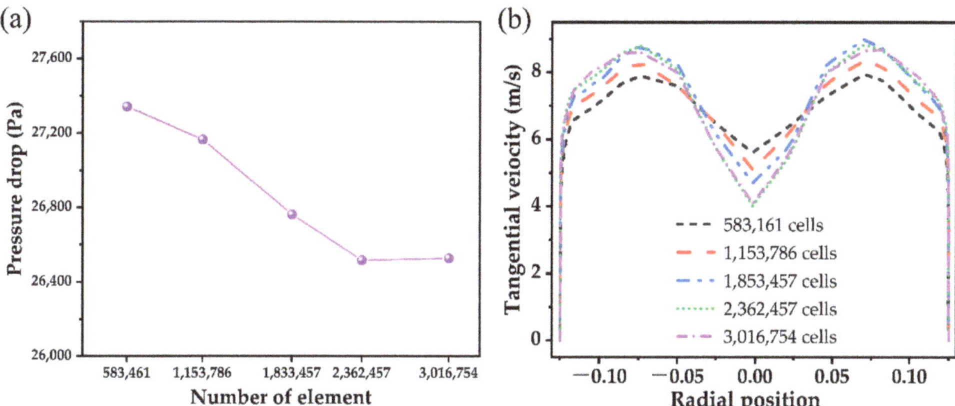

Figure 4. Grid independence verification for (**a**) pressure drop verification and (**b**) tangential velocity verification.

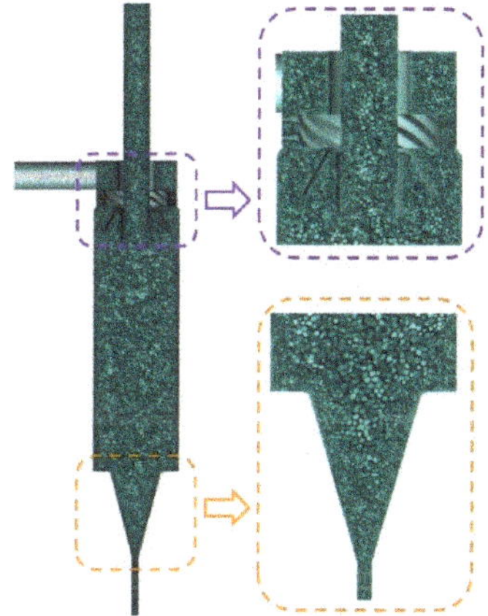

Figure 5. Grid division of the cyclone separator.

2.4. Verification of Simulation

In order to verify the accuracy of the simulation calculation model, data were collected from the production site and the oligomers collected by B/C and $D_{1~4}$ were quantified in the discharge tank. The quantitative data from the two production lines shown in Figure 1 were compared with the simulation data to verify the accuracy of the simulation calculations.

When the discharge tank was filled, the on-site workers removed the oligomers, and the mass of the oligomers was 200 L each time. The number of times the oligomers were extracted from the two production lines in a particular year and quarter were recorded. In that year, the two production lines needed to have the oligomers extracted 38 times and 35 times, respectively, so the total volumes of oligomers separated by the cyclone separators throughout the year at each production line were 7600 L and 7000 L; similarly, the volumes of oligomers for a quarter were 2000 L and 1800 L. The six cyclone separators were simulated using the calculated boundary conditions shown in Table 2.

The separation efficiency was defined as the ratio of the mass flow rate of the droplets captured at the bottom flow port to the inlet mass flow rate from the Fluent post-processing results.

$$\eta = \frac{\sum n_k \frac{\pi d_k^3}{6} \rho_o}{\sum m_k} \tag{6}$$

where n_k is the number of k-sized particles captured per unit of time, m_k is the mass flow rate of k-sized oligomers in the feed, d_k is the particle diameter, and ρ_o is the oligomer density.

Six cyclone separators in series with a mixed media flow rate of 7.5 kg/s and an oligomer mass concentration of 0.0033% were calculated to obtain an oligomer mass flow rate (Q_m) of 2.48×10^{-4} kg/s for cyclone separator B, which was substituted into Fluent for simulation. The oligomer mass flow rate not separated out by each cyclone separator went to the next cyclone separator for separation. For cyclone separator B, according to the particle size distribution in Figure 3a and using Equation (6), the exhaust port is directly connected to the inlet of cyclone separator C through a short pipeline, so the mass flow rate of the liquid phase at the inlet of cyclone separator C is the mass flow rate that was not separated in cyclone separator B. The particle size distribution obtained using Fluent post-processing can be used as the inlet boundary condition for cyclone separator C. The exhaust port of cyclone separator C is connected to cyclone separator D after the pipeline and cooler, and the calculation for cyclone separator D used the particle size distribution shown in Figure 3b. The cooler between cyclone separators $D_{1\sim4}$ has a long pipeline, so it was difficult to judge the actual droplet size and distribution after collision and aggregation in the pipeline. Therefore, it was necessary to assume that the droplet size distribution between the cyclone separator and the next cyclone separator in this process returned to the particle size distribution in this state. After processing by the six cyclone separators, the mass flow rate of the remaining unseparated oligomers was 0.56×10^{-4} kg/s.

From the oligomer density (ρ_o), the oligomer volume flow rate was calculated using the following equation:

$$Q_V = \frac{Q_m \times 1000 \times 3600}{\rho_o} = 1.204 \text{ L/h} \tag{7}$$

The simulated separation efficiency and the calculated oligomer content throughput of the six separators are shown in Table 3. According to the actual operation of the cyclone separator, it operated for an average of 2000 h in a quarter and 8000 h in a year. The oligomer content generated in one quarter of operation of separator B is 2408 L. The calculated oligomer content and separation volume of the other separators are shown in Table 3.

Table 3. Separation conditions of the cyclone separator.

Device Number	B	C	D_1	D_2	D_3	D_4
Separation efficiency (%)	36.73	28.24	23.28	14.11	13.49	12.87
Import mass flow rate (kg/s)	2.48×10^{-4}	1.57×10^{-4}	1.12×10^{-4}	0.86×10^{-4}	0.74×10^{-4}	0.64×10^{-4}
Separating mass flow rate (kg/s)	9.09×10^{-5}	4.42×10^{-5}	2.62×10^{-5}	1.22×10^{-5}	1.00×10^{-5}	0.82×10^{-5}
Separation volume flow (L/h)	0.44	0.22	0.13	0.06	0.05	0.04
One quarter's worth of oligomer content (L)	2408	1523.58	1093.35	838.82	720.44	623.23
One quarter's worth of separation volume (L)	884.42	430.23	254.53	118.38	97.21	80.2

The volume of oligomers that can be separated by all six separators in one quarter is 1864.97 L, as calculated from Table 3, which is equivalent to 7459.88 L in one year. In comparison, the volumes of oligomers separated in one quarter in the field were 1800 L and 2000 L with a maximum relative error of 7.24%, and the volumes of oligomers separated in one year in the field were 7000 L and 7600 L with a maximum relative error of 6.57%. Due to the uncertainty of the simulated results and the differences in the values of physical parameters such as density, viscosity, particle size distribution, etc., the relative error between the simulated results and the actual field results is generally acceptable, as it falls within 10%. Therefore, the simulation results are considered verified by the field data and can be considered for use in engineering applications. The numerical model established herein can thus well predict the gas–liquid separation behavior within and the hydrodynamic characteristics of the cyclone separator.

3. Experimental Design

The experimental design for the function of the D_1 separator in the high-pressure polyethylene process was carried out using the boundary conditions of D_1 as stated in Table 2 and Figure 3b. Due to the interactions between the various structural factors of the cyclone separator, its performance cannot be analyzed based on a single structure. The size matching and restriction relationship between the various structures should be considered comprehensively in order to design a better cyclone separator. The fastest and most effective method for establishing the restriction and matching relationship is through the experimental method, but for multiple factors, wherein each factor has a number of levels, it is necessary to consider multiple test factors at the same time. If this full test design was adopted, the workload would be unmanageable, so the experiment utilized the Taguchi design method [54], which is often used to improve the quality of products in manufacturing. This method is mainly used in the field of engineering design and industrial engineering to achieve improvements in the performance of an existing system by optimizing the design parameters [55]. The purpose of this experimental design is to reduce and control the changes in process or design parameters to improve the system's performance characteristics. Through analysis of the experimental results, the significance of the factors and the analysis of variance can be obtained. The influence of each parameter on performance can be determined, the optimal combination can be calculated, and the linear regression equation can be optimized.

In the high-pressure production environment, the size and shape of the structure will be subject to many restrictions. The diameters of the inlet and outlet are affected by the treatment capacity and are not considered to be design variables. Preliminary analysis shows that the diameter of the umbrella structure has a small effect on the separation effect and is also not used as a design variable. The variables selected for the design of experiments (DOE) were cylinder diameter (D), cylinder height (H_{cy}), cone diameter (D_c), cone height (H_c), guide vane height (h_v), guide vane angle (α_v), exhaust pipe extension length (h_e), and umbrella structure height (h_{um}). The above eight factors were studied at three levels and an experimental design was carried out using a three-level orthogonal table with eight structural parameters of $L_{27}(13^3)$. The three levels of the above factors are listed in Table 4.

Table 4. Levels of the factors.

No.	D (mm)	H_{cy} (mm)	D_c (mm)	H_c (mm)	h_v (mm)	α_v (°)	h_e (mm)	h_{um} (mm)
1	220	641	100	202	45	15	91	58
2	250	741	150	242	60	30	111	68
3	280	841	200	282	75	45	131	78

An orthogonal table is a fractional factorial design used to form a design matrix for multiple combinations of design factors, and the response values were predicted experimentally or theoretically based on the different combinations of design factors. The orthogonal table gave 27 combinations of design parameters, and based on these combinations, the separation efficiency and pressure drop responses were calculated using Fluent. The separation efficiency is the ratio of the mass flow rate between the bottom flow port and the inlet of the cyclone separator, as shown in Equation (6), and the pressure drop is the static pressure difference between the inlet and exhaust ports. In this Taguchi design, the signal-to-noise ratios (SNRs) of separation efficiency and pressure drop were calculated. The separation efficiency was optimized for both reactions by using the maxims of "larger is better" and "smaller is better", respectively. The SNRs for pressure drop and separation efficiency were calculated according to these equations [55]:

Larger is better:

$$\frac{S}{N} = -10\log_{10}\frac{1}{Y_i^2} \tag{8}$$

Smaller is better:

$$\frac{S}{N} = -10\log_{10}Y_i^2 \tag{9}$$

where Y is the response variable. Table 5 lists the pressure drop, separation efficiency, and their respective SNRs for each simulated operation.

Table 5. L_{27} orthogonal array values of the signal-to-noise ratio (SNR) and response parameters.

No.	D (mm)	H_{cy} (mm)	D_c (mm)	H_c (mm)	h_v (mm)	α_v (°)	h_e (mm)	h_{um} (mm)	η (%)	SNR for η	ΔP (Pa)	SNR for ΔP	σ	SNR for σ
1	220	641	100	202	45	15	91	58	25.84	28.25	31,007	−89.83	0.83	−1.58
2	220	641	100	202	60	30	111	68	29.43	29.38	28,309	−89.04	1.04	0.34
3	220	641	100	202	75	45	131	78	33.34	30.46	25,808	−88.24	1.29	2.23
4	220	741	150	242	45	15	91	68	29.13	29.29	29,254	−89.32	1.00	−0.03
5	220	741	150	242	60	30	111	78	33.54	30.51	26,668	−88.52	1.26	2.00
6	220	741	150	242	75	45	131	58	31.53	29.97	22,749	−87.14	1.39	2.84
7	220	841	200	282	45	15	91	78	33.24	30.43	27,399	−88.75	1.21	1.68
8	220	841	200	282	60	30	111	58	31.73	30.03	23,407	−87.39	1.36	2.65
9	220	841	200	282	75	45	131	68	35.64	31.04	20,817	−86.37	1.71	4.68
10	250	641	150	282	45	30	131	58	29.53	29.41	30,653	−89.73	0.96	−0.32
11	250	641	150	282	60	45	91	68	26.84	28.58	24,404	−87.75	1.10	0.83
12	250	641	150	282	75	15	111	78	39.93	32.03	28,752	−89.17	1.39	2.86
13	250	741	200	202	45	30	131	68	29.14	29.29	25,893	−88.26	1.13	1.03
14	250	741	200	202	60	45	91	78	25.63	28.18	19,739	−85.91	1.30	2.27
15	250	741	200	202	75	15	111	58	34.44	30.74	22,421	−87.01	1.54	3.73
16	250	841	100	242	45	30	131	78	24.93	27.93	29,324	−89.34	0.85	−1.40
17	250	841	100	242	60	45	91	58	17.14	24.68	21,657	−86.71	0.79	−2.03
18	250	841	100	242	75	15	111	68	30.23	29.61	25,805	−88.23	1.17	1.38
19	280	641	200	242	45	45	111	58	22.44	27.02	23,556	−87.44	0.95	−0.42
20	280	641	200	242	60	15	131	68	35.53	31.01	27,903	−88.91	1.27	2.10
21	280	641	200	242	75	30	91	78	33.34	30.46	21,738	−86.74	1.53	3.72
22	280	741	100	282	45	45	111	68	18.23	25.22	27,180	−88.69	0.67	−3.46
23	280	741	100	282	60	15	131	78	32.14	30.14	31,496	−89.97	1.02	0.18
24	280	741	100	282	75	30	91	58	24.03	27.62	23,547	−87.44	1.02	0.18
25	280	841	150	202	45	45	111	78	17.84	25.03	22,422	−87.01	0.80	−1.98
26	280	841	150	202	60	15	131	58	25.83	28.24	25,157	−88.01	1.03	0.23
27	280	841	150	202	75	30	91	68	23.64	27.47	19,010	−85.58	1.24	1.90

4. Results and Discussion

4.1. Analysis of the Control Factors

Based on the results in Table 5, further analysis was performed to analyze the effect of each factor on the separation efficiency and pressure drop using Taguchi's SNR response table and to analyze the optimal combination of structural parameters. The results of the simulations with different combinations of structural parameters at different levels were calculated and analyzed, and the average SNR responses of separation efficiency and pressure drop are shown in Table 6. The table shows the SNR ratio at each factor level and the trend of each factor as it changes from level 1 to level 3. The greater the change, the greater the influence of the factors. The degree of influence was different for the eight structural parameters. For the evaluation index of separation efficiency, the influencing factors are, in order of priority: guide vane angle (α_v) > guide vane height (h_v) > cone diameter (D_c) > cylinder diameter (D) > exhaust pipe extension length (h_e) > cylinder length (H_{cy}) > umbrella structure height (h_{um}) > cone height (h_c). For the evaluation index of pressure drop, the degree of influence of structural parameters was: guide vane angle (α_v) > guide vane height (h_v) > cone diameter (D_c) > cylinder length (H_{cy}) > exhaust pipe extension length (h_e) > cone height (h_c) > cylinder diameter (D) > umbrella structure height (h_{um}). The optimal experimental parameters for the eight factors corresponding to the two responses and the trend of the effect of each factor on the response values can be determined from the SNR main effects plot in Figure 6. The horizontal axis of the plot shows the values of each parameter at three different levels, and the vertical axis shows the response values (average SNR). The optimal parameter levels for separation efficiency were taken as 220 mm, 641 mm, 200 mm, 282 mm, 75 mm, 15°, 131 mm, and 78 mm. The optimal parameter levels for pressure drop were taken as 280 mm, 841 mm, 200 mm, 202 mm, 75 mm, 45°, 91 mm, and 58 mm.

Table 6. The average SNR response table for separation efficiency and pressure drop.

	Level	D (mm)	H_{cy} (mm)	D_c (mm)	H_c (mm)	h_v (mm)	α_v (°)	h_e (mm)	h_{um} (mm)
Separation efficiency	1	29.93	29.62	28.14	28.56	27.98	29.97	28.33	28.44
	2	28.94	28.99	28.95	28.94	28.97	29.12	28.84	28.99
	3	28.02	28.27	29.80	29.39	29.93	27.80	29.72	29.46
	Delta	1.91	1.35	1.66	0.83	1.95	2.17	1.39	1.02
	Rank	3	6	4	8	2	1	5	7
Pressure drop	1	−88.29	−88.54	−88.61	−87.65	−88.71	−88.80	−87.56	−87.86
	2	−88.01	−88.03	−88.03	−88.04	−88.02	−88.01	−88.06	−88.02
	3	−87.75	−87.49	−87.42	−88.36	−87.33	−87.25	−88.44	−88.18
	Delta	0.53	1.05	1.19	0.71	1.38	1.55	0.88	0.33
	Rank	7	4	3	6	2	1	5	8

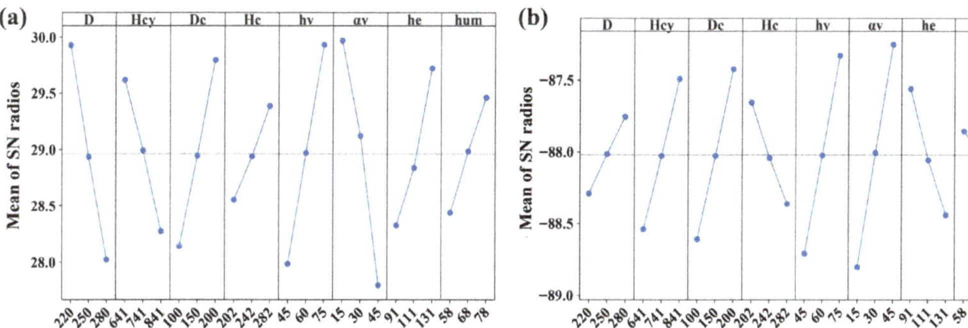

Figure 6. The mean of SNR vs the geometrical parameters of (**a**) separation efficiency and (**b**) pressure drop.

4.2. Analysis of Variance

The significance of the developed response linear mathematical equations was verified using analysis of variance (ANOVA). P-tests were performed based on the response results and geometric factors. The sources of variation, degrees of freedom (DF), sum of squares (SS), mean square (MS), F-value (F), and p-value (p) of the response results for different size parameters are shown in Table 7.

Table 7. Analysis of variance (ANOVA) for factors on SNR of separation efficiency and pressure drop.

	Source	DF	SS	MS	F	p
Separation efficiency	D	2	16.3	8.17	100	2.43×10^{-7}
	H_{cy}	2	8.16	4.08	50.0	6.20×10^{-6}
	D_c	2	12.4	6.19	75.8	9.06×10^{-7}
	H_c	2	3.09	1.54	18.9	3.98×10^{-4}
	h_v	2	17.1	8.54	104	1.97×10^{-7}
	α_v	2	21.6	10.8	132	6.34×10^{-8}
	h_e	2	8.96	4.48	54.9	4.05×10^{-6}
	h_{um}	2	4.72	2.36	28.9	6.94×10^{-5}
	Residual error	10	0.816	0.0820		
	Total	26	93.2			
Pressure drop	D	2	1.28	0.640	59.1	2.89×10^{-6}
	H_{cy}	2	4.96	2.48	229	4.47×10^{-9}
	D_c	2	6.35	3.17	293	1.33×10^{-9}
	H_c	2	2.25	1.13	104	2.04×10^{-7}
	h_v	2	8.62	4.31	398	2.94×10^{-10}
	α_v	2	10.8	5.42	500	9.49×10^{-11}
	h_e	2	3.52	1.76	162	2.39×10^{-8}
	h_{um}	2	0.485	0.242	22.4	2.04×10^{-4}
	Residual error	10	0.108	0.0110		
	Total	26	38.4			

For ANOVA, if the p-value of any term is less than the confidence level, the term is said to have a significant effect on the response, and the p-value was calculated at a 99% confidence interval (significance level $\alpha = 0.01$). The ANOVA results show that the eight factors, cylinder diameter (D), cylinder length (H_{cy}), cone diameter (D_c), cone height (H_c), guide vane height (h_v), guide vane angle (α_v), exhaust pipe extension length (h_e), and umbrella structure height (h_{um}), all had significant effects on separation efficiency ($p < 0.01$) and all eight factors also had significant effects on pressure drop ($p < 0.01$), indicating that the stability of the established model was at 95% of the confidence limit.

4.3. Regression Analysis

Regression analysis was performed using SPSS statistical tools. An important goal of regression analysis is to create regression equations for optimization. In addition, the quality of the established regression equation can be verified via regression analysis [55]. The regression analysis considered a total of eight independent variables, D, H_{cy}, D_c, H_c, h_v, α_v, h_e, and h_{um}, and two dependent variables, η and ΔP. In order to determine the regression coefficients, the mean of each variable must be calculated. Then, the sum of squares of each variable was predicted; the sum of squares is the square of the difference between each variable and its overall mean. After, the multiplication cross of each independent variable and the dependent variable was determined. Based on the secondary data resources obtained from these calculations, the regression coefficients were predicted using the least squares method. The regression equation for each dependent variable was derived from this regression coefficient. Subsequently, the standard error (SE) was predicted from the ratio of the standard deviation to the square root of the total sample. The T-value of each variable was predicted from the ratio of the regression coefficient to the SE. The F-value was calculated from the ratio of the mean squared deviation to the residuals of the ANOVA. A p-value of less than 0.01 indicated a significant effect on the response value. Similarly, for p-values greater than 0.01, the results for the response values were insignificant. The regression analysis is reported in Table 8, where all eight factors were found to have a significant effect on the separation efficiency and pressure drop.

Table 8. Regression analysis for factors on separation efficiency and pressure drop.

	Predictor	Coef	SE Coef	T	p
Separation efficiency	Constant	20.6	9.89×10^{-1}	20.8	4.90×10^{-14}
	D	-9.33×10^{-2}	2.02×10^{-3}	-46.3	3.58×10^{-20}
	H_{cy}	-2.00×10^{-2}	6.05×10^{-4}	-33.1	1.42×10^{-17}
	D_c	5.09×10^{-2}	1.21×10^{-3}	42.1	1.96×10^{-19}
	H_c	3.64×10^{-2}	1.51×10^{-3}	24.0	3.90×10^{-15}
	h_v	0.207	4.03×10^{-3}	51.3	5.80×10^{-21}
	α_v	-0.214	4.03×10^{-3}	-53.0	3.21×10^{-21}
	h_e	0.108	3.02×10^{-3}	35.6	3.81×10^{-18}
	h_{um}	0.175	6.05×10^{-3}	28.9	1.58×10^{-16}
Pressure drop	Constant	4.43×10^{4}	302	146	3.83×10^{-29}
	D	-24.8	0.616	-40.3	4.25×10^{-19}
	H_{cy}	-15.1	0.185	-81.6	1.41×10^{-24}
	D_c	-34.7	0.370	-94.0	1.11×10^{-25}
	H_c	24.8	0.462	53.8	2.47×10^{-21}
	h_v	-133	1.23	-108	8.59×10^{-27}
	α_v	-151	1.23	-123	8.99×10^{-28}
	h_e	61.2	0.924	66.3	5.85×10^{-23}
	h_{um}	51.1	1.85	27.6	3.42×10^{-16}

Another important factor in regression analysis is R^2 (coefficient of determination), which is defined as the ratio of the variance of the independent variables that have been accounted for by all the independent variables in the model to the total variance of the independent variables. When R^2 converges to the unit value of 1, it indicates a good fit between this parameter and the response. The R^2 values of the response parameters (η and ΔP) were 0.998 and 0.999, respectively, indicating that the established linear equation had a high optimization effect. Regression analysis was performed using SPSS software to obtain a mathematical model based on eight parameters, D, H_{cy}, D_c, H_c, h_v, α_v, h_e, and h_{um}, and to predict the separation efficiency and pressure drop. The regression equations are shown in Equation (10) and Equation (11), respectively. To further verify the accuracy of the regression equation, the predicted values of the regression equation were compared with the simulated values shown in Figure 7. The comparison results showed that the predicted values were in good agreement with the experimental values, and in the correspondence between the residual values and the predicted values of the regression equation, the residual values showed an irregular distribution, which indicated that the calculation results were good. As shown in Table 9, the calculated separation efficiency and pressure drop of the optimal separation efficiency and optimal pressure drop combination with the original structure were compared with the predicted values of the regression equation. The results show that the separation efficiency of the optimal separation efficiency combination was improved by 98.7% compared with the original structure, and the maximum prediction error of the regression equation was 3.1%. The optimal pressure drop combination reduced the pressure drop by 42.8% compared to the original structure, and the maximum prediction error of the regression equation was 3.6%.

$$\eta = 20.575 - 0.093D - 0.020H_{cy} + 0.051D_c + 0.036H_c + 0.207h_v - 0.214\alpha_v + 0.108h_e + 0.175h_{um} \tag{10}$$

$$\Delta P = 44264 - 24.831D - 15.073H_{cy} - 34.733D_c + 24.846H_c - 133.490h_v - 151.340\alpha_v + 61.236h_e + 51.070h_{um} \tag{11}$$

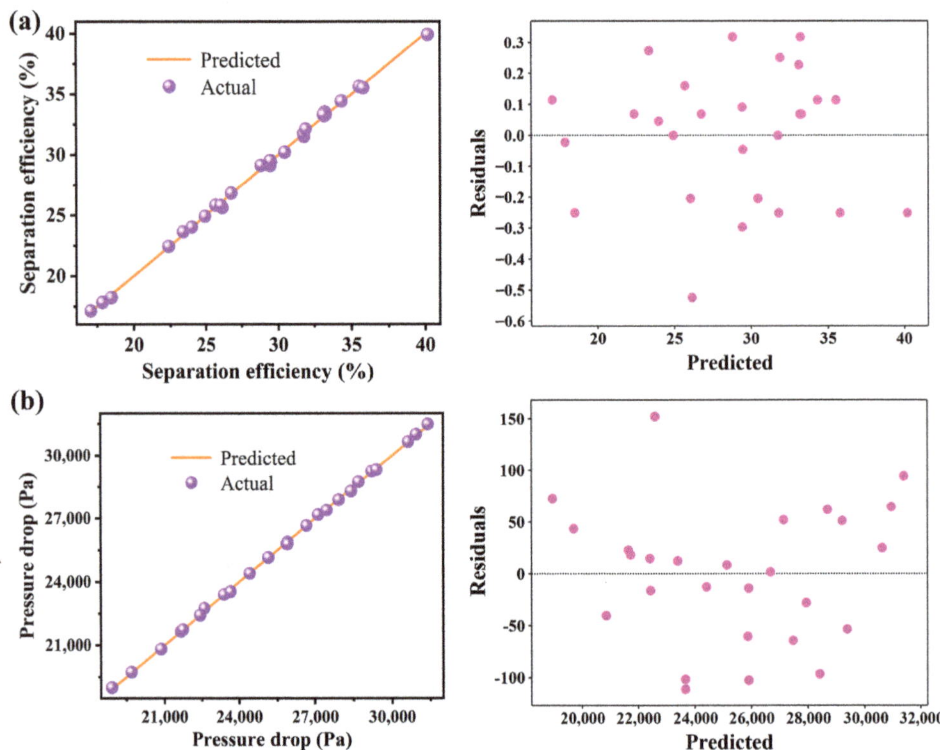

Figure 7. From left to right: comparison of the simulated response values of separation efficiency and the predicted values of the regression equation and correspondence between residuals and predicted values of the regression equation for (**a**) separation efficiency and (**b**) pressure drop.

Table 9. Optimal combination of separation efficiency and pressure drop.

	Simulation Value of Separation Efficiency (%)	Prediction Value of Separation Efficiency (%)	Simulation Value of Pressure Drop (Pa)	Prediction Value of Pressure Drop (Pa)
The original structure	23.28	23.42	26,157	26,188
Optimal structure for separation efficiency	46.25	47.68	28,294	28,923
Optimal structure for pressure drop	21.61	20.96	14,951	14,420

The parameters of the guide vane and the diameters of the cylinder section and the cone section, which have a greater influence on the separation performance and pressure drop, were selected for analysis. The performance parameters and the variable surface are shown in Figure 8. As the guide vane angle increases, the separation efficiency of the cyclone separator becomes smaller and the pressure drop decreases; as the guide vane height increases, the separation efficiency increases and the pressure drop decreases; as the cylinder diameter increases, the separation efficiency decreases and the pressure drop also decreases; as the cone diameter increases, the separation efficiency increases and the pressure drop decreases.

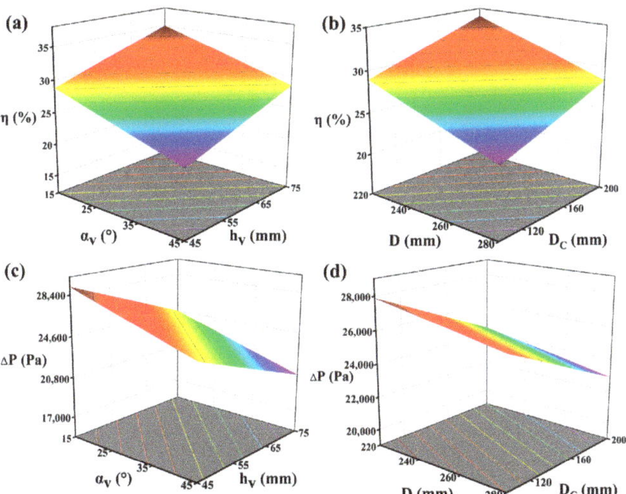

Figure 8. The influence of guide vane and cone diameter on response parameters.

4.4. Multi-Objective Optimization

The previous paper is based on statistical principles, using the trend and variance analysis of the influence of factors on a single inspection target to obtain a preliminary optimization scheme. This analysis method is more systematic and scientific. However, in the actual multi-objective optimization problem, it is often necessary to account for changes in the objectives and conduct a systematic analysis to obtain the highest possible benefits with the lowest possible comprehensive consumption, and then determine a more relevant optimization scheme. The larger the ratio between the efficiency class objective and the energy consumption class objective, the better the optimization scheme. The concept of relative separation efficiency was introduced, which was expressed by η', as shown in the following equation:

$$\eta' = \frac{\eta}{\eta_0} \tag{12}$$

where η is the separation efficiency obtained from the test and η_0 is the separation efficiency of the original structure.

The concept of relative pressure drop was also introduced and expressed by $\Delta P'$, as shown in the following equation:

$$\Delta P' = \frac{\Delta P}{\Delta P_0} \tag{13}$$

where ΔP is the pressure drop obtained from the simulation and ΔP_0 is the pressure drop of the original structure.

Among the two objectives to be investigated in the comprehensive analysis, the relative separation efficiency η' compared to the original structure belongs to the benefit target and the relative pressure drop $\Delta P'$ belongs to the energy consumption target. This was expressed in terms of the energy-efficiency factor, σ, as shown in the following equation:

$$\sigma = \frac{\eta'}{\Delta P'} \tag{14}$$

The σ values were calculated separately for each test group, and the SNRs are shown in Table 5. From Table 10, it can be seen that the three factors that have the greatest influence on the σ value are the guide vane height (h_v), the cone diameter (D_c) and the cylinder diameter (D). The surface plot of the performance parameters and variables analyzing the influence of the three factors on the σ value is shown in Figure 9. As the height of the guide vanes (h_v) and the cone diameter (D_c) increase, the σ value increases accordingly; as the cylinder diameter (D) decreases, the σ value also decreases accordingly.

Table 10. The average SNR response table for σ.

Level	D (mm)	H_{cy} (mm)	D_c (mm)	H_c (mm)	h_v (mm)	α_v (°)	h_e (mm)	h_{um} (mm)
1	1.646	1.086	−0.462	0.910	−0.720	1.174	0.773	0.589
2	0.929	0.972	0.926	0.907	0.954	1.122	0.789	0.975
3	0.274	0.790	2.384	1.031	2.613	0.552	1.286	1.285
Delta	1.372	0.296	2.846	0.124	3.333	0.622	0.513	0.696
Rank	3	7	2	8	1	5	6	4

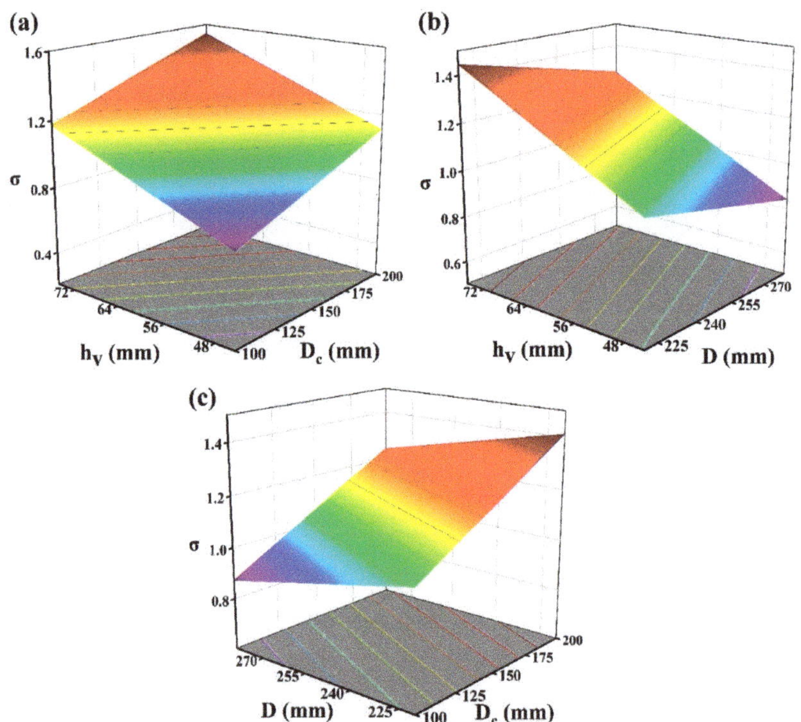

Figure 9. The influence of guide vane height (h_v), cone section diameter (D_c), and cylinder section diameter (D) on σ.

The regression analysis was performed using SPSS software, and a mathematical model based on the parameters D, H_{cy}, D_c, H_c, h_v, α_v, h_e and h_{um} was obtained with an R^2 of 0.974. The regression analysis is reported in Table 11. D, D_c, h_v, α_v, h_e, and h_{um} had a significant effect on the response values, and H_{cy} and H_c did not have a significant effect on the response values. The established regression equation is shown in Equation (15). As shown in Figure 10, the predicted values agree well with the experimental values, and the residual values are scattered and show irregular distribution. The established linear equation has a high optimization effect and can thus be used to predict the separation efficiency and pressure drop.

$$\sigma = 1.19 \times 10^{-2} - 2.87 \times 10^{-3}D - 1.20 \times 10^{-4}H_{cy} + 3.68 \times 10^{-3}D_c + 3.53 \times 10^{-4}H_c + 1.44 \times 10^{-2}h_v$$
$$-1.71 \times 10^{-3}\alpha_v + 1.72 \times 10^{-3}h_e + 4.36 \times 10^{-3}h_{um} \tag{15}$$

Table 11. Regression analysis for factors on σ.

Predictor	Coef	SE Coef	T	p
Constant	1.19×10^{-2}	1.57×10^{-1}	0.0757	9.40×10^{-1}
D	-2.86×10^{-3}	3.21×10^{-4}	-8.95	4.82×10^{-8}
H_{cy}	-1.20×10^{-4}	9.62×10^{-5}	-1.25	2.27×10^{-1}
D_c	3.68×10^{-3}	1.92×10^{-4}	19.1	2.06×10^{-13}
H_c	3.53×10^{-4}	2.40×10^{-4}	1.47	1.60×10^{-1}
h_v	1.44×10^{-2}	6.41×10^{-4}	22.4	1.30×10^{-14}
α_v	-1.71×10^{-3}	6.41×10^{-4}	-2.66	1.59×10^{-2}
h_e	1.72×10^{-3}	4.81×10^{-4}	3.58	2.14×10^{-3}
h_{um}	4.36×10^{-3}	9.62×10^{-4}	4.53	2.58×10^{-4}

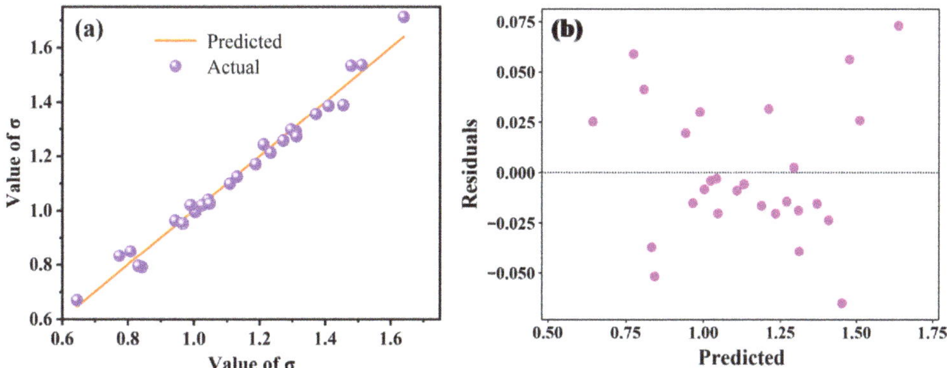

Figure 10. (**a**) The relationship between simulated response value and predicted value and (**b**) the relationship between residual and predicted value.

The optimal value of the regression equation was solved using the linprog function in MATLAB R2018a software. The larger the value of σ, the better. The lower and upper limits were set to 220, 641, 100, 202, 45, 15, 91, 58 and 280, 841, 200, 282, 75, 45, 131, 78, respectively. The optimal solution was calculated to be $\sigma = 1.7595$, at which time the values of the independent variables were 220, 641, 200, 282, 75, 15, 131, 78. The comprehensive optimal combination was the same as the optimal separation efficiency combination.

4.5. Improved and Original Structure Comparison

The comprehensive optimal combination parameters were applied to the cyclone separator, and the optimized model is shown in Figure 11, where the cyclone separator is narrower and smaller, the cylinder and cone joints are smoother, and the guide vane angle is inclined and the height is taller. The simulated value of the cyclone separator was in good agreement with the predicted value of the equation, with an error of 5.5%. Under the optimal structural parameters, the improved structure increases the separation efficiency by 98.7% compared to the original structure, while the pressure drop only increases by 8.2%. The calculation results for substituting this comprehensive optimal combination into the high-pressure polyethylene process are shown in Table 12. Separators $D_{1\sim4}$ were all improved, resulting in a separation efficiency which indicated that 2194 L of oligomers could be separated in one quarter and 8775 L could be separated in one year. The total separation efficiency of the six cyclone separators could thus reach 91.1%, which is 17.7% higher than the separation efficiency achieved by the original structure, which was 77.4%. Additionally, the improved structure could effectively reduce the required frequency of filter cleaning.

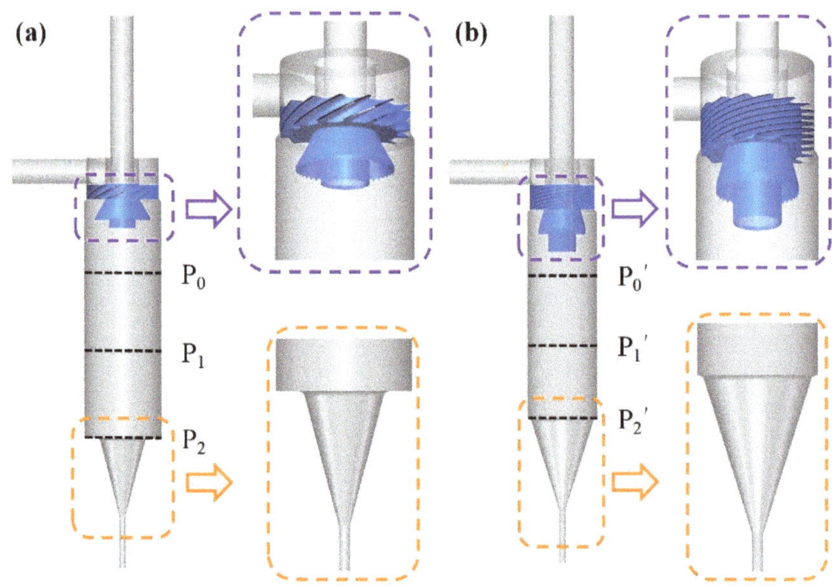

Figure 11. Structure comparison: (**a**) original structure and (**b**) improved structure.

Table 12. Separation status after optimization of the high-pressure polyethylene process.

Device Number	B	C	D_1	D_2	D_3	D_4
Separation efficiency (%)	36.73	28.24	47.68	33.23	26.56	23.66
Import mass flow rate (kg/s)	2.48×10^{-4}	1.57×10^{-4}	1.12×10^{-4}	5.88×10^{-5}	3.93×10^{-5}	2.88×10^{-5}
Separating mass flow rate (kg/s)	9.09×10^{-5}	4.42×10^{-5}	5.36×10^{-5}	1.95×10^{-5}	1.04×10^{-5}	6.82×10^{-6}
Separation volume flow (L/h)	0.44	0.22	0.26	0.10	0.05	0.03
One quarter's worth of oligomer content (L)	2408	1523.58	1093.35	572.04	381.95	280.51
One quarter's worth of separation volume (L)	884.42	430.23	521.31	190.09	101.45	66.37

Figure 12 shows the pressure contour of the original structure and that of the improved structure. The results show that the pressure was distributed in a V-shape along the radial direction, and the cyclone separator showed good symmetry in terms of overall pressure distribution, with a low-pressure zone at the center. This is due to the presence of a highly intensified forced vortex due to the high vortex velocity. The maximum pressure of the original structure of the cyclone separator was located in the cylinder wall area, and the pressure decreased gradually from the cylinder wall to the exhaust pipe. The maximum pressure of the improved cyclone separator structure was located in the channel in front of the guide vane, followed by the wall area, and the pressure decreased gradually from the cylinder wall to the exhaust pipe. This is because when the mixed medium enters the cyclone tangentially, it is subject to the reaction force of the circular wall of the cyclone separator and generates vortex. As the angle of the guide vanes of the improved structure becomes smaller, the axial force generated when its feed fluid hits the wall of the guide vanes also becomes smaller, resulting in the highest pressure in the channel, which was also the main reason for the high pressure drop of the improved structure compared with that of the original structure. After the feed fluid passes through the guide vane, the pressure distribution in the cyclone separator was similar to that of the original structure, and the pressure decreased from the cylinder wall to the center of the cyclone separator. The decrease of the cylinder diameter and the increase of the cone diameter smoothed out the area where the cylinder and cone made contact, which helped to reduce the pressure drop.

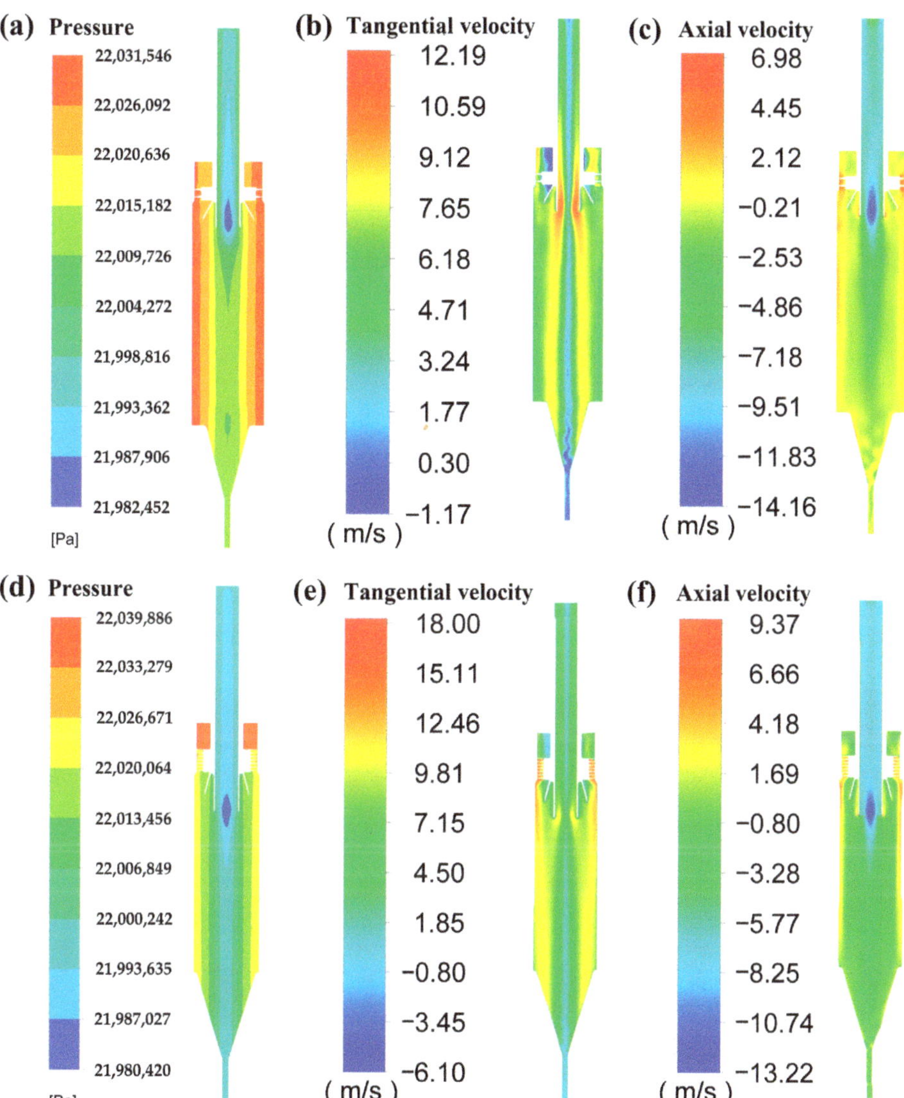

Figure 12. The contour of the original structure and the improved structure. From top to bottom: original structure and improved structure. From left to right: pressure contour, tangential velocity contour, and axial velocity contour.

Figure 12 shows the tangential and axial velocity distribution contours for the two cyclone separators. The radial distribution of tangential velocity and axial velocity is shown in Figure 13, where R is the radius of the cylinder of the corresponding structure and r is the radial position. The positions of the three lines selected in the radial direction are shown in Figure 11, where the positions P_0 (P_0') and P_1 (P_1') trisect the height of the cylinder, the position P_2 is the contact area of the cylinder and cone, and the lengths of all three lines are the diameters of the cylinders. Since the wall was set as a no-slip wall boundary condition (the fluid has zero relative velocity at the wall), and the original structure and the improved structure were different here, the velocity is zero in some areas on Figure 13c,f. In a centrifugal separator, the separation efficiency is mainly affected by the centrifugal force, which is mainly related to the tangential velocity, and the centrifugal force generated in the cyclone separator is greater when there is a larger tangential velocity. The larger centrifugal force allows more droplets to hit against the wall, thus increasing the collection of the liquid phase. Therefore, the tangential velocity magnitude in the cyclone separator is very important for separation efficiency. The maximum tangential velocity of the original structure occurs in the bottom area of the exhaust pipe of the cyclone separator, and the maximum tangential velocity of the improved structure occurs in the flow channel of the guide vanes. This is because the guide vane angle and guide vane height are optimized so that the mixed media has a higher tangential velocity after flowing through the flow channel, and thus the mixed media has a higher tangential velocity for separation in the cylinder. The tangential velocities of the two forms were in good symmetry. From the tangential velocity distribution of the two cyclone separators in Figure 13, both tangential velocities have an M-shaped distribution, with lower tangential velocity at the center gradually increasing tangential velocity in the region of $0 \leq |R'| < 2/3R$, and decreasing tangential velocity in the region of $2/3R \leq |R'| < R$ for both.

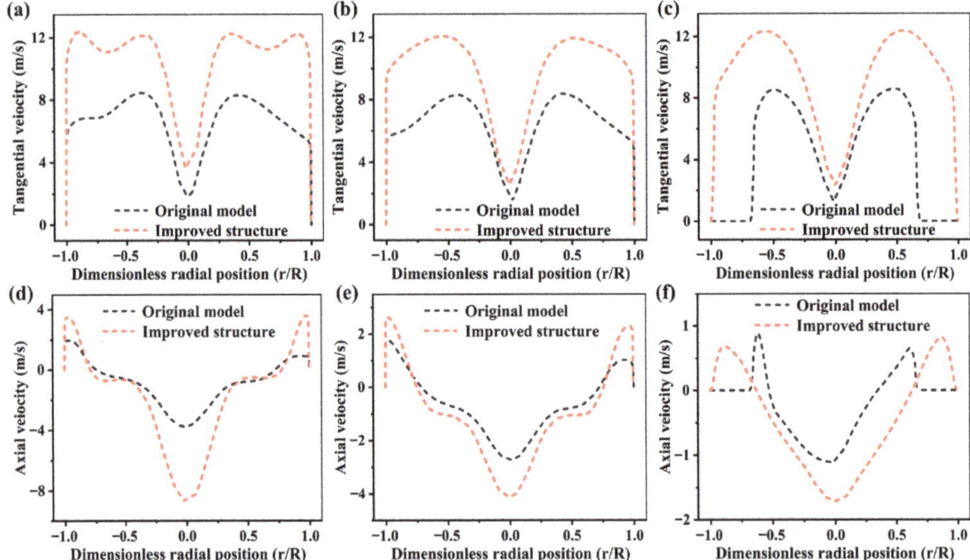

Figure 13. The tangential velocity and axial velocity distribution of the original structure and the improved structure. From top to bottom: tangential velocity and axial velocity. From left to right: P_0, P_1, and P_2.

In order to compare the axial velocity of the different structure forms, it was specified that the axial velocity values are positive along the axis downward with the positive direction and negative along the axis upward with the negative direction. From the axial velocity contour in Figure 12, it can be seen that both cyclone structures show relatively good symmetry, and the maximum axial velocity of both structures is in a small area near the outer wall of the exhaust pipe, which is because a small amount of mixed medium will escape directly from here along the exhaust pipe. From Figure 13, the axial velocity is smaller from the wall of the cylinder to the position near the central axis, and it has an M-shaped distribution. At the same cross-sectional position, the axial velocity increased and then decreased along the radial direction from the center to the wall. This is due to the existence of upstream flow and downstream flow in the separator. Numerically, the velocity of the upstream flow was greater than that of the downstream flow, which is due to the upstream flow flowing to the exhaust pipe. When the fluid flows through the exhaust pipe, the flow area decreases, resulting in an increase in its velocity. In the downflow region, the maximum axial velocity was close to the side wall, and its size gradually decreased as the axial position decreased. In the original structure, the phenomenon of large axial velocity and asymmetric distribution of axial velocity in the cone section was also observed in some areas of the cone wall, as evidenced in Figure 13f, indicating that the structural ratio of the cylinder section to the cone section affected the stability of the axial velocity in this area. The improved structure did not have this phenomenon, which indicates that the structural ratio of the cylinder and cone sections of the improved structure was more conducive to the smooth flow of fluid in this area.

5. Conclusions

The Taguchi design method was used to optimize the structural design of the cyclone separator, and the best mathematical equations for the two response parameters (η and ΔP) and the equation for the energy-efficiency factor, σ, were fitted. The numerical simulation method, Taguchi method, and multi-objective optimization method were applied to investigate the effects of eight structural parameters on the separation performance of the high-pressure polyethylene cyclone separator. The conclusions are as follows:

(1) For the separation efficiency, the influencing factors were, in order of priority: guide vane angle > guide vane height > cone diameter > cylinder diameter > exhaust pipe extension length > cylinder height > umbrella structure height > cone height. The optimal parameter levels for separation efficiency were taken as: cylinder diameter at 220 mm, cylinder height at 641 mm, cone diameter at 200 mm, cone height at 282 mm, guide vane height at 75 mm, guide vane angle at 15°, exhaust pipe extension length at 131 mm, and umbrella structure height at 78 mm. The optimal combination for separation efficiency improved the separation efficiency by 98.7% compared with that of the original structure. The mathematical equations for separation efficiency and the eight parameters were also obtained and predicted with a maximum error of 3.1%.

(2) For the pressure drop, the influencing factors were, in order of priority: guide vane angle > guide vane height > cone diameter > cylinder length > exhaust pipe extension length > cone height > cylinder diameter > umbrella structure height. The optimal parameter levels for pressure drop were taken as: cylinder diameter at 280 mm, cylinder height at 841 mm, cone diameter at 200 mm, cone height at 202 mm, guide vane height at 75 mm, guide vane angle at 45°, exhaust pipe extension length at 91 mm, and umbrella structure height at 58 mm. The optimal pressure drop combination reduced the pressure drop by 42.8% compared with that of the original structure. The mathematical equations for pressure drop and the eight parameters were also obtained and predicted, and the maximum error of the predicted value was 3.6%.

(3) An energy-efficiency factor was defined and used to reduce the multi-objective optimization problem and achieve a higher separation efficiency and lower pressure drop. The guide vane height, cone diameter, and cylinder diameter had the greatest influence on the comprehensive performance of the cyclone separator and the comprehensive optimal combination was the same as the optimal separation efficiency combination. The improved structure increased the separation efficiency by 98.7% compared to the original structure, while the pressure drop increased by only 8.2%. When the improved structure was applied in the high-pressure polyethylene process, the separation efficiency reached 91.1%, which was 17.7% higher than that of the original structure, which could thus effectively reduce the frequency of filter cleaning and greatly improve the production efficiency. The mathematical equations of the energy-efficiency factor and the eight parameters were also obtained and predicted, and the simulated values of the improved structure were in good agreement with the predicted values, with a relative error within 5.5%.

Although the proposed improved structure is effective in the relevant operating and physical parameters of the cyclone separator in the high-pressure polyethylene process, it may not be capable of achieving this effect under other process conditions. The cyclone separator behaves differently under different physical and operating conditions. Therefore, more work can be done on the applicability of the cyclone separator equipment to study the range of process parameters to which the cyclone separator can be adapted.

Author Contributions: Conceptualization, B.H. and B.G.; Methodology, B.H.; Software, S.L.; Validation, C.W. and B.G.; Formal Analysis, B.H.; Investigation, S.L.; Resources, C.W.; Data Curation, S.L.; Writing—Original Draft Preparation, S.L.; Writing—Review & Editing, B.H. and B.G.; Visualization, S.L.; Supervision, B.G.; Project Administration, B.G.; Funding Acquisition, B.G. All authors have read and agreed to the published version of the manuscript.

Funding: This research received no external funding.

Data Availability Statement: Data sharing not applicable.

Acknowledgments: The authors acknowledge constructive comments of Baoling Han from Beijing Yanhua Engineering Construction Company and the partial calculation work done by Suli Yang from Hebei University of Technology.

Conflicts of Interest: The authors declare no conflict of interest.

Nomenclature

General symbols

d	Inlet diameter (mm)
D	Cylinder diameter (mm)
D_b	Bottom flow pipe diameter (mm)
D_c	Cone diameter (mm)
D_e	Exhaust pipe diameter (mm)
D_{ij}	Stress diffusion
d_k	Particle diameter (μm)
H	Total height of the cyclone separator (mm)
H_b	Bottom flow pipe height (mm)
H_c	Cone height (mm)
H_{cy}	Cylinder height (mm)
h_e	Exhaust pipe extension length (mm)
h_{um}	Umbrella structure height (mm)
h_v	Guide vane height (mm)
m_k	Mass flow rate of k-sized oligomers in the feed
n_k	Number of k-sized particles captured per unit of time
P_{ij}	Shear production
Q_m	Mass flow rate of oligomers
Q_v	Volume flow rate of oligomers
R	Cylinder radius of the corresponding structure
r	Radial position
u_i	Time-averaged fluid velocity i (m/s)
u'_i	Fluctuating velocity to direction i (m/s)
$\overline{u'}$	Sub-grid scale velocity (m/s)
Y	Response variable

Greek symbols

α_v	Guide vane angle (°)
β_{um}	Umbrella structure angle (mm)
δ_{ij}	Kronecker delta
ΔP	Pressure drop (Pa)
$\Delta P'$	Relative pressure drop (Pa)
ε_{ij}	Source term
η	Separation efficiency
η'	Relative separation efficiency
μ	Viscosity (kg/m.s)
ρ	Gas density (kg/m^3)
ρ_o	Oligomer density (kg/m^3)
σ	Energy-efficiency factor
μ_t	Eddy viscosity (kg/m.s)
Φ_{ij}	Pressure-strain

References

1. Mazyan, W.I.; Ahmadi, A.; Brinkerhoff, J.; Ahmed, H.; Hoorfar, M. Enhancement of Cyclone Solid Particle Separation Performance Based on Geometrical Modification: Numerical Analysis. *Sep. Purif. Technol.* **2018**, *191*, 276–285. [CrossRef]
2. Vieira, L.G.M.; Silva, D.O.; Barrozo, M.A.S. Effect of Inlet Diameter on the Performance of a Filtering Hydrocyclone Separator. *Chem. Eng. Technol.* **2016**, *39*, 1406–1412. [CrossRef]
3. Chi, Y.; Meng, X.; Zhang, R.; Liu, H.; Liu, Z. Progress in numerical simulation of liquid-liquid cyclone separator. *Chin. J. Process Eng.* **2021**, *21*, 1132–1141.
4. El-Emam, M.A.; Shi, W.; Zhou, L. CFD-DEM Simulation and Optimization of Gas-Cyclone Performance with Realistic Macroscopic Particulate Matter. *Adv. Powder Technol.* **2019**, *30*, 2686–2702. [CrossRef]
5. Yang, L.; Zhang, J.; Ma, Y.; Xu, J.; Wang, J. Experimental and Numerical Study of Separation Characteristics in Gas-Liquid Cylindrical Cyclone. *Chem. Eng. Sci.* **2020**, *214*, 115362. [CrossRef]
6. Katare, P.; Krupan, A.; Dewasthale, A.; Datar, A.; Dalkilic, A.S. CFD Analysis of Cyclone Separator Used for Fine Filtration in Separation Industry. *Case Stud. Therm. Eng.* **2021**, *28*, 101384. [CrossRef]
7. Bymaster, A.; Olson, M.; Grave, E.; Svedeman, S.J.; Viana, F.; Akdim, M.R.; Mikkelsen, R. High Pressure Gas-Liquid Separation: An Experimental Study on Separator Performance of Natural Gas Streams at Elevated Pressures. In Proceedings of the Offshore Technology Conference, Houston, TX, USA, 2–5 May 2011.
8. Dueck, J.; Farghaly, M.; Neesse, T. The Theoretical Partition Curve of the Hydrocyclone. *Miner. Eng.* **2014**, *62*, 25–30. [CrossRef]
9. He, F.; Wang, H.; Wang, J.; Li, S.; Fan, Y.; Xu, X. Experimental Study of Mini-Hydrocyclones with Different Vortex Finder Depths Using Particle Imaging Velocimetry. *Sep. Purif. Technol.* **2020**, *236*, 116296. [CrossRef]
10. Jiang, L.; Liu, P.; Yang, X.; Zhang, Y.; Li, X.; Zhang, Y.; Wang, H. Experimental Research on the Separation Performance of W-Shaped Hydrocyclone. *Powder Technol.* **2020**, *372*, 532–541. [CrossRef]
11. Kozić, M.; Ristic, S.; Puharic, M.; Linic, S. CFD Analysis of the Influence of Centrifugal Separator Geometry Modification on the Pulverized Coal Distribution at the Burners. *Trans. FAMENA* **2014**, *38*, 25–36.
12. Ye, J.; Xu, Y.; Song, X.; Yu, J. Novel Conical Section Design for Ultra-Fine Particles Classification by a Hydrocyclone. *Chem. Eng. Res. Des.* **2019**, *144*, 135–149. [CrossRef]
13. Fadaei, M.; Ameri, M.J.; Rafiei, Y.; Ghorbanpour, K. A Modified Semi-Empirical Correlation for Designing Two-Phase Separators. *J. Pet. Sci. Eng.* **2021**, *205*, 108782. [CrossRef]
14. Hwang, K.-J.; Hwang, Y.-W.; Yoshida, H. Design of Novel Hydrocyclone for Improving Fine Particle Separation Using Computational Fluid Dynamics. *Chem. Eng. Sci.* **2013**, *85*, 62–68. [CrossRef]
15. Tian, J.; Ni, L.; Song, T.; Shen, C.; Yao, Y.; Zhao, J. Numerical Study of Foulant-Water Separation Using Hydrocyclones Enhanced by Reflux Device: Effect of Underflow Pipe Diameter. *Sep. Purif. Technol.* **2019**, *215*, 10–24. [CrossRef]
16. Yohana, E.; Tauviqirrahman, M.; Yusuf, B.; Choi, K.-H.; Paramita, V. Effect of Vortex Limiter Position and Metal Rod Insertion on the Flow Field, Heat Rate, and Performance of Cyclone Separator. *Powder Technol.* **2021**, *377*, 464–475. [CrossRef]
17. Chu, K.; Chen, J.; Yu, A.B.; Williams, R.A. Numerical Studies of Multiphase Flow and Separation Performance of Natural Medium Cyclones for Recovering Waste Coal. *Powder Technol.* **2017**, *314*, 532–541. [CrossRef]
18. Pei, B.; Yang, L.; Dong, K.; Jiang, Y.; Du, X.; Wang, B. The Effect of Cross-Shaped Vortex Finder on the Performance of Cyclone Separator. *Powder Technol.* **2017**, *313*, 135–144. [CrossRef]
19. Shukla, S.K.; Shukla, P.; Ghosh, P. The Effect of Modeling of Velocity Fluctuations on Prediction of Collection Efficiency of Cyclone Separators. *Appl. Math. Model.* **2013**, *37*, 5774–5789. [CrossRef]
20. Qiu, S.; Wang, G. Effects of Reservoir Parameters on Separation Behaviors of the Spiral Separator for Purifying Natural Gas Hydrate. *Energies* **2020**, *13*, 5346. [CrossRef]

21. Gao, X.; Chen, J.; Feng, J.; Peng, X. Numerical Investigation of the Effects of the Central Channel on the Flow Field in an Oil–Gas Cyclone Separator. *Comput. Fluids* **2014**, *92*, 45–55. [CrossRef]
22. Kępa, A. The Efficiency Improvement of a Large-Diameter Cyclone—The CFD Calculations. *Sep. Purif. Technol.* **2013**, *118*, 105–111 [CrossRef]
23. Chuah, T.G.; Gimbun, J.; Choong, T.S.Y. A CFD Study of the Effect of Cone Dimensions on Sampling Aerocyclones Performance and Hydrodynamics. *Powder Technol.* **2006**, *162*, 126–132. [CrossRef]
24. Zhang, H.; Dewil, R.; Degrève, J.; Baeyens, J. The Design of Cyclonic Pre-Heaters in Suspension Cement Kilns. *Int. J. Sustain. Eng.* **2014**, *7*, 307–312. [CrossRef]
25. Wang, Y.; Chen, J.; Yang, Y.; Han, M.; Zhou, Y.; Ye, S.; Yan, C.; Yue, T. Experimental and Numerical Performance Study of a Downward Dual-Inlet Gas-Liquid Cylindrical Cyclone (GLCC). *Chem. Eng. Sci.* **2021**, *238*, 116595. [CrossRef]
26. Lim, J.-H.; Oh, S.-H.; Kang, S.; Lee, K.-J.; Yook, S.-J. Development of Cutoff Size Adjustable Omnidirectional Inlet Cyclone Separator. *Sep. Purif. Technol.* **2021**, *276*, 119397. [CrossRef]
27. Safikhani, H.; Mehrabian, P. Numerical Study of Flow Field in New Cyclone Separators. *Adv. Powder Technol.* **2016**, *27*, 379–387 [CrossRef]
28. Liu, L.; Bai, B. Numerical Study on Swirling Flow and Separation Performance of Swirl Vane Separator. *Interfacial Phenom. Heat Transf.* **2017**, *5*, 9–21. [CrossRef]
29. Elsayed, K.; Lacor, C. The Effect of Cyclone Vortex Finder Dimensions on the Flow Pattern and Performance Using LES. *Comput. Fluids* **2013**, *71*, 224–239. [CrossRef]
30. Hamdy, O.; Bassily, M.A.; El-Batsh, H.M.; Mekhail, T.A. Numerical Study of the Effect of Changing the Cyclone Cone Length on the Gas Flow Field. *Appl. Math. Model.* **2017**, *46*, 81–97. [CrossRef]
31. Shastri, R.; Brar, L.S. Numerical Investigations of the Flow-Field inside Cyclone Separators with Different Cylinder-to-Cone Ratios Using Large-Eddy Simulation. *Sep. Purif. Technol.* **2020**, *249*, 117149. [CrossRef]
32. Misiulia, D.; Elsayed, K.; Andersson, A.G. Geometry Optimization of a Deswirler for Cyclone Separator in Terms of Pressure Drop Using CFD and Artificial Neural Network. *Sep. Purif. Technol.* **2017**, *185*, 10–23. [CrossRef]
33. Zhou, F.; Sun, G.; Han, X.; Zhang, Y.; Bi, W. Experimental and CFD Study on Effects of Spiral Guide Vanes on Cyclone Performance. *Adv. Powder Technol.* **2018**, *29*, 3394–3403. [CrossRef]
34. Misiulia, D.I.; Kuz'min, V.V.; Markov, V.A. Developing an Untwisting Device for Cyclones and Estimating Its Parameters. *Theor. Found. Chem. Eng.* **2013**, *47*, 274–283. [CrossRef]
35. Wasilewski, M.; Singh Brar, L.; Ligus, G. Effect of the Central Rod Dimensions on the Performance of Cyclone Separators—Optimization Study. *Sep. Purif. Technol.* **2021**, *274*, 119020. [CrossRef]
36. Ficici, F.; Ari, V. Optimization of the Preheater Cyclone Separators Used in the Cement Industry. *Int. J. Green Energy* **2013**, *10*, 12–27. [CrossRef]
37. Safikhani, H.; Hajiloo, A.; Ranjbar, M.A. Modeling and Multi-Objective Optimization of Cyclone Separators Using CFD and Genetic Algorithms. *Comput. Chem. Eng.* **2011**, *35*, 1064–1071. [CrossRef]
38. Sankar, P.S.; Prasad, R.K. Process Modeling and Particle Flow Simulation of Sand Separation in Cyclone Separator. *Part. Sci. Technol.* **2015**, *33*, 385–392. [CrossRef]
39. Mariani, F.; Risi, F.; Grimaldi, C.N. Separation Efficiency and Heat Exchange Optimization in a Cyclone. *Sep. Purif. Technol.* **2017**, *179*, 393–402. [CrossRef]
40. Venkatesh, S.; Suresh Kumar, R.; Sivapirakasam, S.P.; Sakthivel, M.; Venkatesh, D.; Yasar Arafath, S. Multi-Objective Optimization Experimental and CFD Approach for Performance Analysis in Square Cyclone Separator. *Powder Technol.* **2020**, *371*, 115–129 [CrossRef]
41. Babaoğlu, N.U.; Hosseini, S.H.; Ahmadi, G.; Elsayed, K. The Effect of Axial Cyclone Inlet Velocity and Geometrical Dimensions on the Flow Pattern, Performance, and Acoustic Noise. *Powder Technol.* **2022**, *407*, 117692. [CrossRef]
42. Shastri, R.; Singh Brar, L.; Elsayed, K. Multi-Objective Optimization of Cyclone Separators Using Mathematical Modelling and Large-Eddy Simulation for a Fixed Total Height Condition. *Sep. Purif. Technol.* **2022**, *291*, 120968. [CrossRef]
43. Elsayed, K.; Parvaz, F.; Hosseini, S.H.; Ahmadi, G. Influence of the Dipleg and Dustbin Dimensions on Performance of Gas Cyclones: An Optimization Study. *Sep. Purif. Technol.* **2020**, *239*, 116553. [CrossRef]
44. Brar, L.S.; Elsayed, K. Analysis and Optimization of Cyclone Separators with Eccentric Vortex Finders Using Large Eddy Simulation and Artificial Neural Network. *Sep. Purif. Technol.* **2018**, *207*, 269–283. [CrossRef]
45. Yao, X.; Zuo, P.; Lu, C.; E, C.; Liu, M. Characteristics of Flow and Liquid Distribution in a Gas–Liquid Vortex Separator with Multi Spiral Arms. *Particuology* **2022**, *68*, 101–113. [CrossRef]
46. Baltrėnas, P.; Chlebnikovas, A. Removal of Fine Solid Particles in Aggressive Gas Flows in a Newly Designed Multi-Channel Cyclone. *Powder Technol.* **2019**, *356*, 480–492. [CrossRef]
47. Zhou, W.; E, C.; Fan, Y.; Wang, K.; Lu, C. Experimental Research on the Separation Characteristics of a Gas-Liquid Cyclone Separator in WGS. *Powder Technol.* **2020**, *372*, 438–447. [CrossRef]
48. Elsayed, K. Optimization of the Cyclone Separator Geometry for Minimum Pressure Drop Using Co-Kriging. *Powder Technol.* **2015**, *269*, 409–424. [CrossRef]
49. Elsayed, K. Design of a Novel Gas Cyclone Vortex Finder Using the Adjoint Method. *Sep. Purif. Technol.* **2015**, *142*, 274–286 [CrossRef]

30. Singh, P.; Couckuyt, I.; Elsayed, K.; Deschrijver, D.; Dhaene, T. Shape Optimization of a Cyclone Separator Using Multi-Objective Surrogate-Based Optimization. *Appl. Math. Model.* **2016**, *40*, 4248–4259. [CrossRef]
31. Cengel, Y.; Cimbala, J. *Fluid Mechanics Fundamentals and Applications (Si Units)*; McGraw-Hill Higher Education: Boston, MA, USA, 2010.
32. Fathizadeh, N.; Mohebbi, A.; Soltaninejad, S.; Iranmanesh, M. Design and Simulation of High Pressure Cyclones for a Gas City Gate Station Using Semi-Empirical Models, Genetic Algorithm and Computational Fluid Dynamics. *J. Nat. Gas Sci. Eng.* **2015**, *26*, 313–329. [CrossRef]
33. Oh, J.; Choi, S.; Kim, J. Numerical Simulation of an Internal Flow Field in a Uniflow Cyclone Separator. *Powder Technol.* **2015**, *274*, 135–145. [CrossRef]
34. Kui, C. *Design of Experiments and Analysis*; Tsinghua University Press: Beijing, China, 1996.
35. Roy, R.K. *Design of Experiments Using the Taguchi Approach: 16 Steps to Product and Process Improvement*; John Wiley & Sons: New York, NY, USA, 2001.

Article

Research on Multiple Constraints Intelligent Production Line Scheduling Problem Based on Beetle Antennae Search (BAS) Algorithm

Yani Zhang [1,2], Haoshu Xu [3,*], Jun Huang [1,2] and Yongmao Xiao [1,2,*]

1   School of Computer and Information, Qiannan Normal University for Nationalities, Duyun 558000, China
2   Key Laboratory of Complex Systems and Intelligent Optimization of Guizhou Province, Duyun 558000, China
3   Office of Academic Affairs, Qiannan Broadcast Television University, Duyun 558000, China
*   Correspondence: zyn19800126@sina.com (H.X.); xym198302@163.com (Y.X.)

Abstract: Aiming at the intelligent production line scheduling problem, a production line scheduling method considering multiple constraints was proposed. Considering the constraints of production task priority, time limit, and urgent task insertion, a production process optimization scheduling calculation model was established with the minimum waiting time and minimum completion time as objectives. The BAS was used to solve the problem, and a fast response mechanism for emergency processing under multiple constraints was established. Compared with adaptive particle swarm optimization (APSO) and non-dominated sorting genetic algorithm-II (NSGA-II) operation, this algorithm showed its superiority. The practical application in garment processing enterprises showed that the method was effective and can reduce the completion time and waiting time.

Keywords: production line scheduling; multiple constraints; beetle antennae search; multi-objective

1. Introduction

The optimization of production line scheduling is the key to realizing high-efficiency production in modern manufacturing systems. The production line scheduling problem is a complex optimization problem. The state of modern production lines is constantly changing, so the production line needs to have the ability to quickly adjust [1–3]. As the pillar industry of economy and social development, the manufacturing industry is facing severe tests under the background of economic globalization. It is very necessary to upgrade related industries' informatization and intelligence [4,5]. Due to the great differences in the production process, raw materials, and the production equipment of different products produced in the workshop, the complexity of production scheduling is a problem [6–8]. Therefore, it is necessary for enterprises to use the control system to schedule the production line to ensure the stability and continuity of production and improve the production capacity.

The production line scheduling problem is also typical due to its variable production environment, diverse research objects, complex constraints, and other factors. Its main characteristics are complexity, randomness, multivariable, and multi-objective. Therefore, the difficulty of solving shop floor scheduling will increase exponentially with the passage of time and the accumulation of tasks, which is recognized as a NP problem. Therefore, in the research on the production line scheduling problem, the improvement and perfection of the solution method has become the key research field to solve the problem. For the production line scheduling problem, scholars at home and abroad have carried out corresponding research. Hu et al. [9] proposed a production scheduling model based on grey prediction, which predicted the product demand and planned the inventory production capacity through the grey prediction method. The method is verified by a case study of a glass manufacturing enterprise. Zhang et al. [10] proposed a new firefly

Citation: Zhang, Y.; Xu, H.; Huang, J.; Xiao, Y. Research on Multiple Constraints Intelligent Production Line Scheduling Problem Based on Beetle Antennae Search (BAS) Algorithm. *Processes* **2023**, *11*, 904. https://doi.org/10.3390/pr11030904

Academic Editor: Sergey Y. Yurish

Received: 2 February 2023
Revised: 9 March 2023
Accepted: 14 March 2023
Published: 16 March 2023

algorithm based on Levy Flight, aiming at problems such as hunger and blockage in the intelligent production line scheduling process. Compared with GSO (glowworm swarm optimization), SGSO (glowworm swarm optimization of scene understanding), and CGSO (chaos glowworm swarm optimization), it has better solving accuracy, convergence, and stability. Ma et al. [11] built an enterprise production line simulation model based on Agent modeling technology, which can effectively simulate the actual production process of the factory. Jia et al. [12] proposed a Petri net model suitable for the study of flexible production line scheduling and verified the method by taking the machine tool seat-type flexible production line as an example. Wu et al. [13] proposed a data-driven semiconductor production line scheduling framework. Based on scheduling optimization data samples and a machine learning algorithm, the framework can determine the approximate optimal scheduling strategy in real time according to its current production state. Jin et al. [14] developed a central control and scheduling system based on the logistics scheduling method of discrete production lines. Eroglu et al. [15] carried out research on large-scale loom scheduling, focusing on the series-related setting time and scheduling model with machine resource constraints, and proposed an improved hybrid genetic algorithm that could solve machine resource constraints. Wang et al. [16] improved and optimized the knitting shop scheduling model and used the improved genetic algorithm to solve the above model. Zhang et al. [17] proposed a production optimization method based on deep reinforcement learning and applied it to reservoir models. This approach maximized the net present value throughout the lifecycle and enabled real-time adjustments to the good control solution. Duan et al. [18] proposed a fixed-time, time-varying output formation-containment (FT-TV-OFC) control system for heterogeneous universal multi-agent systems and verified the system through a case study. Wang et al. [19] proposed a generalized growth-oriented remanufacturing services (GGRMS) method, which can maximize the residual value of retired products and reduce process consumption and resource waste. The method was verified by gearbox remanufacturing. Cao et al. [20] constructed a multi-objective optimization model for multi-segment heterogeneous vehicles, which took into account four objective functions: total cost, maximum time, carbon emission, and load utilization. A meme algorithm based on Two_Arch2 is proposed to deal with the model.

The above literature studies the production line scheduling and puts forward some optimization methods. However, the research on the constraints of the production line is not comprehensive enough and seldom considers the characteristics of multiple constraints. The multi-constraint intelligent production line is a process in which multiple modules and different devices run at the same time under multiple constraints. Aiming at the complexity and multi-constraint characteristics of the intelligent production line scheduling problem, this paper determines the optimization model of the scheduling problem by analyzing the scheduling process of the production line. The scheduling model takes the minimum capacity completion time and the minimum station waiting time as the optimization objectives, and the processing time, worker skill proficiency, process, and equipment as constraints. The model is solved by BAS, and the comparison of BAS with APSO and NSGA-II algorithms shows that BAS is faster and more accurate. The model is oriented to the scheduling problem of the manufacturing production line, and the problem of untimely mass production scheduling of the garment production line is analyzed as an example.

2. Multi-Constraint Intelligent Production Process

The main body of the production line scheduling problem mainly includes production tasks and plans, schedulable resources, scheduling constraints, and production optimization indicators. For general production line scheduling problems, production tasks and plans generally refer to jobs, and the schedulable resources are generally the production and processing equipment of enterprises [21]. The key to production line scheduling is to allocate tasks reasonably and efficiently to the schedulable resources. The optimization index of production line scheduling, that is, the optimization objective, has diversity and variability according to the different strategies of the production enterprise. The determination of the optimization objective mainly depends on the production cost and product quality [22]. For most enterprises, common optimization objectives include: minimum sum of production material cost, processing cost, and logistics cost, maximum product output per unit production time, minimum and maximum completion time of production tasks, timely delivery of urgent orders and high-priority production tasks without delay, high utilization rate of workshop equipment, and low energy consumption of key production and processing equipment. The constraints of production line scheduling problems mainly include: product process constraints, transport path constraints of material handling equipment such as AGVs, operating load constraints of processing machine tools, and constraints of limited resources, such as the number of personnel, equipment, and tools and fixtures [23,24].

2.1. Multi-Constraint Intelligent Production Line Scheduling Process Analysis

The multi-constraint intelligent production line process mainly includes a scheduling control module, transportation module, loading and unloading module, execution module, and an auxiliary equipment module. The multi-constraint intelligent production line scheduling control module sends out instructions based on the detected location of the workpiece and material.

The workpiece and material should be timely delivered to the pre-processing area of the corresponding station, as required. The operator in the processing area will execute the processing order, and then transfer to the next area for further processing or return to the unloading area of the material center [25,26].

2.2. Scheduling Requirements

The operation process of the multi-constraint intelligent production line is a process in which multi-functional modules run at the same time with different equipment [27]. The production task of each equipment is formulated according to the demand. Multiple single equipment runs at the same time to complete specific tasks, such as transportation production and assembly, loading and unloading, etc., and the equipment and processing time of product production during the operation process may be adjusted. The operation process of the multi-constraint intelligent production line is a combination of equipment, personnel, and materials formed to complete a specific task under specific constraints. Therefore, the operation process of the multi-constraint intelligent production line should meet the following requirements [28–30]:

(1) Multi-constraint intelligent production line operation requires that under specific constraints, multiple devices in the intelligent production line work cooperatively to complete the production task. Real-time dynamic priority division can be conducted on the urgency of production tasks according to the actual situation of multi-constraint production, the production sequence can be arranged according to the level, and production tasks can be rescheduled during the process of task execution. The balance rate of the intelligent production line is optimized to improve the overall utilization rate of resources and reduce the waiting time to achieve optimal scheduling objectives.

(2) Each process of the multi-constraint intelligent production line needs different modules to cooperate. The function of the same module is the same, and the theoretical maximum production capacity of each process is the same. If there are multiple execution combinations in the same process, the distribution of production tasks shall be coordinated according to the total amount of tasks, and the task shall be equally distributed when the maximum production capacity of all combinations is met. If a production task does not meet the maximum capacity, allocate devices based on production requirements.

(3) Multi-constraint intelligent production line product production under constraints requires multiple sets or groups of equipment to work together. According to the different process route to establish different constraints.

(4) The multi-constraint intelligent production line can set the overall ideal production balance rate, and schedule and adjust the production process according to the actual production balance rate.

(5) The higher the daily production capacity, the better, and it is necessary to try to reduce the number of scheduling adjustments in the production process.

2.3. Scheduling Model

The production process, raw materials, and production equipment of different enterprises greatly differ, but the scheduling objectives mainly include time objectives, resource objectives, carbon emission objectives, efficiency objectives, etc. [31,32]. The scheduling of the multi-constraint intelligent production line is similar. Completion time is the time it takes to complete a job or an entire process. It is the time consumed by workers to directly process the object of labor and complete each process operation. It is the basic component of the quota time. The waiting time is the idle time between the completion of the last process and the start of the next process. Minimizing the waiting time can effectively improve the device utilization. In this paper, the minimum production capacity completion time and the minimum waiting time of the station are taken as the optimization objectives. The completion time and station waiting time are targeted for intelligent production line operation scheduling under multiple constraints [33–35]. The objective function is shown as follows:

$$\mathrm{min}f_1 = \mathrm{min}\sum_{i=1}^{n} = F_{ijm,x} \tag{1}$$

$$\mathrm{min}f_2 = \sum_{i=1}^{n}\sum_{j}^{n_i} \left(T_{ijk} - C_{i(j-1)}\right) \tag{2}$$

$$C_{ijk} = \max\{C_{i(j-1)k}, T_{ijk}\} + P_{ijkh}, k \in M_{ij} \tag{3}$$

where T_{ijk} is the processable time of process j of order i on equipment k, and $C_{i\,(j-1)}$ is the completion time of process $j-1$ of order i. The objective function f_1 is the total completion time, where $F_{ijm,x}$ is the completion time of the assumed process on machine m. The objective function f_2 is the station waiting time, and it is the difference between the processing time of the next process of the order $T_{ijk}, C_{i(j-1)}$ and the completion time of the last process.

The objective function of intelligent production line is subject to the following constraints:

$$P_{ijkh} = \frac{P_{ijk}}{a_h} \tag{4}$$

$$S_{ijk} \geq t_i, k \in M_{ij} \tag{5}$$

$$\sum_{i=1}^{n} X_{ijk} = 1, k \in M_{ij} \tag{6}$$

$$C_{ijk} = \max\left\{C_{i(j-1)k}, T_{ijk}\right\} + P_{ijkh}, k \in M_{ij} \tag{7}$$

$$C_{ijk} \leq S_{i(j+1)k}, k \in M_{ij} \tag{8}$$

where, P_{ijkh} is the processing time constraint of the process, P_{ijk} is the processing time of process j of order i on equipment k, a_h is the production efficiency of employees with different skill levels, the skill level of personnel is A, B, and C, and the production efficiency ratio of each level is $\alpha_A : \alpha_B : \alpha_C = 1.2 : 1 : 0.8$. X_{ijk} is the discriminant condition of process j of order i processed on equipment k, and when equipment k is selected for processing, X_{ijk} is 1, otherwise it is 0. S_{ijk} and C_{ijk} are the actual start time and completion time of the j process of order i on equipment k, respectively. Formula (4) indicates that the actual processing time of an operation is affected by the production efficiency of employees. For the same operation, the actual processing time of efficient employees is shorter. Formula (5) indicates that the order can be put into production only after the user places the order. Formula (6) indicates that a process can only be processed once on one equipment. Formula (7) indicates the completion time constraint of the order. Formula (8) represents the processing order constraint of the order. In addition to the above constraints, the following assumptions are specified in this study: one machine can only process one process, the processing order of different orders has no sequence constraints, etc.

3. Steps of BAS

3.1. Intelligent Production Scheduling Process under Multiple Constraints

The intelligent production scheduling process under multiple constraints has the following steps [36–38]:

1. Constraint parameter expression.
2. Calculate the number of stations required by each process according to the actual situation of the process.
3. Select all stations suitable for each process according to the configuration information of each station.
4. Form a preliminary pipeline distribution plan according to the log-on status of employees at each station and the historical production data of employees and implement the distribution.
5. Measure the balance rate of the production line according to the actual production capacity of each process and judge the rationality of the current production line process allocation.
6. Generally, the balance rate of the production line is used to measure the balance of the production line. When the balance rate of the production line is greater than 85%, it indicates that the load is distributed evenly. If the balance rate of the production line is >85%, proceed to the next step, if not, return to the previous step.
7. If it is judged that the current balance rate of the production line is lower than the present value, the working procedure shall be arranged again according to the actual production efficiency and station configuration information of each station, and the production data of each station shall be recorded as reference data for the next intelligent production scheduling.
8. Check whether the site memory of each station reaches the site threshold of this process.
9. Select the appropriate number of stations according to the production capacity of each process, site configuration information, and employee production data to help process this process.
10. Check whether the balance rate reaches the maximum value. If yes, proceed to the next step. If no, return to the previous step.
11. Stop emergency dispatching.

Figure 1 shows the process flow of multi-constraint intelligent production line scheduling, where Y is the discriminant condition satisfied and N is the condition not satisfied.

3.2. Algorithm Analysis

Many methods have been proposed to solve the job shop scheduling problem. BAS is a heuristic optimization algorithm proposed by Jiang et al. [39]. This algorithm simulates the foraging behavior of beetles to build a mathematical model and solve complex problems. The beetle does not know the location of the target food point during the foraging process, but uses the two whiskers on its head to collect the strength of the smell. When the scent is stronger on the left than on the right, the next step is to the left, and then the next step is to the right. The beetles are efficient at finding food based on this principle. Compared with other traditional algorithms, BAS requires only one individual to search, and the computation is greatly reduced. It is easy to realize, and the convergence speed is fast, so BAS was chosen as the computational solution of this study [40,41].

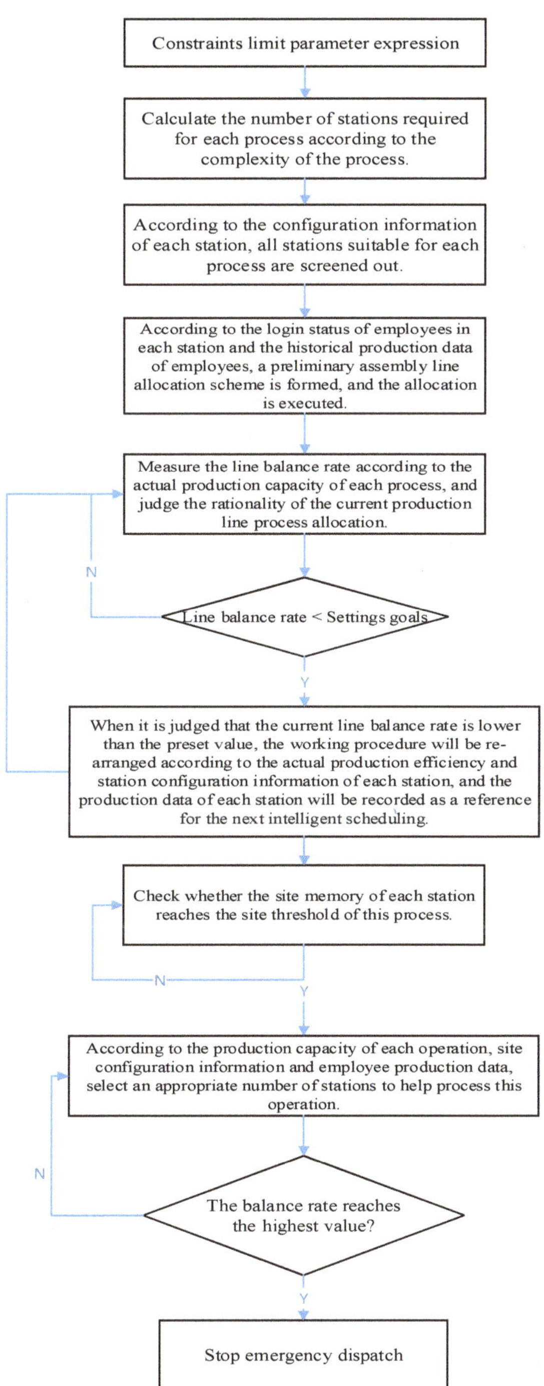

Figure 1. Multi-constraint intelligent production line scheduling process flow chart.

The idea of the BAS algorithm is that the specific location of food is equivalent to the maximum value point of the objective function, and beetles move step-by-step towards the location with the strongest smell of food. Compared with the artificial bee colony algorithm and the ant colony algorithm, the BAS can perform optimization calculation without an explicit objective function and gradient vector. BAS is different from other optimization algorithms as BAS only needs one individual, namely a beetle, and this algorithm uses a single individual to solve the optimization problem, with a low computational complexity and fast optimization speed [42–44]. The body steps are as follows:

(1) Randomly generate and standardize the direction vector:

$$b = \frac{rands(N,1)}{||rands(N,1)||} \tag{9}$$

where, N is the spatial dimension of variables, and $rands(\cdot)$ is a random function.

(2) Calculate the coordinates of the left and right whiskers:

$$x_l - x_r = 2d^t b \tag{10}$$

x_l, x_r can be represented by the center of mass, as follows:

$$x_l = x^t + d^t b \tag{11}$$

$$x_r = x^t - d^t b \tag{12}$$

where, x_t is the coordinate of ceramide at time t, and d^t is the horizontal projection search distance from the center of mass to the whisker at time t.

(3) To determine the odor intensity of the left and right antennae, $f(x_l)$ and $f(x_r)$ are used to substitute the left and right positions, and $f(x)$ is the objective function.

(4) According to the corresponding smell of the two whiskers, determine the next movement position of the longicorn:

$$x^{t+1} = x^t - \delta^t b\, sign(f(x_r) - f(x_l)) \tag{13}$$

where, δ^t is the step size at time t.

(5) Search distance and step update:

$$d^{t+1} = \eta_d d^t + d_0 \tag{14}$$

$$\delta_t = \eta_\delta \delta^{t-1} \tag{15}$$

where, d_0 represents the artificially set minimum step size, and η_d and η_δ are the attenuation coefficients of search distance and step size renewal, respectively.

Figure 2 shows the workflow of the beetle whisker search algorithm, and Figure 3 shows the algorithm workflow combination process diagram, where Y is the discriminant condition satisfied and N is the condition not satisfied.

The traditional BAS algorithm is only suitable for the optimization of continuous functions, so it needs to be discretized to solve the scheduling problem. In this paper, a non-fixed point discrete division method was adopted. Control vector parameterization is a direct method for numerically solving optimal control problems. This method transforms the original infinite-dimensional dynamic optimization problem into a finite-dimensional static optimization problem via discretization. The parameterization process of the control vector requires a finite number of parameters to approximate the control vector $u(t)$, which varies continuously with time. Time intervals are usually divided into n subintervals. In dynamic optimization, the time interval is divided by equal division and unequal division. Unequal partition of time intervals involves initializing a set of parameters in the time domain $\left[t_0,\ t_f\right]$ according to some rules, namely $(t_1,\ t_2,\ \ldots,\ t_n \in (t_0, t_f))$, and the division formula of unequal division is as follows [45]:

$$t_i = t_0 + \frac{(t_f - t_0) \times \sum\limits_{k=1}^{i} \tau_k}{\sum\limits_{i=1}^{n} \tau_i}, i = 1, 2, \ldots, n \tag{16}$$

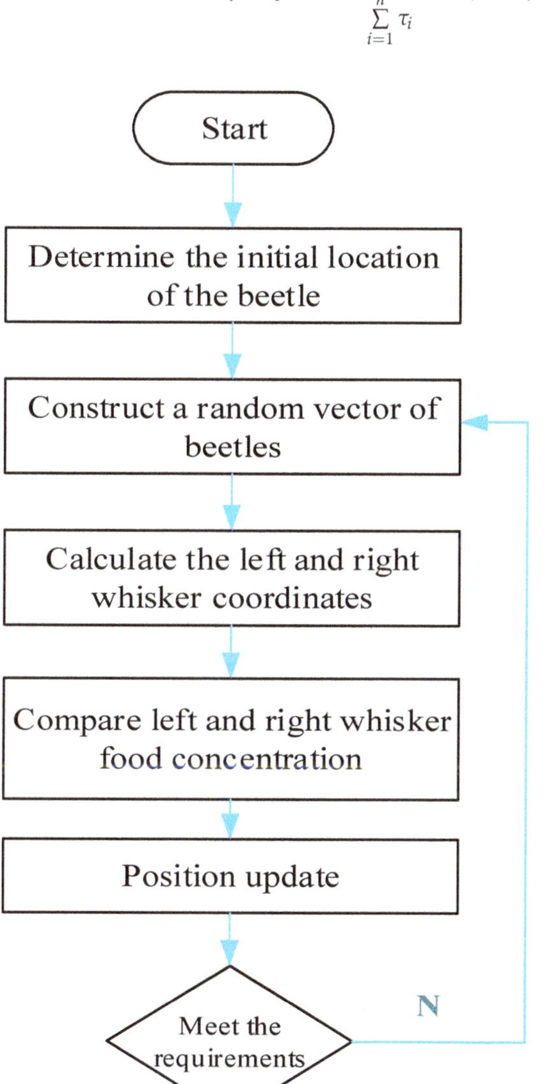

Figure 2. BAS algorithm flow.

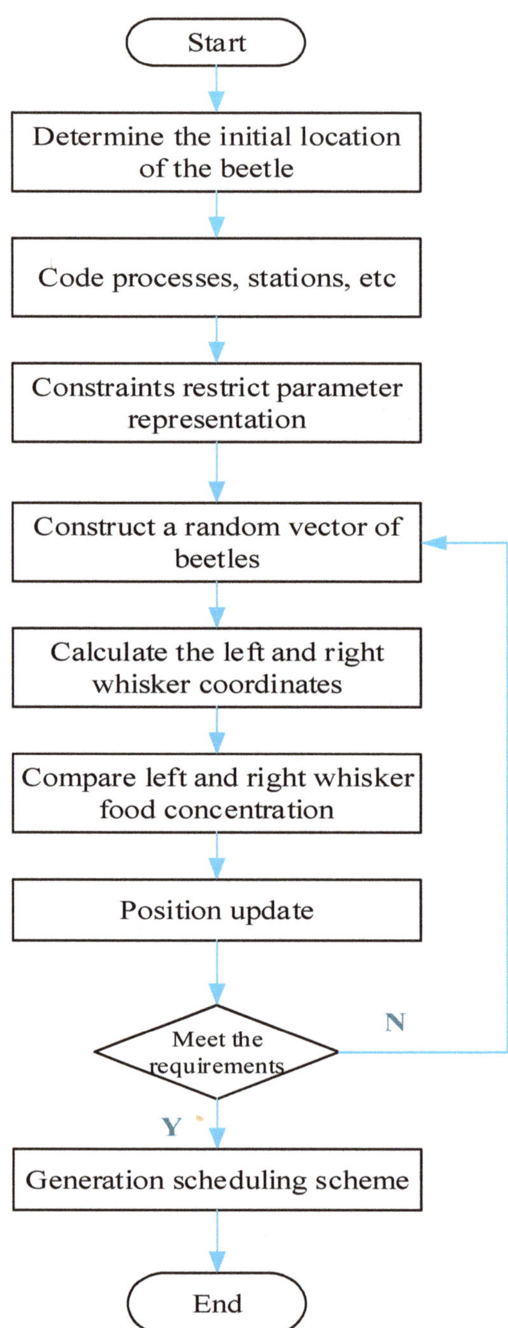

Figure 3. Algorithm flow diagram combined with the process diagram.

The non-fixed point discrete division method is a new method to determine tim nodes in the time domain. It is based on the control parameterization idea of the piecewis constant method. The method is to randomly select n time interval points, t_i, $i = 1, 2, \ldots, r$ within the time interval $\left[t_0,\ t_f\right]$ between the beginning and the end of the reaction proces

The distance between two different adjacent points can be longer or shorter, the position of nodes can be sparse or dense, and $[t_i, t_{i+1}]$ is calculated in turn to obtain the control trajectory. Compared with the method of equal division and unequal division, the non-fixed point discrete division method can determine time nodes more randomly. The non-fixed point discrete division method can refine the control process and produce a more accurate control trajectory.

In this paper, the production line was scheduled by using the longicorn whisker search algorithm. Combining the longicorn whisker search algorithm and the multi-constraint production line scheduling model, the minimum capacity completion time and the shortest waiting time of the station are the optimization objectives. The algorithm and scheduling process are shown in Figure 3. The parameters in the Tendon search algorithm were initialized, including the space dimension k, the distance between the left and right whiskers d, the initial step length, and the number of iterations, and the position of Tendon was initialized and the direction was random. The production line capacity completion time, station waiting time, station information, equipment, etc., were coded. The beetle random vector was constructed, and the next moving direction of longicorn was determined by comparing the signal size of longicorn's left and right whiskers, and then the position of longicorn was updated. The fitness of the solution was iteratively calculated, the individuals with better fitness were selected, and the iteration was completed after continuous screening until the conditions were met.

4. Cases

The intelligent production line is widely used in modern enterprises, and different enterprises have different characteristics. There are many technological processes in the garment manufacturing industry, and the intelligent production line is long, and each stage has its own characteristics. The garment intelligent production line is more complex than the ordinary mechanical product processing intelligent production line, and may have several kinds of ordinary mechanical product production line problems. Intelligent production line scheduling can be used to solve the problem of garment production, and the method can be applied to many industries.

Garment hanging production systems have been developed since the 1970s. It is a technological crystallization from traditional manual manufacturing in order to improve production efficiency and management levels based on the development of an industrial base and garment production practice. The garment hanging system belongs to the production and transmission mode of the garment assembly line, which is famous for its automatic high-tech method. The hanging system requires the combination of manpower and equipment to form a set of rigorous and mechanized production transmission and management modes. In the production process, the corresponding hardware system is used to directly transport the semi-finished product cutting pieces to the side of each processing machine, and the length of the line can be adjusted at any time. The processing chain and production mode are not limited, with good flexibility. It is convenient for garment mass production and saves time. Due to the impact of order arrival times, style differences, insert order and return order, and other factors under the clothing mass personalized customization mode, the production scheduling is complicated.

Taking multi-constraint garment hanging production system scheduling as an example, the garment hanging production system of many enterprises currently has a large gap in personnel quality, management level, and operational skills, which limits the roles of the system functions, and thus results in the factories equipped with hanging production lines not improving the production efficiency much compared with before. At present, the scheduling function of the garment hanging production system is only to put forward the processes to be scheduled and find feasible scheduling schemes in the production process, and the final scheduling decision is still up to the line leader. However, artificial scheduling cannot guarantee the optimal balance rate of the production line, which may lead to the existence of bottleneck processes and frequent scheduling of the production line.

4.1. Advantages of Multi-Constraint Intelligent Production Line

Figure 4 shows the flow chart of scheduling, where (a) is the ordinary scheduling and (b) is the multi-constraint intelligent scheduling. Multi-constraint intelligent production line has the following advantages.

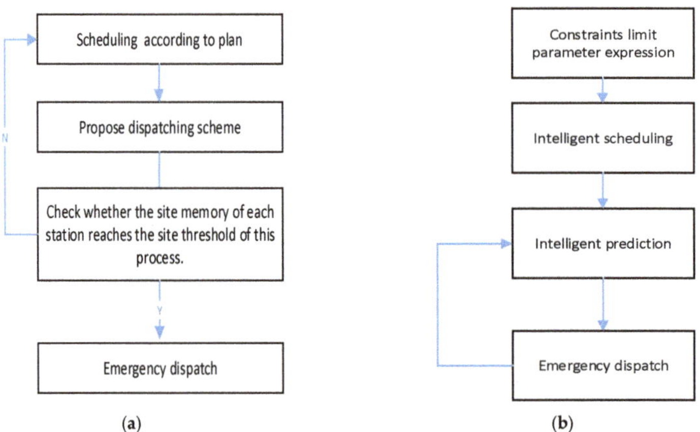

Figure 4. Scheduling flow diagram: (**a**) general scheduling and (**b**) multi-constraint intelligent scheduling.

(1) In the general production system, the production process is greatly affected by the level of operational skills. The production process of the multi-constraint intelligent production line is adjusted according to the emergency situation of production tasks and equipment conditions, which is not affected by the skill level of the operators.

(2) In general, the scheduling power of the production system is largely determined by the operator. The scheduling function of the production system only puts forward the processes to be scheduled and feasible scheduling schemes in the production process, and the final scheduling decision is made by the line manager according to the production situation. Multi-constraint intelligent production line scheduling can predict the production situation according to the production plan and automatically schedule the production.

(3) The overall utilization rate of the equipment in the production process of the general production system is difficult to be used as a good indicator. The multi-constraint intelligent production line scheduling can take the equipment utilization rate as a separate indicator for assessment, which can effectively ensure the equipment utilization rate.

(4) Due to the large gaps in personnel quality, management level, and operational skills in the general production system, it is difficult to exert the production efficiency of the system function to a large extent. The multi-constraint intelligent production line can effectively improve production efficiency according to the real-time scheduling of production tasks and equipment.

4.2. Case Study

Taking the actual clothing processing and production as an example, the general processing and production process of clothing enterprises is as follows: discharge → cutting → making bags → sewing → keyhole nail → ironing → garment inspection → packaging → warehousing or shipping. The production process is arranged according to the demand. The number of people in the shirt hanging production line is 13, with 1 person per 1 station, and each station has 1 or 2 equipment. According to the time ratio of each processing process of shirts, 16 sets of equipment are set, among which the flat sewing machine, ironing table, five-line car, chain machine, keyhole machine, and nailing machine configuration ratio is 7:4:2:1:1.

4.3. Result

To verify the effectiveness of the algorithm, the APSO algorithm, NSGA-II algorithm, and BAS algorithm were selected for comparison. With the minimum waiting time as the goal, the APSO, NSGA-II, and BAS algorithms were compared. MAtlab2016 was used for programming, and the BAS algorithm, APSO algorithm, and NSGA-II algorithm parameters were set as follows, to ensure the maximization of the population size and the maximum number of iterations as far as possible. The population size was set as N = 100 and the maximum number of iterations as 300 generations. The crossover rate was $P_c = 0.8$, and the variation rate was $P_m = 0.02$. BAS parameters were set as follows: dimension n was 3, the coefficient η_δ between the distance of two whiskers and step size was 5, the initial step size of each beetle was 0.3, and the maximum number of iterations was 300. The algorithm comparison results are shown in Figure 5. The optimal value of the APSO algorithm was 1.5 s, of the NSGA-II algorithm was 1.3 s, and of the BAS algorithm was 1.1 s. The results after the BAS algorithm optimization were minimal, and BAS had good convergence. As the number of iterations increased, the waiting time gradually decreased until it became stable. It can be seen from the iteration curves of the three algorithms that the BAS algorithm had the lowest number of iterations, followed by NSGA-II and APSO.

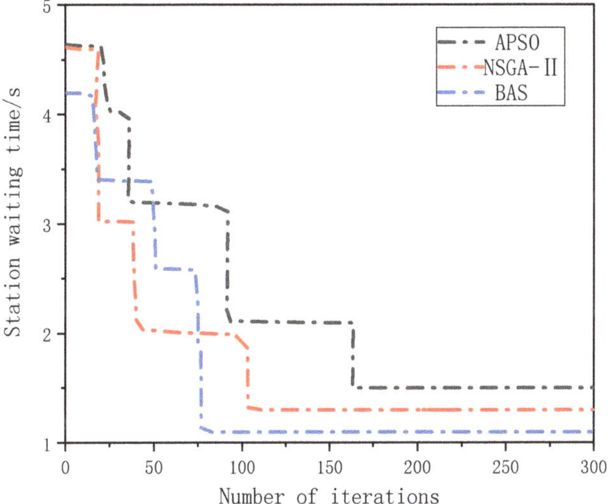

Figure 5. Iteration diagram of station waiting time.

Objective 2 was obtaining the minimum completion time, and the maximum number of iterations was 300. The comparison results of the three algorithms are shown in Figure 6, and the optimal value was 327.2 s. The optimal value of the APSO algorithm was 347.5 s, of the NSGA-II algorithm was 341.3 s, and of the BAS algorithm was 327.2 s. The completion time after the BAS algorithm optimization was minimum, and the algorithm had good convergence. As can be seen from the figure, as the number of iterations increased, the completion time gradually decreased until it became stable. From the iteration curves of the three algorithms, it can be seen that the BAS algorithm had the lowest number of iterations, followed by NSGA-II and APSO.

Figures 7–9 show the scheduling results. Different colors in the figure represent different processes, and the corresponding colors of the process numbers have been given in the figure.

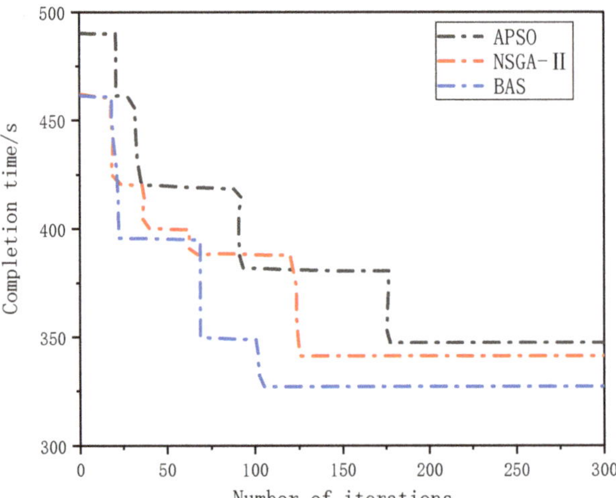

Figure 6. Completion time iteration diagram.

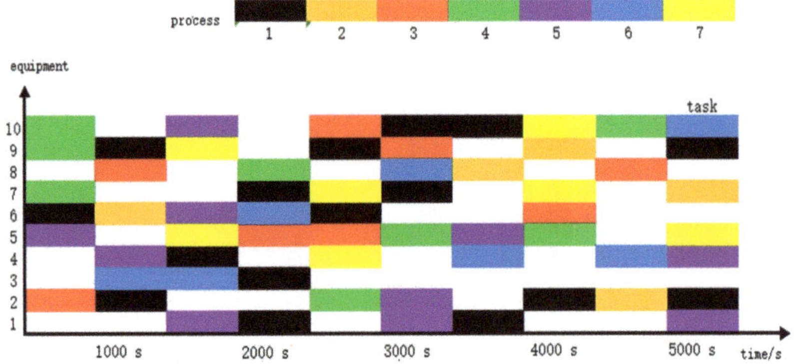

Figure 7. Production plan before scheduling.

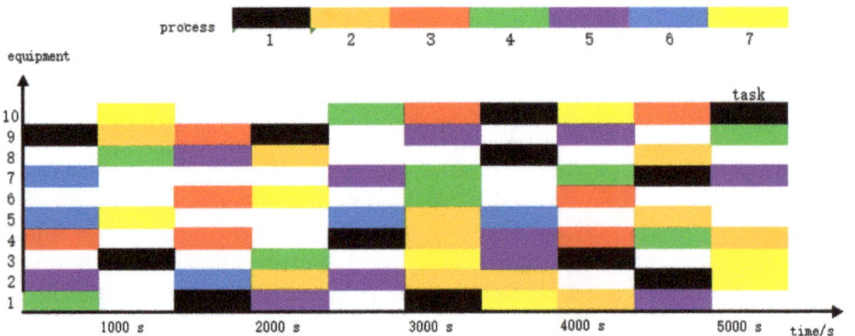

Figure 8. Scheduling scheme for complete human cooperation.

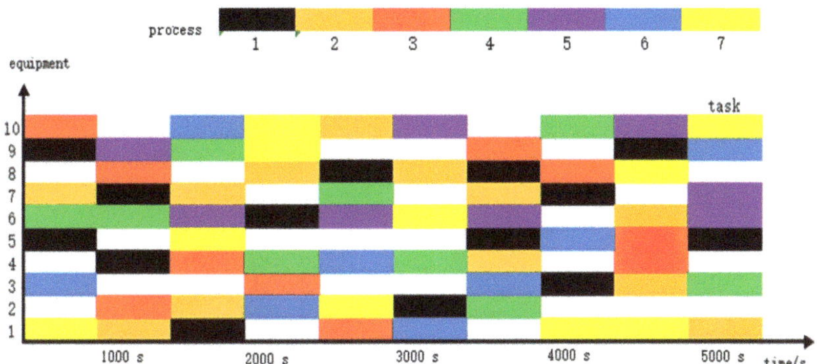

Figure 9. Scheduling scheme for multi-constrained intelligent scheduling.

The experimental results showed that the production time and equipment utilization rate required to complete 500 pieces of clothing were 23,027 s and 61% in the production plan before scheduling, 20,156 s and 64% in the scheduling under the assumption of complete human cooperation, and 18,936 s and 71% in the multi-constrained intelligent scheduling. After multi-constraint intelligent scheduling, the shortest completion time was 18,936 s, which was 82% of that before optimization. The device utilization rate of multi-constraint intelligent scheduling was 10% higher than that before optimization and the waiting time was greatly reduced.

4.4. Comparison with Related Research

Some methods to reduce the completion time and waiting time of job-shop scheduling have been reported at home and abroad [9–26], and these methods and algorithms are effective for saving time. However, these studies are all about job-shop scheduling without considering production constraints. In this case, because the production limit is not precisely limited, it is difficult to ensure accurate and effective scheduling. In the stage of production planning, whether the scheduling has pertinence depends on the limitation of the basic background to a great extent. On the basis of multi-constraint production, this paper optimized the process parameters with the aim of minimizing the completion time and the waiting time. The results are more practical by using the relatively advanced BAS.

5. Conclusions

A multi-constraint intelligent production line scheduling method based on BAS was proposed to solve the problem of untimely scheduling of manufacturing production, and we used the garment mass production line as an example to verify. This method considers production line scheduling under various constraints and uses BAS to solve the problem, aiming at the order import of the multi-constraint intelligent production line, based on the existing production conditions, using the BAS for real-time production scheduling, to maximize the utilization rate of equipment and to achieve the efficiency of the production line. Experimental results showed that compared with traditional scheduling, the completion time of multi-constraint intelligent production line scheduling based on BAS was reduced by 7.1%, and the waiting time of equipment was reduced by 16.2%.

In the future, the research will mainly focus on two points: first, more typical multi-constraint intelligent production line scheduling problems will be studied, and second, the typical process scheduling of the multi-constraint intelligent production line for the same product will be studied.

Processes **2023**, 11, 904

Author Contributions: Conceptualization, Y.Z. and Y.X.; methodology, Y.Z.; software, Y.Z.; validation, H.X.; formal analysis, H.X.; investigation, H.X.; data curation, J.H.; writing—original draft preparation, J.H. and Y.X.; writing—review and editing, Y.X.; funding acquisition, J.H. All authors have read and agreed to the published version of the manuscript.

Funding: This study was supported by the development project of young scientific and technological talents in colleges and universities of Guizhou Province ([2020]195), and the program of Qiannan Normal University for Nationalities under Grant No. QNSY 2018025.

Institutional Review Board Statement: Not applicable.

Informed Consent Statement: Not applicable.

Data Availability Statement: The study did not report any data.

Conflicts of Interest: The authors declare no conflict of interest.

References

1. Ebrahimipour, V.; Najjarbashi, A.; Sheikhalishahi, M. Multi-objective modeling for preventive maintenance scheduling in a multiple production line. *J. Intell. Manuf.* **2015**, *26*, 111–122. [CrossRef]
2. Li, H.Y.; Gui, C.; Xiao, K. Simulation of multivariate scheduling optimization for open production line based on improved genetic algorithm. *Int. J. Simul. Model.* **2018**, *17*, 347–358. [CrossRef]
3. Xiao, Y.; Zhou, J.; Wang, R.; Zhu, X.; Zhang, H. Energy-Saving and Low-Carbon Gear Blank Dimension Design Based on Business Compass. *Processes* **2022**, *10*, 1859. [CrossRef]
4. Qu, C.; Shao, J.; Cheng, Z. Can embedding in global value chain drive green growth in China's manufacturing industry? *J. Clean. Prod.* **2020**, *268*, 121962. [CrossRef]
5. Yu, P.; Ma, H.; Yang, Y.; Tao, L.; Mba, D. A Novel Grey Incidence Decision-making Method Based on Close Degree and Its Application in Manufacturing Industry Upgrading. *J. Grey Syst.* **2020**, *32*, 1–15.
6. Xiao, Y.; Zhao, R.; Yan, W.; Zhu, X. Analysis and Evaluation of Energy Consumption and Carbon Emission Levels of Products Produced by Different Kinds of Equipment Based on Green Development Concept. *Sustainability* **2022**, *14*, 7631. [CrossRef]
7. Mota, B.; Gomes, L.; Faria, P.; Ramos, C.; Vale, Z.; Correia, R. Production Line Optimization to Minimize Energy Cost and Participate in Demand Response Events. *Energies* **2021**, *14*, 462. [CrossRef]
8. Yu, H.; Han, S.; Yang, D.; Wang, Z.; Feng, W. Job Shop Scheduling Based on Digital Twin Technology: A Survey and an Intelligent Platform. *Complexity* **2021**, *2021*, 8823273. [CrossRef]
9. Hu, O. Research on Scheduling Model of Production Line Based on Grey Prediction and Its Application. *J. Wuhan Univ. Technol.* **2020**, *42*, 106–112.
10. Zhang, J.; Li, X. Research on Intelligent Production Line Scheduling Problem Based on LGSO Algorithm. *Comput. Sci.* **2021**, *48*, 668–672.
11. Ma, W.; Zhang, Y. Scheduling batches in shampoo industry based on simulation method. *Comput. Integr. Manuf. Syst.* **2022**, *28*, 3403–3420.
12. Jia, Y.; Zhang, K. Scheduling model of machine tool seat flexible production line based on Petri net. *J. Mech. Electr. Eng.* **2022**, *39*, 7.
13. Wu, Q.; Ma, Y. Data-driven dynamic scheduling method for semiconductor production line. *Control. Theory Appl.* **2015**, *32*, 1233–1239.
14. Jin, Y.; Yang, W. Introduction of logistics scheduling system for a discrate automatic. *Manuf. Autom.* **2021**, *43*, 32–36.
15. Eroglu, D.Y.; Ozmutlu, H.C. Solution method for a large-scale loom scheduling problem with machine eligibility and splitting property. *J. Text. Inst.* **2017**, *108*, 2154–2165. [CrossRef]
16. Wang, J.A.; Pan, R.R.; Gao, W.D.; Wang, H. An automatic scheduling method for weaving enterprises based on genetic algorithm. *J. Text. Inst.* **2015**, *106*, 1377–1387. [CrossRef]
17. Zhang, K.; Wang, Z.; Chen, G.; Zhang, L.; Yang, Y.; Yao, C.; Wang, J.; Yao, J. Training effective deep reinforcement learning agents for real-time life-cycle production optimization. *J. Pet. Sci. Eng.* **2022**, *208*, 109766. [CrossRef]
18. Duan, J.; Duan, G.; Cheng, S.; Cao, S.; Wang, G. Fixed-time time-varying output formation–containment control of heterogeneous general multi-agent systems. *ISA Trans.* **2023**. In Press. [CrossRef]
19. Wang, L.; Zhao, H.; Liu, X.; Zelin, Z.; Xu-Hui, X.; Evans, S. Optimal Remanufacturing Service Resource Allocation for Generalized Growth of Retired Mechanical Products: Maximizing Matching Efficiency. *IEEE Access* **2021**, *9*, 89655–89674.
20. Cao, B.; Zhang, W.; Wang, X.; Zhao, J.; Gu, Y.; Zhang, Y. A memetic algorithm based on two_Arch2 for multi-depot heterogeneous vehicle capacitated arc routing problem. *Swarm Evol. Comput.* **2021**, *63*, 100864. [CrossRef]
21. Dong, Y. Terminal inventory level constraints for online production scheduling. *Eur. J. Oper. Res.* **2021**, *295*, 102–117. [CrossRef]
22. Zhang, H.; Li, X.; Kan, Z.; Zhang, X.; Li, Z. Research on optimization of assembly line based on product scheduling and just-in-time feeding of parts. *Assem. Autom.* **2021**, *41*, 577–588. [CrossRef]

3. Munoz, E.; Capon-Garcia, E. Systematic approach of multi-label classification for production scheduling. *Comput. Chem. Eng.* **2019**, *122*, 238–246. [CrossRef]
4. Spindler, J.; Kec, T.; Ley, T. Lead-time and risk reduction assessment of a sterile drug product manufacturing line using simulation. *Comput. Chem. Eng.* **2021**, *152*, 107401. [CrossRef]
5. Kundakc, N.; Kulak, O. Hybrid genetic algorithms for minimizing makespan in dynamic job shop scheduling problem. *Comput. Ind. Eng.* **2016**, *96*, 31–51. [CrossRef]
6. Wang, S.; Zhang, C.; Liu, Q.; Rao, Y.Q.; Yin, Y. Flexible job shop dynamic scheduling under different reschedule periods. *Comput. Integr. Manuf. Syst.* **2014**, *20*, 2470–2478.
7. Chen, H.; Chen, Y.; Ding, J.; Liu, Y.L. Analysis of garment hanging production line organization. *J. Silk* **2012**, *49*, 30–32.
8. Trevino-Martinez, S.; Sawhney, R.; Sims, C. Energy-carbon neutrality optimization in production scheduling via solar net metering. *J. Clean. Prod.* **2022**, *380*, 134627. [CrossRef]
9. Xiao, Y.; Jiang, Z.; Gu, Q.; Yan, W.; Wang, R. A novel approach to CNC machining center processing parameters optimization considering energy-saving and low-cost. *J. Manuf. Syst.* **2021**, *59*, 535–548. [CrossRef]
10. Music, G. Petri Net based solution supervision and local searchfor Job Shop scheduling. *IFAC-Pap.* **2021**, *54*, 665–670.
11. Hu, L.; Liu, Z.; Hu, W.; Wang, Y.; Tan, J.; Wu, F. Petri-net-based dynamic scheduling of flexible manufacturing system via deep reinforcement learning with graph convolutional network. *J. Manuf. Syst.* **2020**, *55*, 1–14. [CrossRef]
12. Crh, M.; Ribeiro, C.C. Shop scheduling in manufacturing environments: A review. *Int. Trans. Oper. Res.* **2022**, *29*, 3237–3293.
13. Lv, Y.; Weitong, C.; Zhong, B. The Invention Relates to an Intelligent Production Scheduling Method for Garment Hanging Production Line. CN112009975A, 1 December 2020.
14. Ziang, X.; Jinsong, D.; Guohua, Z. Adaptive dynamic scheduling of shirt hanging pipeline. *J. Text. Res.* **2001**, *41*, 144–149.
15. Tamssaouet, K.; Dauzère-Pérès, S.; Knopp, S.; Bitar, A.; Yugma, C. Multiobjective optimization for complex flexible job-shop scheduling problems. *Eur. J. Oper. Res.* **2022**, *296*, 87–100. [CrossRef]
16. Abuhamdah, A.; Alzaqebah, M.; Jawarneh, S.; Althunibat, A.; Banikhalaf, M. Moth optimisation algorithm with local search for the permutation flow shop scheduling problem. *Int. J. Comput. Appl. Technol.* **2021**, *65*, 189. [CrossRef]
17. Park, M.J.; Choi, B.C.; Min, Y.; Kim, K.M. Two-Machine Ordered Flow Shop Scheduling with Generalized Due Dates. *Asia-Pac. J. Oper. Res.* **2020**, *37*, 1950032. [CrossRef]
18. Jia, L.; Baz, D.E. A Dual Heterogeneous Island Genetic Algorithm for Solving Large Size Flexible Flow Shop Scheduling Problems on Hybrid multi-core CPU and GPU Platforms. *Math. Probl. Eng.* **2019**, *2019*, 1–13.
19. Zhao, H.; Yao, H.; Jiao, Y.; Wang, Y. An Improved Beetle Antennae Search Algorithm Based on Inertia Weight and Attenuation Factor. *Math. Probl. Eng.* **2022**, *2022*, 1–20.
20. Khan, A.T.; Cao, X.; Li, S.; Katsikis, V.N.; Brajevic, I.; Stanimirovic, P.S. Fraud detection in publicly traded U.S firms using Beetle Antennae Search: A machine learning approach. *Expert Syst. Appl.* **2022**, *191*, 116148. [CrossRef]
21. Wang, S. Numerical Analysis and Parameter Optimization of Wear Characteristics of Titanium Alloy Cross Wedge Rolling Die. *Metals* **2021**, *11*, 1998.
22. Khan, A.T.; Cao, X.; Li, S. Dual Beetle Antennae Search system for optimal planning and robust control of 5-link biped robots. *J. Comput. Sci.* **2022**, *60*, 101556. [CrossRef]
23. Medvedeva, M.; Katsikis, V.; Mourtas, S.; Simos, T.E. Randomized time-varying knapsack problems via binary beetle antennae search algorithm: Emphasis on applications in portfolio insurance. *Math. Methods Appl. Sci.* **2020**, *44*, 2002–2012. [CrossRef]
24. Wang, J.; Situ, C.; Yu, M. The post-disaster emergency planning problem with facility location and people/resource assignment. *Kybernetes* **2019**. *ahead-of-print*. [CrossRef]
25. Lyu, Y.; Mo, Y.; Lu, Y.; Liu, R. Enhanced Beetle Antennae Algorithm for Chemical Dynamic Optimization Problems' Non-Fixed Points Discrete Solution. *Processes* **2022**, *10*, 148. [CrossRef]

Article

Optimization of Low-Carbon and Highly Efficient Turning Production Equipment Selection Based on Beetle Antennae Search Algorithm (BAS)

Yongmao Xiao [1,2,3], Guohua Chen [4,*], Hao Zhang [5] and Xiaoyong Zhu [6]

[1] School of Computer and Information, Qiannan Normal University for Nationalities, Duyun 558000, China
[2] Key Laboratory of Complex Systems and Intelligent Optimization of Guizhou Province, Duyun 558000, China
[3] Key Laboratory of Complex Systems and Intelligent Optimization of Qiannan, Duyun 558000, China
[4] School of Mechanical Engineering, Hubei University of Arts and Science, Xiangyang 441053, China
[5] College of Mechanical and Electrical Engineering, Xinjiang Agricultural University, Wulumuqi 830052, China
[6] School of Economics & Management, Shaoyang University, Shaoyang 422099, China
* Correspondence: 10863@hbuas.edu.cn; Tel.: +86-13-016-410-611

Abstract: Reducing carbon emission and raising efficient production are the important goals of modern enterprise production process. The same product can be produced by a variety of equipment, and the carbon emissions and processing time of different equipment vary greatly. Choosing suitable production equipment is an important method for manufacturing enterprises to achieve the efficient emission reduction of production process. However, the traditional production equipment selection mode only gives qualitative results, and it is difficult to provide effective advice for enterprises to choose suitable equipment under the needs of carbon neutrality. To solve this problem, this paper systematically analyzes carbon emission and the time of the turning production process, and a unified calculation model for carbon emission and efficient production of multi-type processing equipment is established. The important point of the article is to research the diversity among between carbon emissions and efficiency levels of the same product produced by different devices. The carbon emissions and efficiency levels of different kinds of equipment can be calculated by the BAS algorithm. By turning a shaft part as an example, the results show that this method can calculate the optimal value of carbon emissions and efficiency of the same product produced by different equipment and can provide suggestions for enterprises to select appropriate equipment for low-carbon and efficient production. This paper provides a reference for further research on the quantitative calculation model for the selection of high-efficiency and low-carbon production equipment.

Keywords: production process; equipment selection; low carbon; highly efficient; BAS

Citation: Xiao, Y.; Chen, G.; Zhang, H.; Zhu, X. Optimization of Low-Carbon and Highly Efficient Turning Production Equipment Selection Based on Beetle Antennae Search Algorithm (BAS). *Processes* **2023**, *11*, 911. https://doi.org/10.3390/pr11030911

Academic Editor: Sergey Y. Yurish

Received: 8 February 2023
Revised: 13 March 2023
Accepted: 14 March 2023
Published: 16 March 2023

1. Introduction

The rapid development of manufacturing industry has consumed a lot of resources and caused great damage to the environment [1,2]. Under the background of climate warming, how to reduce carbon emissions in the production field has become the focus of the development of manufacturing industry. Modern manufacturing industry needs to develop in a green and sustainable direction [3]. As the main source of energy consumption and carbon emission in manufacturing process, the use of production equipment has been widely concerned. Production equipment optimization selection is one of the effective ways to promote green manufacturing [4]. There are many kinds of production equipment and a product can often be realized by different equipment, such as in the processing of surface processing equipment, including lathes, milling machines, and so on. Different equipment has a different impact on the processing quality, production energy consumption, production environment, equipment utilization rate, carbon emissions, and so on. On the basis of existing equipment resources, optimizing the selection of machine tool equipment is an important means to achieve efficient emission reduction production process in manufacturing industry, and an effective way to transform the traditional manufacturing mode to green manufacturing [5–8]. Therefore, it is very valuable to study the method of production equipment optimization selection.

Production equipment is the basic processing equipment in mechanical manufacturing industry. The variety, performance, and parameters of production equipment determine the carbon emission and efficient of the process. The optimal selection of equipment is a multi-objective and multi-scheme evaluation decision problem which has been studied by many scholars. At present, the analytic hierarchy process (AHP), fuzzy comprehensive evaluation method, and so on, have been well-applied in some fields. Li et al. [9] established a multi-criteria mixed decision model to comprehensively evaluate machine tool equipment resources. The experimental results verified the feasibility and effectiveness of the multi-criteria mixed decision model. Zhou et al. [10] proposed an equipment selection method based on the combination of fuzzy analytic hierarchy Process (FAHP) and entropy weight ideal point method and took resource consumption and environmental impact into comprehensive consideration. The method was verified by camshaft machining. Yan et al. [11] proposed a method of machine tool equipment selection based on the combination of analytic hierarchy process (AHP) and grey correlation method and verified the method through machine tool selection for gear machining. Zheng et al. [12] proposed the model algorithm of FAHP and fuzzy comprehensive evaluation (FCA) based on triangular fuzzy number for machine tool equipment optimization. Combined with the case of machine tool equipment optimization for blade milling in an aviation manufacturing enterprise, the feasibility and effectiveness of the method was verified. Yan et al. [13] applied Reference Ideal Method (RIM) to solve the optimization and evaluation model of intelligent production equipment. The optimization of machine tool equipment in an aviation enterprise was taken as an example to verify. Zhang et al. [14] made use of fuzzy mathematics theory, established a theoretical model of machine tool optimization selection according to geometric features of parts, machine tool parameters, and other factors, and realized reasonable automatic selection of machine tools with computer-aided process planning (CAPP). Liu et al. [15] proposed an efficient machine tool selection method based on energy efficiency evaluation, which calculated energy efficiency by modeling the features of each alternative machine tool and parts to be processed. Zanuto et al. [16] evaluated the whole life cycle of different processing technologies to determine the equipment with the least impact on the environment. Nguyen et al. [17] proposed a hybrid method for fuzzy multi-attribute decision making in machine tool evaluation. Comparison with other methods shows the effectiveness of this method. Karmiris et al. [18] processed 60CrMoV18-5 Steel by electric discharge (EDM) and used Taguchi experimental design for parameter control to compare the machining performance and surface quality. Based on grey relation analysis, multi-objective optimization of evaluation index number was carried out. Benardos et al. [19] quantified the performance (training and generalization) of a neural network based on genetic algorithms and their complexity, which are applied to practical engineering problems. Danil et al. [20] analyzed the application trends, advantages and disadvantages of resource conservation, optimization and cooling in economical sustainable manufacturing. Karkalos et al. [21] conducted mechanical machining experiments of Ti-6Al-4V using abrasive water jet under different process conditions and carried out sustainability analysis with the help of grey relation analysis (GRA). Kuntoglu et al. [22] analyzed cutting parameters of AISI 5140 steel using a response surface method to obtain minimum vibration and surface roughness.

In the production process, a product can be produced by a variety of devices. How to rationally and reasonably choose the production equipment that meets the processing requirements is very important. Experts have put forward many methods to solve the problem of production equipment selection. However, in most studies, qualitative analysis was used to solve the evaluation model. Although qualitative analysis can simplify the calculation, there is a disconnect between the consistency of the constructed judgment matrix and the consistency of the actual situation. In this paper, a method is proposed to quantitatively compare the carbon emission levels and efficiency of various equipment, which can quantify the carbon emission values and processing time of different equipment under given conditions. The equipment selection model of multi-equipment unified carbon emission and efficient production was established. With minimum carbon emission and minimum processing time as optimization objectives, BAS, a recently proposed algorithm with good computational speed, was used to solve the problem. Compared with other qualitative methods of equipment selection, this paper innovatively proposes a qualitative evaluation method of processing efficiency and carbon emission of equipment production, which can accurately analyze equipment data. The BAS algorithm was used to solve the problem with higher accuracy. The method was verified by turning a shaft part as an example, which provides reference for the selection of production equipment in enterprises, helps enterprises to achieve low carbon and efficient production mode, and contributes to the government's emission reduction policy.

This paper presents a quantitative analysis and selecting method for low-carbon and highly efficient processing equipment. The first chapter introduces the literature about the selection and evaluation of production equipment for the production process and puts forward the existing problems in the current research. Then, the analysis and evaluation of processing time and carbon emission level in the process of processing a product with different equipment were put forward. In Section 2, a unified calculation model for processing time and carbon emission of various equipment was established. Section 3 introduces and improves the BAS algorithm. Three algorithms, BAS, PSO, and GA, were compared and analyzed. In Section 4, the processing characteristics of a product using different equipment were analyzed through case analysis. Three algorithms, BAS, PSO, and GA, were used to optimize the processing time and carbon emission level of different processing equipment, and the results were analyzed. Section 5 is the conclusion of the paper.

2. Materials and Methods

Production equipment selection is the key to the planning of green manufacturing process elements, mainly through a variety of optional production equipment scheme comparative analysis, evaluation, and decisions to obtain the optimal production equipment scheme, so that the overall performance of the parts processing process is the best, especially the performance of resource consumption and environmental impact. With growing concern about global warming, many researchers have focused on manufacturing activities that consume a lot of energy and emit carbon into the atmosphere. Low-carbon manufacturing, which aims to reduce carbon intensity, is becoming a hot topic.

In the selection of machining parameters, it is often encountered to make multiple objective functions in a given area to achieve the best optimization problem, which is called the multi-objective optimization method. In the production process of equipment, the selection of process parameters has an important impact on the processing efficiency, the total cost of processing, the quality of the workpiece, and the amount of carbon dioxide emitted to the environment. The selection of appropriate cutting parameters is of great significance to the government, enterprises, and users [23–25]. The important point of the article is to research the diversity between carbon emissions and efficiency levels of the same product produced by different device. The following calculation takes processing time and carbon emission as the objectives.

2.1. Time Function

The working hours of a working procedure include cutting time, tool change time, and working procedure auxiliary time. The quantity of the shortest machining time can achieve the highest production efficiency. The mathematical model of processing time function can be expressed as [26]:

$$T_P = t_m + t_{ct}\frac{t_m}{T} + t_{ot} \tag{1}$$

$$t_m = \frac{L_w \Delta}{n f a_{sp}} = \frac{\pi d_0 L_w \Delta}{1000 v_c f a_{sp}} \tag{2}$$

The Taylor generalized tool durability calculation formula is

$$T = \frac{C_T}{v_c^x f^y a_{sp}^z} \tag{3}$$

where, t_m is the working procedure cutting time, t_{ct} is the time used for a tool change, t_{ot} is other auxiliary time in addition to the tool change, T is the tool life, L_w is the machining length, Δ is the machining allowance, n is the spindle speed, d_0 is the workpiece diameter, v_c is the cutting speed, f is the feed, a_{sp} is the cutting depth, C_T is the constant related to the cutting conditions, x, y, z are the tool life coefficient. Following that, the processing time function is

$$T_P = \frac{\pi d_0 L_w \Delta}{1000 v_c f a_{sp}} + \frac{t_{ct} \pi d_0 L_w \Delta v_c^{x-1} f^{y-1} a_{sp}^{z-1}}{1000 C_T} + t_{ot} \tag{4}$$

2.2. Carbon Emission Function

The sources of carbon emissions in the production and processing of machine tools mainly include five parts: raw material consumption C_m, electric energy consumption in production and processing C_e, tool wear C_t, cutting fluid loss of machine tools C_c, and the disposal of processing waste materials C_s [27–29]. Figure 1 shows the carbon emission composition of manufacturing process. Raw material consumption (i.e., material resource utilization) is determined to a large extent by the process design stage, and the post-treatment of waste generated during the process is generally carried out after the completion of the process. Therefore, the processing process has limited efforts to optimize the carbon emission C_m caused by raw material consumption and the carbon emission C_s from waste disposal. Therefore, the carbon emissions generated by cutting production are [23,24]:

$$C = C_e + C_t + C_c \tag{5}$$

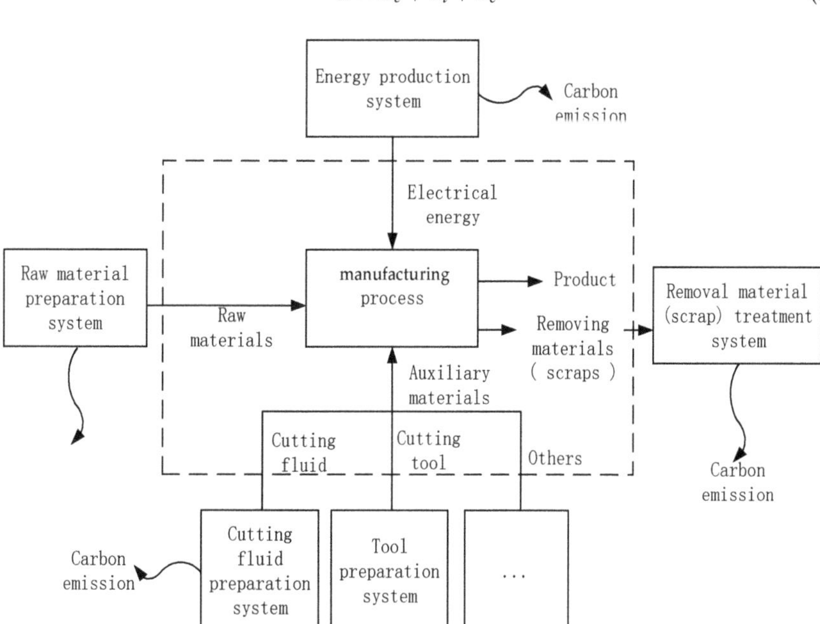

Figure 1. The chart of carbon emissions from the manufacturing process.

(1)   Carbon emissions from electric energy consumption

In the process of machining, it needs to consume a lot of electric energy. The carbon emissions generated by electricity consumption can be expressed as

$$C_e = F_e E_e \tag{6}$$

where, F_e, E_e are carbon emission factor and power consumption of electric energy respectively. Electric energy carbon emission factor F_e has a close relationship with the structure of the power grid, and different power grids have different carbon emission factors. In this paper, the average emission factor of several major power grids in China was used as the carbon emission factor of electric energy, F_e = 0.674 7 kgCO$_2$/kg [23].

(2)   Carbon emissions from tool wear

In cutting production, direct CO_2 emissions from tool wear are less, but mostly indirect CO_2 emissions, that is, CO_2 emissions from tool production, are evenly distributed in the actual cutting process. Therefore, the tool CO_2 emission calculation is based on the calculation method, which is converted into the production process according to the processing time within the tool service time.

$$C_t = \frac{t_m}{T_t} F_t W_t \tag{7}$$

F_t, W_t represent the carbon emission factor and tool quality of the tool, respectively. To determine the carbon emission factor of the tool, it is necessary to know the energy consumption of the tool preparation process. For the energy consumption of the tool preparation process, only the tool manufacturing process is considered in the calculation in this paper, and the carbon emission factor of the tool is 29.6 kgCO$_2$/kg [18].

Tool life T_t refers to the cutting time experienced by a new tool until it is retired, which may include multiple instances of regrinding (represented by N), so tool life is equal to the product of tool life and (N+1).

$$T_t = (N+1)T \tag{8}$$

T_t, N, T are the tool life, grinding tool number, and tool durability, respectively.

(3)   Carbon emissions from cutting fluid consumption

CO_2 emissions from cutting fluid consumption are mainly composed of two components: CO_2 emissions from the manufacture of mineral oils C_o and CO_2 emissions from the disposal of waste fluid after the use of cutting fluid C_w. The replacement time of cutting fluid in production is relatively long. According to the specific conditions in actual production, the CO_2 emissions generated by cutting fluid consumption should be converted into the same time as the CO_2 emissions from turning tools, that is, the carbon emissions generated by cutting fluid consumption is

$$C_c = \frac{T_p}{T_c}(C_o + C_w) \tag{9}$$

$$C_o = F_o(C_C + A_C) \tag{10}$$

$$C_w = F_w[(C_C + A_C)/\delta] \tag{11}$$

where, F_o and F_w are the carbon emission factor of mineral oil and the carbon emission factor of cutting fluid waste treatment, respectively. C_c, A_c are, respectively, the initial amount and additional amount of cutting fluid. δ, T_c are, respectively, cutting fluid concentration and replacement cycle.

The carbon emission factor of cutting fluid consumption is mainly composed of two types: one is the carbon emission factor of pure mineral oil production and processing

F_o, and the other is the carbon emission factor of waste disposal of cutting fluid F_w. The carbon emission factor of pure mineral oil production can be expressed as

$$F_o = E_{Eo} E_{Co} \times \frac{44}{12}$$ (12)

where E_{Eo}, E_{Co} are the internal energy value of mineral oil (GJ/L) and carbon content of mineral oil (kgC/GL), respectively. The contained energy value of mineral oil is between 41,868 and 42,705 kJ/kg. In this paper, 42,287 kJ/kg and CO_2 emission factor 20 kgC/GJ were selected for optimization. At room temperature, the density of mineral oil is generally between 0.86 and 0.98 g/cm3, and 0.92 was adopted in this paper. Based on the above parameters, it can be seen that the CO_2 emission factor of mineral oil is 285 kgCO$_2$/L [24]. As for the carbon emission factor F_w, the main component of waste cutting fluid is water. To facilitate calculation, the carbon emission factor of waste cutting fluid treatment can be replaced by carbon emission factor of waste cutting fluid treatment. The carbon emission factor of wastewater treatment is 0.2 kgCO$_2$/L [30].

2.3. Constraints

In the production process, the selection of process parameters is mainly limited by the machine tool stiffness, the limited range of machine tool cutting parameters, the machine tool power, and the workpiece surface quality requirements.

(1)   Power constraints of machine tools

In metal cutting production, the cutting power cannot exceed the maximum power of the machine tool spindle motor P_{max}, that is

$$\frac{F_c v_c}{1000\eta} \le P_{\mathrm{max}}$$ (13)

η is machine tool efficiency.

(2)   Cutting force constraints

In metal cutting production, the cutting force generated cannot be greater than the rated cutting force of the machine tool, that is

$$F_c \le F_{\mathrm{max}}$$ (14)

(3)   Constraints on spindle speed

$$\frac{\pi d_0 n_{\mathrm{min}}}{1000} \le v \le \frac{\pi d_0 n_{\mathrm{max}}}{1000}$$ (15)

(4)   Feed constraint

$$f_{\mathrm{min}} \le f \le f_{\mathrm{max}}$$ (16)

(5)   Workpiece processing quality constraints

In the process of processing, the machining quality of the workpiece should be guaranteed, and the surface roughness should meet the processing requirements.

$$R_a \le R_{\mathrm{max}}$$ (17)

R_{max} is the maximum surface roughness.

3. Beetle Antennae Search Algorithm

BAS is a new intelligent optimization algorithm for the biological performance of beetles, which was proposed in 2017. Its development was inspired by the foraging principle of beetles [31,32]. The bionic search principle of the algorithm is that beetles search for food according to the intensity of the smell given off. As a single search algorithm, the beetle antennae search algorithm has the advantages of simple principle, few parameters and less computation [33–37]. It has great advantages when dealing with low-dimensional optimization targets. In the early stage, beetles do not need to know the specific location of food, but search for the intensity of food smell through two tentacles [38,39]. If the intensity of food smell received by the left whiskers is greater than that received by the right whiskers, the beetle will move to the left for some distance and make the next food hunt [40–43]. The cycle continues until the beetles find the most flavorful spot to finish their foraging. Different from other heuristic algorithms with a large population, the BAS algorithm only needs one beetle, so its calculation amount will be greatly reduced, making the search speed faster. According to this foraging principle of beetles, the beetle antennae search algorithm can be obtained, as shown in Figure 2. Figure 3 shows the algorithm pseudocode. The specific steps are as follows:

(1) Suppose that the beetle forages in an n-dimensional space, its center of mass is set as zx the left whisker of the beetle is al, the right whisker is ar, the initial distance between the two whiskers is $d0$, and the coefficient between the distance between the two whiskers and the first step length $step$ of the beetle is c. The most important thing to pay attention to is that the initial distance between the two whiskers $d0$ and the setting value of the first step size $step0$ of the beetle should be fully considered to skip out of the local optimal value and ensure the normal optimization of the beetle in the later stage.

(2) Since the head orientation of the beetle each time it forages is random, let us say the head orientation of the beetle is $\vec{b}$,

$$\vec{b} = \frac{rand(n,1)}{||rand(n,1)||} \tag{18}$$

where, $rand(n,1)$ represents the randomly generated n dimension vector.

(3) According to the beetle head orientation established above, the coordinates of the left and right whiskers of the beetle can be represented

$$al = zx^t - \vec{b} \times d^t \tag{19}$$

$$ar = zx^t + \vec{b} \times d^t \tag{20}$$

where, zx^t is the position of the centroid corresponding to the t-th foraging of the beetle, d^t is the distance between the two whiskers corresponding to the t-th foraging of the beetle, its value will decrease with the increase of foraging times. The attenuation coefficient is eta_bc, that is $d^t = eta\_bc \times d^{t-1}$, usually the value of eta_bc ranges from 1 to 0.95.

(4) Fit values $fitnessl$ and $fitnessr$ are obtained by using the coordinates of the left and right whiskers of the beetle, and the difference between the two values was used to influence the position of the centroid of the next beetle,

$$zx^t = zx^{t-1} + step^t \times \vec{b} \times sign(fitnessr - fitnessl) \tag{21}$$

where, $sign$ is the symbolic function, $step^t$ is the t time foraging step, and its value is related to the distance between the two whiskers.

$$step^t = c \times d^t \tag{22}$$

(5) Determine whether the above process can find the optimal value of the function or reach the number of iterations. If the above conditions are not met, the above steps (2)–(4) should be repeated until the established conditions are met and the loop is terminated.

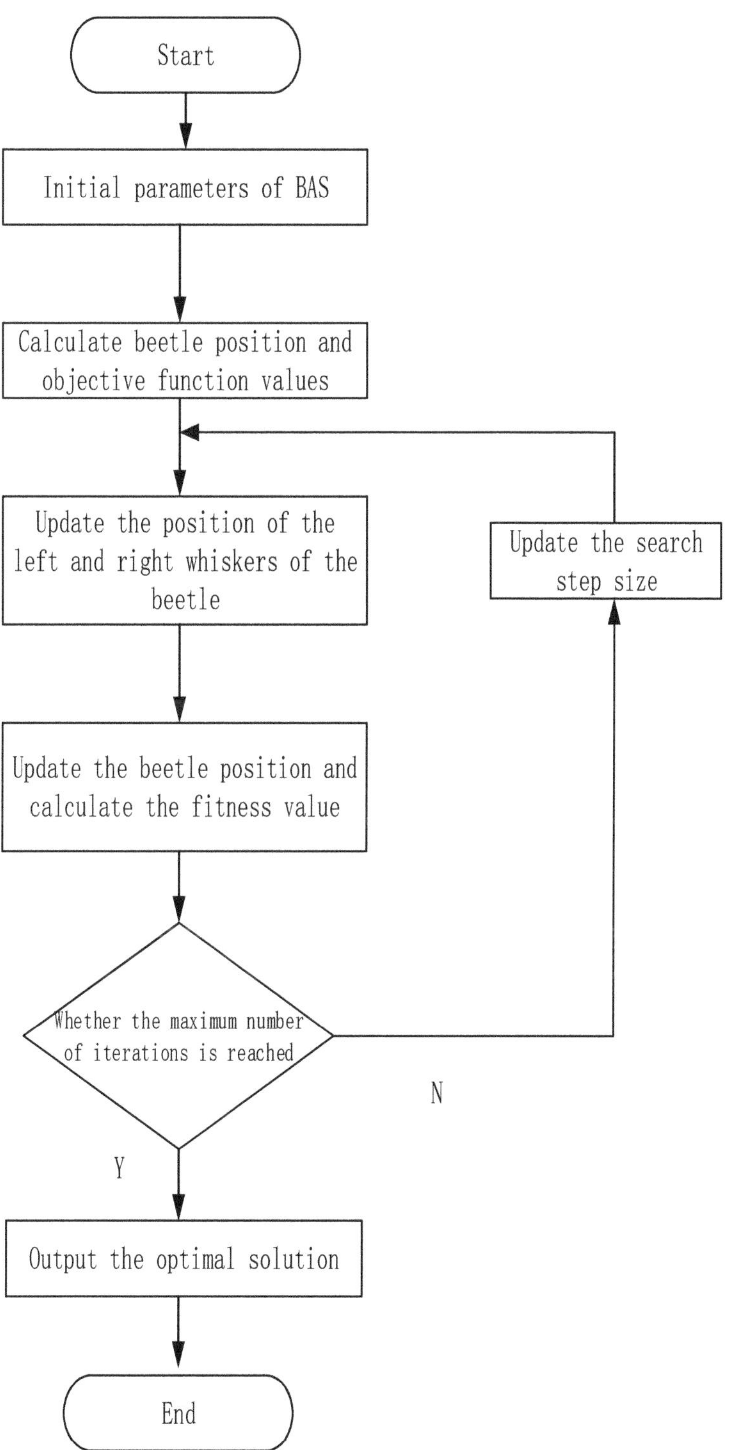

Figure 2. Flow chart of beetle antennae search algorithm.

Algorithm : BAS algorithm

Input: Establish an objective function $f\left(x^t\right)$.

where variable $x^t = \left[x_1,\ldots,x_i\right]^T$.

initialize the parameters x^0, d^0, δ^0.

Output: x_{best}, f_{best}.

while $(t < T_{max})$ or *(stop criterion)* **do**

    Generate the direction vector unit $\vec{b}$ according to (1):

    Search in variable space with two kinds of antennae according to (2):

    Update the state variable x^t according to (3):

    if $f\left(x^t\right) < f_{best}$ **then**

        $f_{best} = f\left(x^t\right), x_{best} = x^t$

    Update sensing diameter d and step size δ with decreasing functions (4) and (5) respectively, which could be further studied by the designers.

return x_{best}, f_{best}.

Figure 3. BAS algorithm pseudocode.

In this paper, the optimization objectives were to minimize the processing time and carbon emission. The process of BAS combined with the calculation model is as follows:

Step 1: Code. In the algorithm, the selection of equipment, cutting tool, and process ordering need to be reasonably reflected in the zero-piece coding method. The encoding mechanism is as follows: each individual in the population has three substrings, namely sequential S_i, device M_i, and tool T_i, whose length is equal to the number of steps of part. The sequential substring Si represents the sequence of operations for machining parts in a continuous list, which takes into account the constraints of processing precedence. The device substring M_i consists of the device number that has been assigned to each operation. The j-th bit on the substring represents the device used to complete step j. The meaning of tool substring is similar to that of equipment substring.

Step 2: Parameter settings include space dimension, distance between left and right whiskers, initial step size, iteration number, etc.

Step 3. Initialize the position of the longicorn with random direction and construct random vector of the beetle.

Step 4. Calculate the fitness value.

Step 5: Compare the signal size of the left and right whiskers of the longicorn to determine the next direction of movement of the longicorn.

Step 6: Update the position of the beetles.

Step 7: Check whether the termination condition is met. If yes, output results. If no, go to Step 4.

Step 8: Output the optimal solution and end the algorithm.

The algorithm runs on the MATLAB 2016 b, the dimension n is 2, the initial step size of the beetle *step* is 0.3, the distance between the two whiskers of the beetle d is 5, and the number of iterations is 300.

In order to verify the optimization performance of BAS algorithm proposed in this paper, the BAS algorithm was compared with the PSO algorithm and the GA algorithm, respectively. The test function is $f(x) = \sum_{i=1}^{n} x_i^2, -100 \leq x_i \leq 100$ [44]. Based on the parameter settings of the above standard functions, the search individual was set to 50, the maximum number of iterations was set to 1000, and each test function was run 30 times to generate statistical results. The results are shown in Table 1. The results obtained by BAS are more accurate and the running time is shorter. Different from PSO algorithm, BAS algorithm is a single search algorithm, with simple principle, fewer parameters, less computation, and other advantages. It can also be seen from the results that the convergence speed of BAS algorithm is faster and the optimization accuracy is higher. Therefore, the BAS algorithm was chosen as the solution method of the model in this paper.

Table 1. Test algorithm optimization results.

Algorithm	Mean Value	Mean Square Error	Mean Running Time
BAS	0	1.1589×10^{-5}	0.4597
PSO	0	2.1857×10^{-6}	0.5153
GA	0.0037	1.8954×10^{-3}	2.3324

4. Case Study

4.1. Experimental Conditions

The workpiece is a shaft with a length of 180 mm and a diameter of 100 mm. The parts diagram is shown in Figure 4. The material is 45# carbon steel. There are two kinds of lathes in the workshop, and their parameters are shown in Table 2. The parameters of the tools used in this optimization are shown in Table 3

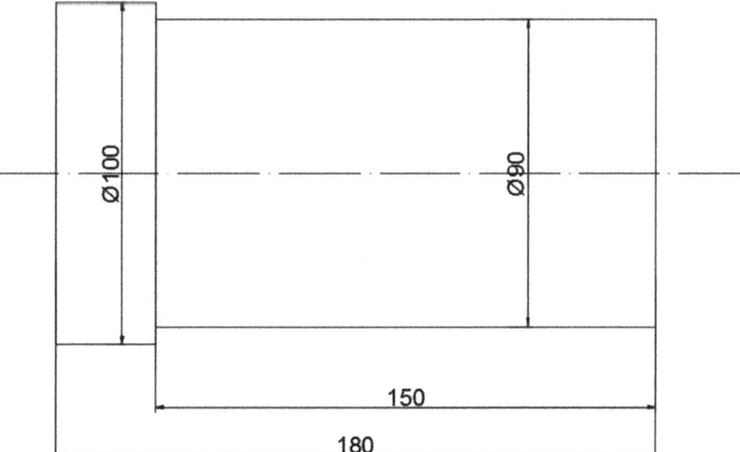

Figure 4. Part size drawing.

Table 2. Machine parameters.

Lathe	n_{min}/(r/min)	n_{max}/(r/min)	f_{min}/(mm/r)	f_{max}/(mm/r)	F_{max}/N	P_{max}/kW	η
M1	100	1400	0.1	2.5	1700	8.0	0.85
M2	80	1400	0.1	3.5	9000	15	0.8

Table 3. Parameters of tool.

Tool	Material	Main Cutting Edge Angle (°)	Rake Angle (°)	Inclination Angle (°)	Tip Arc Radius r_θ (mm)
K1	Hard metal alloy	75°	10°	−5°	1
K2	Hard metal alloy	45°	20°	5°	0.8

Tool life and cutting force coefficient are shown in Table 4.

Table 4. Tool life and cutting force coefficient.

x	y	z	C_{Ff}	K_{Ff}	x_{Ff}	y_{Ff}	n_{Ff}
5	1.75	0.75	2880	1	1	0.5	−0.4

The relevant parameters of the calculation model are shown in Tables 5 and 6.

Table 5. Calculation of correlation coefficient 1.

One Knife Change Time t_{ct}/min	Auxiliary Time t_{ot}/min	Cutting Replacement Cycle T_c /Month	Total Tool W_t/g	Initial Cutting Oil C_c/L	Additional Cutting Oil Quantity A_c/L	Minimum No−LOAD Power P_{u0}/kW
0.5	0.8	2	15	8.5	4.5	40.6

Table 6. Calculation of correlation coefficient 2.

Cutting Fluid Concentration δ	Number of Grinding N	Number of Grinding w_1	Carbon Emission Weight Number w_2	Spindle Speed Coefficient A_1	Spindle Speed Coefficient A_2
0.05	1	0.5	0.5	0.227	-0.667×10^{-6}

Table 7 shows the carbon emission factors of related attributes [23,45].

Table 7. Carbon emission factor table of related attributes.

Serial Number	Attribute	Carbon Emission Factor/(kgCO$_2$/kWh)
1	electricity	0.6747
2	steel	5.926
3	aluminum	12.807
4	cast iron	4.445
5	cutting fluid	2.87
6	cutter	29.6

4.2. Optimization Results

The parameters of the beetle antennae search algorithm are set as follows: dimension n is 3, number of beetles is 35, coefficient c between the distance between the two whiskers and the step size is 5, the initial step size of each beetle step is 0.3, and the maximum number of iterations *MAXGEN* is 100. The iterative process of M1 and M2 processing time is shown in Figures 5 and 6, respectively. As can be seen from the figure, with the increase of the number of iterations, the processing time gradually decreases until it becomes stable, and the processing time of products using M1 is obviously less than that of M2. The iterative process of carbon emission of M1 and M2 is shown in Figures 7 and 8, respectively. As can be seen from the Figures 7 and 8, with the number of iterations increases, the carbon emission generated by processing gradually decreases until it becomes stable. The carbon emission of products processed by M1 is obviously less than that of M2.

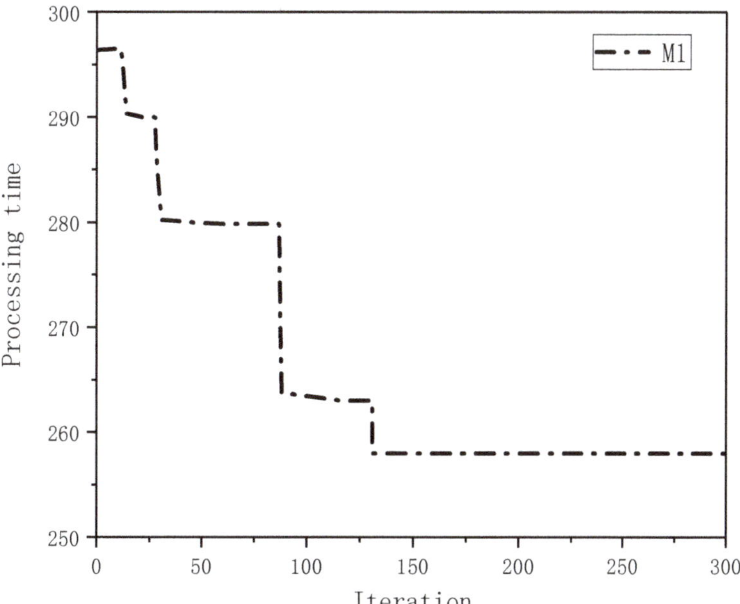

Figure 5. M1 processing time iteration diagram.

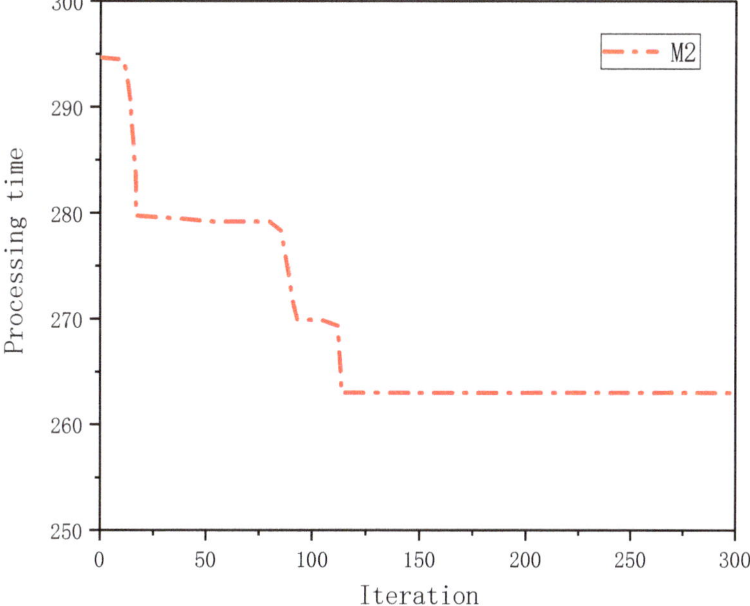

Figure 6. M2 processing time iteration diagram.

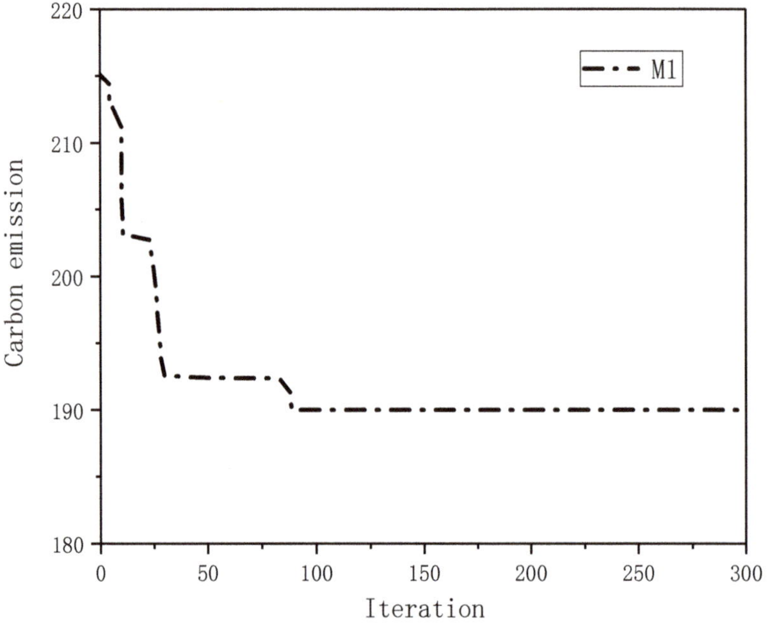

Figure 7. M1 carbon emission iteration diagram.

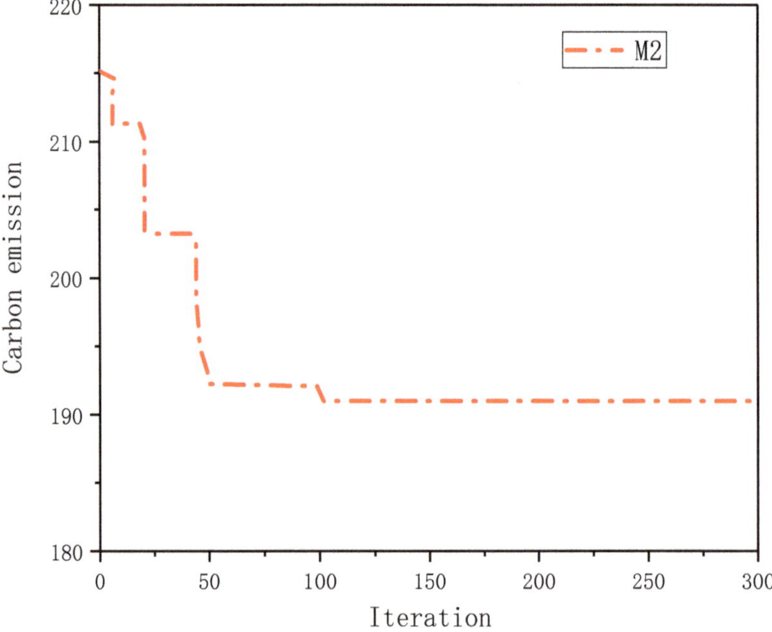

Figure 8. M2 carbon emission iteration diagram.

4.3. Optimization Result Analysis

The optimization results of processing the same product with different equipment are shown in Table 8. As can be seen from the table, the processing time of machine tool M1 is 258, the processing time of machine tool M2 is 263 when processing the above workpiece, and the processing time of M2 is 1.02 times that of M1. The carbon emission of machine tool M1 is 190, the carbon emission of machine tool M2 is 191 when processing the above workpiece, and the carbon emission of machine tool M2 is 1.005 times that of machine tool M1. The processing time and carbon emission of machine tool M1 are reduced compared with that of machine tool M2, so M1 was chosen to process this product. Although there is a small gap between the processing time and carbon emission of a product by using M1 and M2, the use of M1 can save more time and reduce carbon emission for enterprises in mass production.

Table 8. Comparison of different equipment optimization results.

Lathe	Processing Time	Carbon Emission
M1	258	190
M2	263	191

4.4. Comparison of Previous Literatures

In order to achieve efficient and low-carbon production, many scholars studied the optimization of equipment selection in literature [8–16], proposed some optimization methods, and effectively achieved the goal. However, there are some deviations between the theoretical data and the actual data in the traditional equipment selection method for qualitative analysis of evaluation indicators, and the reliability of the data is uneven. Compared with the literature [19,21], the calculation method and the process goal have differences. Therefore, this paper established an efficient and low-carbon production equipment selection model, which was solved by BAS, which can calculate the time and carbon emissions of the different production equipment. In this paper, three algorithms, BAS, PSO, and GA, were used to solve the problem of equipment selection for processing the workpiece in the above cases. The results are shown in Table 9. Among the two kinds of processing and turning equipment M1 and M2, the optimization results of the three algorithms show that the processing time and carbon emission of M1 are less. The optimized processing time of BAS algorithm is 258, which is 0.01% lower than PSO algorithm and 0.05% lower than the GA algorithm. The carbon emission after BAS algorithm optimization is 190, which is 0.03% lower than PSO algorithm and 0.06% lower than GA algorithm. BAS optimized processing time and least carbon emissions. The optimal equipment scheme was obtained by using the beetle antennae search algorithm. The algorithm has low complexity, a strong searching ability, and less computation.

Table 9. Comparison of algorithm optimization results.

Algorithm	Lathe	Processing Time	Carbon Emission
BAS	M1	258	190
	M2	263	191
PSO	M1	260	195
	M2	268	201
GA	M1	271	202
	M2	275	210

5. Conclusions

The selection of production equipment is an indispensable step in machining. Many factors should be considered comprehensively when selecting machine tool equipment so as to make a reasonable choice. This paper put forward a low-carbon and high production equipment selection method which can help producers to quantify the carbon emissions and efficiency levels of different equipment.

1. Systematically analyze the carbon emission and time of the production process and establish the calculation model of the carbon emission and time of the same product produced by different equipment.
2. Use the BAS to solve the model and get the best equipment scheme. The comparison of algorithm optimization results showed BAS is simple and has stronger search ability.
3. By turning a shaft part as an example, the effectiveness of the proposed method is proved.

The results show that this method can analyze and evaluate the processing time and carbon emission level of the same product produced by multiple equipment. It can provide suggestions for enterprises to choose high efficiency and low carbon production equipment. However, this paper only analyzes the processing time and carbon emission of parts processing. In the future, more process objectives and alternative devices will be considered, and the main challenge is that algorithms and devices continue to improve and new problems will arise.

Author Contributions: Conceptualization, Y.X. and G.C.; methodology, Y.X., G.C. and H.Z.; software, Y.X.; validation, G.C., H.Z. and X.Z.; formal analysis, G.C. and X.Z.; investigation, Y.X. and X.Z.; writing—original draft preparation, Y.X. and H.Z.; writing—review and editing, Y.X. All authors have read and agreed to the published version of the manuscript.

Funding: Hubei Provincial Natural Science Foundation for Innovation and Development (2022 CFD081); Xiangyang Science and Technology Plan Project (2022 ABH006436); China Education Department of Hunan Province (Project number: 21B0695); Project of Hunan social science achievement evaluation committee in 2022 (Project number: XSP22YBC081).

Institutional Review Board Statement: Not applicable.

Informed Consent Statement: Informed consent was obtained from all subjects involved in the study.

Data Availability Statement: Not applicable.

Conflicts of Interest: The authors declare no conflict of interest.

References

1. Zhai, X.Q.; An, Y.F. Analyzing influencing factors of green transformation in China's manufacturing industry under environmental regulation: A structural equation model. *J. Clean. Prod.* **2020**, *251*, 119760. [CrossRef]
2. Brodny, J.; Tutak, M. Analysis of the efficiency and structure of energy consumption in the industrial sector in the European Union countries between 1995 and 2019. *Sci. Total Environ.* **2022**, *808*, 152052. [CrossRef] [PubMed]
3. Wang, Q.; Hu, Y.J.; Hao, J.; Lv, N.; Li, T.Y.; Tang, B.J. Exploring the influences of green industrial building on the energy consumption of industrial enterprises: A case study of Chinese cigarette manufactures. *J. Clean. Prod.* **2019**, *231*, 370–385. [CrossRef]
4. Salem, A.; Hegab, H.; Kishawy, H. An integrated approach for sustainable machining processes: Assessment, performance analysis, and optimization. *Sustain. Prod. Consum.* **2021**, *25*, 450–470. [CrossRef]
5. Niroomand, S. Hybrid artificial electric field algorithm for assembly line balancing problem with equipment model selection possibility. *Knowl.-Based Syst.* **2021**, *8*, 106905. [CrossRef]
6. Klink, A.; Arntz, K.; Johannsen, L. Technology-based assessment of subtractive machining processes for mold manufacture. *Procedia Cirp* **2018**, *71*, 401–406. [CrossRef]
7. Xiong, Q.; Wang, J.; Zhou, Q. Prediction model of machining error based on precision and process parameters of machine tools. *Acta Aeronaut. Et Astronaut. Sin.* **2018**, *39*, 421713.
8. Xiao, Y.; Wang, R.; Yan, W.; Ma, L. Optimum Design of Blank Dimensions Guided by a Business Compass in the Machining Process. *Processes* **2021**, *9*, 1286; [CrossRef]
9. Li, H.; Wang, W. Selection method of machine tool equipment resource in cloud manufacturing environment. *Acta Aeronaut. E Astronaut. Sin.* **2020**, *41*, 54–65.
10. Zhou, L.; Yuan, B. The Optimization Selection of Machine Tool for Green Manufacturing. *Modul. Mach. Tool Autom. Manuf. Tech.* **2018**, *2*, 157–160.

1. Yan, P.; Zhao, G. Green Evaluation System of Machine Tool Equipment Based on Manufacturing Process. *Modul. Mach. Tool Autom. Manuf. Tech.* **2021**, *12*, 4.
2. Zheng, Y.; Hu, F.; Wang, M. Research on the machine tool selection optimization method based on fuzzy analytic hierarchy process and fuzzy comprehensive assessment. *Chin. J. Eng. Des.* **2015**, *22*, 405–411.
3. Yan, Q.; Zhang, H. Research on Optimized Selection of Machine Tool Equipment for Intelligent Manufacturing. *Modul. Mach. Tool Autom. Manuf. Tech.* **2020**, *12*, 144–148.
4. Zhang, B.; Guan, S. Machine tool selection based on fuzzy evaluation and optimization of cutting parameter. *J. Meas. Sci. Instrum.* **2015**, *6*, 384–389.
5. Liu, P.; Tuo, J.; Liu, F. A novel method for energy efficiency evaluation to support efficient machine tool selection. *J. Clean. Prod.* **2018**, *191*, 57–66. [CrossRef]
6. Zanuto, R.; Hassui, A.; Lima, F. Environmental impacts-based milling process planning using a life cycle assessment tool. *J. Clean. Prod.* **2019**, *206*, 349–355. [CrossRef]
7. Nguyen, H.; Dawal, S.; Nukman, Y. A hybrid approach for fuzzy multi-attribute decision making in machine tool selection with consideration of the interactions of attributes. *Expert Syst. Appl.* **2014**, *41*, 3078–3090. [CrossRef]
8. Karmiris-Obratański, P.; Papazoglou, E.L.; Leszczyńska-Madej, B.; Karkalos, N.E.; Markopoulos, A.P. An Optimalization Study on the Surface Texture and Machining Parameters of 60CrMoV18-5 Steel by EDM. *Materials* **2022**, *15*, 3559. [CrossRef]
9. Benardos, P.G.; Vosniakos, G.C. Optimizing feedforward artificial neural network architecture. *Eng. Appl. Artif. Intell.* **2007**, *20*, 365–382. [CrossRef]
10. Pimenov, D.Y.; Mia, M.; Gupta, M.K.; Machado, Á.R.; Pintaude, G.; Unune, D.R.; Khanna, N.; Khan, A.M.; Tomaz, Í.; Wojciechowski, S.; et al. Resource saving by optimization and machining environments for sustainable manufacturing: A review and future prospects. *Renew. Sustain. Energy Rev.* **2022**, *166*, 112660. [CrossRef]
11. Karkalos, N.E.; Karmiris-Obratański, P.; Kudelski, R.; Markopoulos, A.P. Experimental Study on the Sustainability Assessment of AWJ Machining of Ti-6Al-4V Using Glass Beads Abrasive Particles. *Sustainability* **2021**, *13*, 8917. [CrossRef]
12. Kuntoğlu, M.; Aslan, A.; Pimenov, D.Y.; Giasin, K.; Mikolajczyk, T.; Sharma, S. Modeling of Cutting Parameters and Tool Geometry for Multi-Criteria Optimization of Surface Roughness and Vibration via Response Surface Methodology in Turning of AISI 5140 Steel. *Materials* **2020**, *13*, 4242. [CrossRef] [PubMed]
13. Li, C.; Cui, L. Multi-objective NC Machining Parameters Optimization Model for High Efficiency and Low Carbon. *J. Mech. Eng.* **2013**, *49*, 87–96. [CrossRef]
14. Xiao, Y.; Zhao, R.; Yan, W.; Zhu, X. Analysis and Evaluation of Energy Consumption and Carbon Emission Levels of Products Produced by Different Kinds of Equipment Based on Green Development Concept. *Sustainability* **2022**, *14*, 7631. [CrossRef]
15. Xiao, Y.; Jiang, Z.; Gu, Q.; Yan, W.; Wang, R. A novel approach to CNC machining center processing parameters optimization considering energy-saving and low-cost. *J. Manuf. Syst.* **2021**, *59*, 535–548. [CrossRef]
16. Xiao, Y.; Zhou, J.; Wang, R.; Zhu, X.; Zhang, H. Energy-Saving and Efficient Equipment Selection for Machining Process Based on Business Compass Model. *Processes* **2022**, *10*, 1846. [CrossRef]
17. Liu, Z.; Sun, D.; Lin, C. Multi-objective optimization of the operating conditions in a cutting process based on low carbon emission costs. *J. Clean. Prod.* **2016**, *124*, 266–275. [CrossRef]
18. Gu, J.; Jiang, T. Low-Carbon Job Shop Scheduling Problem with Discrete Genetic-Grey Wolf Optimization Algorithm. *J. Adv. Manuf. Syst.* **2020**, *19*, 1–14. [CrossRef]
19. Zhang, Y.; Xu, H.; Huang, J.; Xiao, Y. Low-Carbon and Low-Energy-Consumption Gear Processing Route Optimization Based on Gray Wolf Algorithm. *Processes* **2022**, *10*, 2585. [CrossRef]
20. China National Institute of Standardization. *Corporation Green Gas Measuring and Reporting*; China Zhijian Publishing House: Beijing, China, 2011.
21. Du, H.; Ge, Z. Inverse kinematics solution algorithm of electric climbing robot based on improved beetle antennae search algorithm. *Control. Decis.* **2022**, *37*, 2217–2225.
22. Huang, J.; Zhou, M.; Sabri, M. A Novel Neural Computing Model Applied to Estimate the Dynamic Modulus (DM) of Asphalt Mixtures by the Improved Beetle Antennae Search. *Sustainability* **2022**, *14*, 5938. [CrossRef]
23. Song, D. Modified Beetle Annealing Search (BAS) Optimization Strategy for Maxing Wind Farm Power through an Adaptive Wake Digraph Clustering Approach. *Energies* **2021**, *14*, 7326.
24. Khan, A.; Cao, X. Fraud detection in publicly traded U.S. firms using Beetle Antennae Search: A machine learning approach. *Expert Syst. Appl.* **2022**, *191*, 116148.
25. Medvedeva, M.; Katsikis, V. Randomized time-varying knapsack problems via binary beetle antennae search algorithm: Emphasis on applications in portfolio insurance. *Math. Methods Appl. Sci.* **2021**, *44*, 2002–2012. [CrossRef]
26. Wang, J.; Situ, C. The post-disaster emergency planning problem with facility location and people/resource assignment. *Kybernetes Int. J. Syst. Cybern.* **2020**, *49*, 2385–2418. [CrossRef]
27. Sun, Y.; Zhang, J. Determination of Young's modulus of jet grouted coalcretes using an intelligent model. *Eng. Geol.* **2019**, *252*, 43–53. [CrossRef]
28. Khan, A.; Cao, X. Enhanced Beetle Antennae Search with Zeroing Neural Network for online solution of constrained optimization. *Neurocomputing* **2021**, *447*, 294–306. [CrossRef]

39. Wu, Q.; Wang, J. Beetle antennae search for neural network model with application to population prediction: An intelligent optimization algorithm. *Filomat* **2020**, *34*, 4937–4952. [CrossRef]
40. Fu, W.; Fang, P.; Wang, K. Multi-step ahead short-term wind speed forecasting approach coupling variational mode decomposition improved beetle antennae search algorithm-based synchronous optimization and Volterra series model. *Renewable energy* **2021**, *179*, 1122–1139. [CrossRef]
41. Jiang, X.; Lin, Z. Dynamical Attitude Configuration with Wearable Wireless Body Sensor Networks through Beetle Antennae Search Strategy. *Measurement* **2020**, *167*, 108128. [CrossRef]
42. Xie, S.; Chu, X. Ship predictive collision avoidance method based on an improved beetle antennae search algorithm. *Ocean. Eng.* **2019**, *192*, 106542. [CrossRef]
43. Qian, J.; Wang, P. Joint application of multi-object beetle antennae search algorithm and BAS-BP fuel cost forecast network on optimal active power dispatch problems. *Knowl.-Based Syst.* **2021**, *226*, 107149. [CrossRef]
44. Zhang, W. *Research on PSS Parameter Optimization Based on Improved Prony Identification and Beetle Swarm Algorithm*; Liaoning Technical University: Liaoning, China, 2021.
45. Zhang, H.; Wang, Z. Research on Multidimensional-Feature Data-Driven for Carbon Emission Prediction of CNC Turning Process. *Mach. Des. Manuf.* **2022**, *11*, 22–32.

Article

Study on the Optimization of the Material Distribution Path in an Electronic Assembly Manufacturing Company Workshop Based on a Genetic Algorithm Considering Carbon Emissions

Xiaoyong Zhu [1], Lili Jiang [2,*] and Yongmao Xiao [3,4,5]

1 School of Economics & Management, Shaoyang University, Shaoyang 422000, China
2 School of Management, China West Normal University, Nanchong 637009, China
3 School of Computer and Information, Qiannan Normal University for Nationalities, Duyun 558000, China
4 Key Laboratory of Complex Systems and Intelligent Optimization of Guizhou Province, Duyun 558000, China
5 Key Laboratory of Complex Systems and Intelligent Optimization of Qiannan, Duyun 558000, China
* Correspondence: jianglili@cwnu.edu.cn

Abstract: AbstractIn order to solve the problems of high carbon emissions, low distribution efficiency and high costs related to the process of material distribution in manufacturing workshops, a multi-objective workshop material distribution path optimization problem model is established, and the model is solved using an improved genetic algorithm. The problem is processed using Gray code and crossover and variation operations with a genetic algorithm. To improve the search accuracy and convergence speed of the algorithm, an adaptive mutation method is proposed to enhance the diversity of the population and to achieve global optimal path objective finding. The improved algorithm is applied to workshop path multi-station logistics path planning, which effectively solves the transport path optimization and station solving problems in workshop logistics distribution, and the convergence speed and convergence accuracy of the algorithm are significantly improved. Finally, a simulation analysis is carried out on the optimization of the production material distribution of a smart gas meter workshop owned by K Company, which is an electronic assembly manufacturing company. We used MATLAB software for the case company logistics distribution route model for data analysis and solving. Due to the consideration of carbon emissions, we did not consider two kinds of experiments, which were two different cases of the optimal path. The experimental results verify that the distribution optimization scheduling model can meet the demands for immediate material distribution in the production workshop, which is conducive to improving material distribution efficiency, reducing logistics costs and achieving the goal of lowering carbon emissions. This optimization model has a certain utility in that in the current context of aiming for carbon neutral and carbon peaking, early low carbon distribution layout can reduce the environmental cost of the enterprise, making material distribution a more environmental economic path.

Keywords: carbon emission; genetic algorithm; material distribution; optimization model

Citation: Zhu, X.; Jiang, L.; Xiao, Y. Study on the Optimization of the Material Distribution Path in an Electronic Assembly Manufacturing Company Workshop Based on a Genetic Algorithm Considering Carbon Emissions. *Processes* **2023**, *11*, 500. https://doi.org/10.3390/pr11051500

Academic Editor: Sergey Y. Yurish

Received: 2 April 2023
Revised: 6 May 2023
Accepted: 10 May 2023
Published: 15 May 2023

1. Introduction

With the accelerated development of global integration, the competition between assembly and manufacturing companies has become increasingly fierce, and higher operating costs have put the development of companies to the test. In discrete-oriented manufacturing companies, the auxiliary time in production logistics (storage and handling of materials, etc.) accounts for 90 to 95% of the total operation time [1], and so increasing the optimization of the logistics distribution path has become an effective way to reduce operating costs and improve economic efficiency [2]. Therefore, optimizing the level of operation of production plant logistics plays a vital role in the survival and development of enterprises. Assembly and manufacturing enterprises are often faced with handling many types and models of materials, making it difficult to carry out sorting storage and distribution, and material distribution is the intermediate link between storage and production, playing an important connecting role in the whole system of production logistics, which will directly affect the production efficiency and delivery speed of products. At the same time, in the operational aspects of production logistics in contemporary assembly and manufacturing plants, problems such as high transport costs, low distribution efficiency, untimely distribution and poor service satisfaction at workstations are common [3]. As the "third source of profit", logistics plays a vital role in reducing production costs, improving production efficiency and enhancing market competitiveness. With the continuous development of lean, intelligent, collaborative and green production, improving the transport efficiency of existing production logistics in assembly and manufacturing workshops, improving the satisfaction of workplace services and achieving green and low-carbon logistics have become some of the major challenges that assembly and manufacturing enterprises need to solve.

The development of intelligent green logistics in assembly manufacturing enterprises will be conducive to promoting the optimization of the production mode of enterprises, driving the manufacturing industry to achieve industrial structure upgrades, and is now becoming one of the important driving forces for the transformation of the manufacturing industry. In 2015, China released the "Made in China 2025" strategic plan, highlighting that its main goal of taking intelligent and green manufacturing is an important measure for China's manufacturing industry in order for it to shift from low-end manufacturing to high-end manufacturing, and complete the strategic goal of manufacturing power [4]. There has been a long history of research on production logistics systems, but there is not much literature with a focus on intelligent manufacturing; no practical application models for intelligent production logistics systems have so far been proposed, and research on the vehicle routing problem (VRP) has focused on the design of optimization algorithms and their application in sales logistics. Therefore, this paper combines the production characteristics of K Company, a discrete assembly manufacturing company, and introduces the VRP to solve the problem of real-time material distribution in the production logistics process to achieve intelligent and low-carbon logistics distribution and to lay the foundation for intelligent manufacturing in the workshop.

The workshop material distribution path optimization problem is essentially a vehicle routing problem, and VRP was first proposed by Danting and Ramser in 1959 [5]. The vehicle routing problem refers to a distribution routing solution with the goal of minimizing distribution costs under certain constraints. These constraints include, but are not limited to, the following: customer demand, customer demand time, vehicle capacity, etc. The sequence of distribution vehicles visiting customers needs to be reasonably planned so as to develop a vehicle distribution path scheme that maximizes the above objectives [6–8]. However, the optimization of logistics distribution routes is a complex mathematical problem involving modern optimization algorithms, which are highly difficult to solve and computationally intensive. At the beginning of the research period, attempts to solve this type of problem are mainly focused on the shortest driving path, the lowest cost of consumption or the least time spent on single-objective optimization. With the development of technology, though, research hotspots gradually shift to multi-objective optimization considering more diverse factors that are more in line with the actual process of shop material distribution.

At this stage of research on the shop floor material distribution problem in manufacturing enterprises, scholars have focused on two aspects of the vehicle path problem: model building and solution algorithms. In model building, the main focus is the transition from single-objective to multi-objective, from optimization objectives and constraints, etc., so as to improve the model, and the use of a variety of algorithms to solve the algorithmic synthesis. Based on the idea of two-layer planning, Lou Zhenkai established a multi-objective optimization model with fuzzy time windows, considering the number of vehicles used and the total transport mileage, and applied the simulated annealing algorithm to solve the multi-objective optimization problem for distribution [9]. Muller decomposed the multi-objective optimization problem and solved it using a heuristic algorithm, taking into account the soft time window [10]. Murao et al. investigated VRP using soft time window constraints, with fuzzy variables and penalty functions [11]. Xia Y developed a bi-objective open vehicle routing problem (OVRP) model considering soft time windows and satisfaction rates, in which the OVRP was analyzed and an improved taboo search algorithm (ITSA) fusion algorithm with an adaptive penalty mechanism and multi-neighborhood structure was applied to solve the problem [12]. Yan Zhengfeng et al. proposed a distribution path optimization method based on fuzzy soft time windows for complex mechanical assembly workshops in order to solve the problem of uncertain material demand time at work stations in the actual production process, established a material distribution path optimization model with fuzzy soft time windows with the objective of minimizing distribution costs, and used a hybrid algorithm combining a dynamic programming algorithm and a simulated annealing algorithm to solve the model [13]. Li Siguo and Guo Yu et al. established a material distribution model in a real-time environment for a discrete manufacturing workshop with a complex environment and many external disturbing factors, combined with considerations of the material distribution time window requirement and the minimum material distribution cost as the optimization objective, and used an improved genetic algorithm to solve the model [14]. Ferani et al. proposed a green vehicle path problem that optimizes the transportation costs, spoilage costs, and carbon emissions of perishable products with high transportation costs and serious air pollution issues, and solved it using a multi-objective gradient evolutionary algorithm [15]. Asma et al. proposed a new hybrid vehicle path planning algorithm for the capacitated vehicle path problem to improve the solution quality and speed up convergence [16]. After studying the two-stage vehicle route problem (2S-VRP) in logistics distribution, Zhong X et al. applied a hybrid algorithm combining an artificial bee colony and genetic algorithm (ABCGA) to solve it [17]. In addition, some scholars have applied Flexsim, Arena, Witness and other types of system simulation software to model and simulate various aspects of production logistics in order to find the bottlenecks and optimize the system by adjusting parameters; in this way, they sought to achieve goals such as the highest efficiency, lowest cost and best service [18–20]. To this day, the vehicle routing problem is a popular research topic not only in China but also abroad in the field of logistics research. The problem model is mainly studied in terms of three main pairs of considerations: vehicle capacity, time constraint and vehicle class. For the vehicle path problem with capacity constraints, Lee et al. (2010) combined the properties of the simulated annealing algorithm and designed an improved ant colony algorithm, which was experimentally shown to outperform both the original ant colony algorithm and the simulated annealing algorithm [21]. Nishi and Izuno (2014) proposed a column generation-based heuristic algorithm to solve and compared the performance with branch delimitation algorithm and manual operator, and then verified the feasibility and effectiveness of the algorithm [22]. Ahmed (2018) proposed an efficient particle swarm optimization algorithm based on two-layer local search and verified that the proposed algorithm outperforms other particle swarm optimization algorithms [23]. Reihaneh and Ghoniem (2018) developed a branch-and-cut algorithm for solving [24]. Ana Moura et al. (2023) proposed a commodity-flow model and a formulation of the Three-Dimensional Packing Problem to solve a distribution problem of a Portuguese company in the automotive industry as time windows and loading due to the limited capacity of the fleet in terms of weight and volume. [25]. Smiti et al. (2020) addressed the cumulative capacity constrained vehicle path problem, a mathematical model with the shortest arrival time to the customer as the optimization objective was developed and two optimization models were proposed to solve it [26].

For the vehicle path problem with time windows, Vidal et al. (2013) proposed a hybrid genetic search algorithm that effectively solves a variety of large-scale vehicle path problems such as route duration constraints and those involving customer assignment to specific vehicle types [27]. Nalepa and Blocho (2016) proposed an improved modal algorithm to solve an optimization model with the objective of using the minimum number of vehicles and the shortest vehicle travel distance, and verified its effectiveness through extensive experimental studies [28]. Molina et al. (2020) proposed a hybrid ant colony algorithm with local search and verified experimentally that the method has good performance [29]. Bogue et al. (2020) proposed a column generation algorithm and a post-optimization heuristic algorithm for solving [30]. Jalilvand et al. (2021) developed a two-stage stochastic model and proposed a recursive hedging algorithm for a vehicle path problem with a two-level time window allocation and stochastic service times [31]. Tilk et al. (2021) and designed a branch pricing-cut algorithm to solve the model [32]. Hoogeboom et al. (2021) solved the model using a branch-and-cut approach with the objective of minimizing the travel time and the risk of violating the time window [33].

For the problem of multiple paths for pairs of vehicles, Pietrabissa (2016) developed an algorithm for the multi-model vehicle path problem without communication and showed through simulation experiments that the algorithm has better performance [34]. Avc and Topaloglu (2016) studied the heterogeneous vehicle path problem with simultaneous pickup and delivery and proposed a hybrid local search algorithm for solving it [35]. Gholami et al. (2019) used a genetic algorithm to solve a mixed integer nonlinear model with cost minimization as an objective when studying a multi-vehicle path problem considering product transfer between vehicles in a dynamic situation [36]. Wang et al. (2019) developed a mathematical model with the optimization objective of minimizing total carbon emissions when considering an integrated single-vehicle scheduling and multi-vehicle path problem, and then proposed a forbidden search hybrid algorithm to solve it [37]. Behnke et al. (2021) proposed a column generation method for solving the vehicle path problem with heterogeneous vehicles and heterogeneous roads [38]. It is known from the research of foreign scholars that most of the current research in this field is based on vehicle path problems that consider both vehicle capacity and time window or vehicle path problems that consider both vehicle capacity and vehicle type, while relatively few studies consider vehicle capacity, service attitude, carbon emission, time window and vehicle type simultaneously.

How to reduce carbon emissions in the process of vehicle transportation has received the attention of domestic scholars, and a series of studies on low-carbon logistics has been launched. Montoya A et al. proposed a two-stage heuristic algorithm suitable for solving the green vehicle path problem (G-VRP), incorporating the case of a vehicle visiting a gas station on its way to distribution, and finally verified the effectiveness of the algorithm through experiments [39]. Koc C et al. proposed a simulated annealing algorithm based on the branch-and-bound method to improve the efficiency of solving the G-VRP problem [40]. Jabir E and Zhang S combined vehicle transportation paths and greening to construct a vehicle path optimization model incorporating carbon emission parameters followed by an improved ant algorithm and a hybrid artificial bee colony algorithm [41,42]. Niu Yunyun et al. designed a hybrid taboo search algorithm, which yielded a reduction in the total cost of the open route compared to the closed route, with a significant reduction in CO_2 emission cost and fuel consumption cost [43]. Bektase et al. innovated the concept of low-carbon transport by including a fixed operating speed and load of the vehicle in the calculation of carbon emissions [44]. Marcel also studied fixed vehicle speed, an important factor affecting the green vehicle path model, for energy consumption and carbon emission at a fixed speed [45]. Kwon Y et al. investigated the problem of optimizing heterogeneous vehicle paths considering carbon emissions and found that the implementation of carbon emissions trading can significantly reduce carbon emissions without increasing costs [46].

Although the above research has made some progress, there are still some problems such as poor solution quality, ease of falling into the local optimum and the development of a logistics distribution path with little consideration of environmental factors. It can be found that scholars at home and abroad have made certain achievements in the research of discrete workshop workstation logistics distribution path optimization. Scholars at home and abroad have studied the two issues of low carbon logistics and workshop workstation logistics distribution path optimization separately, yet not many combine the two issues

As national carbon emission governance is becoming more and more regulated, carbon tax policy implementation is coming closer and closer, and manufacturing enterprises must integrate the concept of low carbon into the internal logistics activities of the vehicle in the operation process to reduce the pressure brought by environmental costs. This paper proposes a study on the optimization of shop floor material distribution paths with multiple optimization objectives considering carbon emissions, aiming to help enterprises reduce carbon emission costs based on reducing distribution costs, and achieving customer satisfaction, and realizing the unity of economic benefits and environmental protection. In the analysis of existing workshop material distribution path optimization problems, it has been found that in some workshops, there is a high time penalty cost related to the fact that the distribution trolley cannot deliver the required materials within the time window specified by each workstation, resulting in low service satisfaction at the workstations. Furthermore, unreasonable route planning, which entails the use of more trolleys in the distribution process, increases the distribution distance and distribution costs, etc. Therefore, this paper establishes a multi-objective shop floor material distribution route optimization model based on the cost factors and carbon dioxide emissions generated during the distribution process, taking into account the demand for materials produced at each workstation. Based on this, the example of material distribution within a time window in a smart gas meter workshop owned by an electronic assembly manufacturing K enterprise is analyzed. The results show that the established model can enable the enterprise to more effectively control the carbon emissions and costs generated during development, as well as achieving improved service satisfaction at the workstation and shortening the distribution distance.

This article is structured as follows: Section 2 provides related work on the shop floor material distribution path optimization model. Section 3 presents the vehicle path optimization algorithms. Section 4 presents a discrete assembly manufacturing company workshop, which is taken as an example to verify the correctness of the proposed method. Lastly, the paper ends with Section 5, which concludes the research outcome with future work.

2. Shop Floor Material Distribution Path Optimization Model

2.1. Constructing an Optimization Model

This paper constructs a logistics path optimization model for internal logistics distribution in the workshop of a discrete manufacturing enterprise, seeking the lowest total distribution cost and the highest service satisfaction at the workstations as the optimization objectives. The problem of logistics path optimization for the dispatching of material distribution vehicles to each workstation in the workshop of a discrete manufacturing enterprise can be formulated as follows. The mathematical model of the problem is established under the conditions that the model of material distribution of each workstation in the workshop is known and the number of vehicles is sufficient, and the number and demand of workstations in the workshop, the expected delivery time and the acceptable time interval of each workstation are also known. An optimization model is constructed to minimize costs and achieve a high level of service satisfaction at the workstations while providing the quantity demanded at each workstation.

2.1.1. Cost Components of Material Distribution for Discrete Manufacturing Companies

In order to closely represent the actual situation of material distribution on the shop floor of a discrete manufacturing enterprise, a cost analysis of the various factors affecting the distribution of intra-enterprise logistics is carried out. The distribution path cost model constructed in this paper consists of four cost components, as shown below.

(1) Fixed cost C_1. The fixed cost of the vehicle includes the distribution vehicle consumption costs, vehicle maintenance costs, driver wages and other costs. Usually, the fixed cost of the vehicle in the transportation process is determined by the number k of task-execution vehicles distributed by the distribution center, so the fixed cost in this paper can be expressed according to the relationship between the fixed cost of the vehicle

and the number of distribution vehicles as a positive proportional function, as shown in Equation (1):

$$C_1 = \sum_{k=1}^{k} f_k s_k \tag{1}$$

where f_k is the fixed cost of using the k-th vehicle; $s_k = 1$ or 0, with 1 indicating that the k-th vehicle participates in the distribution and 0 indicating that the k-th vehicle does not participate in the distribution.

(2) Vehicle transportation costs C_2. The transportation costs are the distance-related costs incurred by the vehicle as it undertakes material distribution. Transport costs are proportional to the distance traveled by the vehicle, as shown in Equation (2):

$$C_2 = \sum_{k=1}^{k} \sum_{i=0}^{n} \sum_{j=0}^{n} c_p x_{ij}^k d_{ij} \tag{2}$$

where c_p is the transportation cost per unit distance per distribution vehicle; $x_{ij}^k = 1$ or 0, where 1 means the k-th vehicle participates in distribution and 0 means the k-th vehicle does not participate in distribution, while d_{ij} is the distance from station i to station j.

(3) Penalty cost C_3. The workshop logistics distribution is contained by a time window; distribution vehicles that deliver their materials outside the acceptable time window required by the work station will incur a set penalty cost. Untimely distribution will lead to production line stoppages, delay the completion of the production plan, affect the delivery time of the product and lead to customer dissatisfaction. The penalty cost $p_i(t_i)$ corresponding to workstation point i is linear as a function of vehicle arrival time t_i, and is calculated as shown in Equation (3).

$$p_i(t_i) = \begin{cases} y_i^k \mu_1 p_1 q_i (e_i - t_i^k), e_i \le t_i^k \le et_i \\ 0, et_i \le t_i^k \le lt_i \\ y_i^k \mu_2 p_1 q_i (t_i^k - l_i), lt_i \le t_i^k \le l_i \\ M, t_i^k \prec e_i, t_i^k \succ l_i \end{cases} \tag{3}$$

Therefore, the penalty cost C_3 is calculated as shown in Equation (4):

$$C_4 = \sum_{i=1}^{n} p_i(t_i) \tag{4}$$

where y_i^k indicates whether the k-th vehicle delivers material to station i, and it takes the value 0 or 1, where 1 means delivery and 0 means no delivery; p_1 is the unit price; q_i is the material demand at station i; t_i^k is the actual delivery process time for the k-th vehicle to arrive at station i; μ_1 and μ_2 are pre-set penalty factors; M denotes an extreme value $[et_i, lt_i]$ is a mandatory hard time interval; $[e_i, l_i]$ is an acceptable late time interval.

(4) Carbon emission cost C_4. In order to calculate the carbon emission cost, the carbon dioxide emissions generated by the distribution vehicle during the distribution journey must first be accurately calculated. The carbon dioxide emissions produced by vehicles on the way to distribution are mainly carbon emissions generated by the burning of natural energy sources, such as by fuel consumption during the vehicle's journey. The fuel consumption of the vehicle is determined by the load and distance; for the fuel consumption $\rho(x)$ of a vehicle with load x over a fixed distance, the calculation formula is shown in Equation (5). E_1 indicates the carbon emissions generated by the distribution vehicle's material distribution process; the calculation formula is shown in Equation (6). Carbon emission cost C_4 is calculated as shown in Equation (7).

$$\rho(x) = \rho_o + \frac{\rho^* - \rho_o)}{Q} \times x \tag{5}$$

$$E_1 = e_o \rho(x) d \tag{6}$$

$$C_4 = \sum_{k=1}^{k} \sum_{i,j=0}^{n} p_e e_o \rho(x) x_{ij}^k d_{ij} \tag{7}$$

Here, the load of the material distribution vehicle is represented by x; ρ_o is the fuel consumption of the material distribution vehicle when it is not loaded; ρ^* is the fuel consumption of the material distribution vehicle when it is fully loaded; Q is the rated load; e_o is the CO_2 emission factor of the fuel; the distance traveled by the material distribution vehicle is represented by d, and p_e represents the carbon tax price on the carbon emissions trading market.

2.1.2. Workstation Service Satisfaction Model

With the development and progress of science and technology, and an increasingly competitive social environment, slogans such as "service quality" and "service level" can be seen everywhere in major workshops, and these have obviously become important factors as enterprises seek to improve their competitiveness, production efficiency and service quality. However, in the process of material distribution in discrete manufacturing workshops, various uncertainties can lead to fluctuations in the production pace of the workstations, resulting in changes in the material demand time of each workstation. Therefore, in this paper, the satisfaction of service workstations is one of the optimization objectives. Here, the satisfaction of each workstation i can be expressed using the fuzzy affiliation function $SA(t)$, as shown in Equation (8).

$$SA(t) = \sum_{k=1}^{k} \sum_{i=0}^{n} \{\max[(e_i - T_i), 0] + \max[(T_i - l_i), 0]\} \tag{8}$$

2.1.3. Mathematical Modeling

$$Z_1 = \min(C_1 + C_2 + C_3 + C_4) \tag{9}$$

$$Z_2 = \min\left(\sum_{k=1}^{k} \sum_{i=0}^{n} \sum_{j=0}^{n} d_{ij} x_{ij}^k\right) \tag{10}$$

$$Z_3 = \min\left\langle \sum_{k=1}^{k} \sum_{i=0}^{n} \{\max[(e_i - T_i), 0] + \max[(T_i - l_i), 0]\}\right\rangle \tag{11}$$

such that

$$\sum_{i=1}^{n} q_i Y_i^k \leq Q \quad k = 1, 2, 3, \ldots K \tag{12}$$

$$\sum_{k=1}^{k} \sum_{j=1}^{n} x_{ij}^k \leq K \quad i = 0 \tag{13}$$

$$\sum_{k=1}^{k} \sum_{j=0}^{n} x_{ij}^k = 1 \quad i \neq j, i \in U_1 \tag{14}$$

$$\sum_{k=1}^{k} \sum_{j=0}^{n} x_{ji}^k = 1 \quad i \neq j, i \in U_1 \tag{15}$$

$$\sum_{k=1}^{K} Y_i^k = 1 \quad i \in U_1 \tag{16}$$

$$\sum_{j=1}^{n} x_{ij}^{k} = \sum_{j=1}^{n} x_{ji}^{k} \leq 1 \quad i = 0, k = 1,2,3,\ldots K \tag{17}$$

$$x_{ij} = \begin{cases} 1 \\ 0 \end{cases} \quad i,j \in U_1, k = 1,2,3,\ldots K \tag{18}$$

$$Y_i^k = \begin{cases} 1 \\ 0 \end{cases} \quad i \in U_1, k = 1,2,3,\ldots K \tag{19}$$

$$U_1 = \{i = 1,2,3,\ldots n\} \tag{20}$$

In the above multi-objective optimization model, $i,j = 0$ denotes the starting and ending points of the distribution trolley, and d_{ij} denotes the distance between station i and station j. Equation (9) indicates that the objective function of the model is the lowest total distribution cost; Equation (10) indicates that the distance traveled by the distribution trolley is the smallest; Equation (11) indicates that the service satisfaction of the workstation is the largest; Equation (12) represents that when the distribution center assigns the distribution task to the kth vehicle, the total amount to be transported cannot exceed the maximum load weight of the distribution vehicle; Equation (13) indicates that the number of vehicles carrying out the distribution operation cannot exceed the total number of vehicles at the distribution center's disposal; Equations (14) and (15) show that each station point can only dock one vehicle in order to obtain one distribution service; Equation (16) shows that each station point has one vehicle as the target of its distribution service; Equation (17) shows that the distribution vehicle carrying out the transportation task departs from the distribution center, and returns there after the distribution operation is completed; Equations (18)–(20) establish the setting of decision variables.

2.2. Model Conditional Assumptions

The problem studied in this paper concerns the optimization of the distribution path of K company as regards carbon emissions. In order to facilitate the construction of the model and the implementation of the algorithm that follows, the following assumptions are made:

(1) There is only one material distribution center in the workshop, and the location of the distribution center is known. The material inventory of each workstation in the distribution center is sufficient to meet the demand of all workstations;

(2) The material requirements of each workstation i are known, and the sum of the material requirements of each workstation on each distribution path cannot exceed the maximum capacity Q of the distribution trolley;

(3) The distribution vehicles are of the same type. Each vehicle starts from the distribution center and travels at the same speed and uniformity, and all vehicles must return to the distribution center after completing the distribution task. Factors such as the stopping, starting, and loading and unloading times of the distribution trolley, and trolley failure are ignored;

(4) In the process of the vehicle carrying out the distribution task, its corresponding distribution service object remains unchanged—for example, the service customer information, the distribution order, etc., will not change. A delivery trolley can serve multiple workstations at the same time, but each workstation can only be served by one delivery trolley as it undertakes its material distribution activities;

(5) All the information required for the distribution process is known, including all the details of the distribution center and each workstation, etc.;

(6) Each workstation can only be provided with one delivery service by one vehicle. The materials of a workstation cannot be split during delivery, and the vehicle only performs delivery tasks during the delivery process, without pick-up tasks. There is a time window constraint for distribution. For each workstation i, the delivery trolley must perform

the service within [ei, li]. If the delivery trolley arrives earlier than ei, it must wait at the workstation, and if the delivery trolley arrives later than li, the delivery service will be delayed.

3. Vehicle Path Optimization Algorithms

3.1. Algorithm Design Ideas

This paper studies the optimization of material distribution paths for discrete assembly manufacturing enterprises, considering the carbon emissions of shop floor workstations, which is an NP-hard problem with high complexity. Scholars at home and abroad have proposed a variety of algorithms to solve such problems, and the two main categories are exact algorithms and heuristic algorithms. The established exact algorithms are suitable for solving problems of small scale and low complexity, but for large-scale VRP problems, exact algorithms may not be able to obtain the optimal solution due to the complexity of the solution process, so scholars have proposed heuristic algorithms. Genetic algorithms and forbidden search algorithms are the most widely used types of heuristic algorithms. Genetic algorithms are more globally searchable and more computationally efficient than forbidden search algorithms. Genetic algorithms have strong applicability in solving vehicle path problems, and can solve complex VRP problems well; they are thus widely used by scholars at home and abroad due to their good performance. Table 1 shows a summary of the advantages, disadvantages and applicability of five common modern heuristic algorithms. Therefore, in this paper, taking the characteristics of the study case, the genetic algorithm is chosen to solve the model; the genetic algorithm flow chart is shown in Figure 1. While the traditional genetic algorithm can solve the problem proposed in this paper, there are many disadvantages of this algorithm, such as its slow convergence speed, low efficiency of optimization, and other problems. The vehicle path problem requires an algorithm with a strong global search capability in the early stage and a strong local search capability in the later stage. In order to solve the above problems, this paper makes improvements to the standard genetic algorithm. The improvements focus on two aspects: firstly, designing the operator so that the genetic algorithm has spiral characteristic; secondly, improving the crossover and variation probability such that they have adaptive characteristics. Therefore, in this paper, starting from the mutation operation of the genetic algorithm, we adopt the adaptive mutation method to enhance the variance difference [47,48], which can avoid falling into the local optimal solution; meanwhile, we introduce the Metropolis criterion of simulated annealing algorithm [49–51] to judge whether it accepts the individuals generated by the mutation, which can improve the convergence accuracy of the algorithm.

Table 1. Comparison of the characteristics of modern heuristics.

No.	Algorithm Type	Advantage	Disadvantage	Scope of Application
1	Ant colony algorithm	Good positive feedback mechanism and easy association with other algorithms.	Long search time, need to constantly adjust variables, slow solution speed.	It is applicable to multi-objective optimization problems.
2	Simulated annealing algorithm	High robustness, parallel processing at multiple constraints	The accuracy of the results is not high and the running time is long and inefficient.	Applicable to the modification of existing path problems
3	Particle swarm algorithm	The algorithm is simple and fast to compute, with strong global search capability.	It is not applicable to discrete problems and tends to converge prematurely.	Solved in combination with other algorithms.
4	Taboo search algorithm	Strong local search ability, prone to premature convergence.	The solution is complex, computationally inefficient, and dependent on the initial solution obtained.	Solving large-scale problems.
5	Genetic Algorithm	High computational efficiency and strong bureau search capability.	Poor local search capability.	VRP and other complex realities that fit the problem.

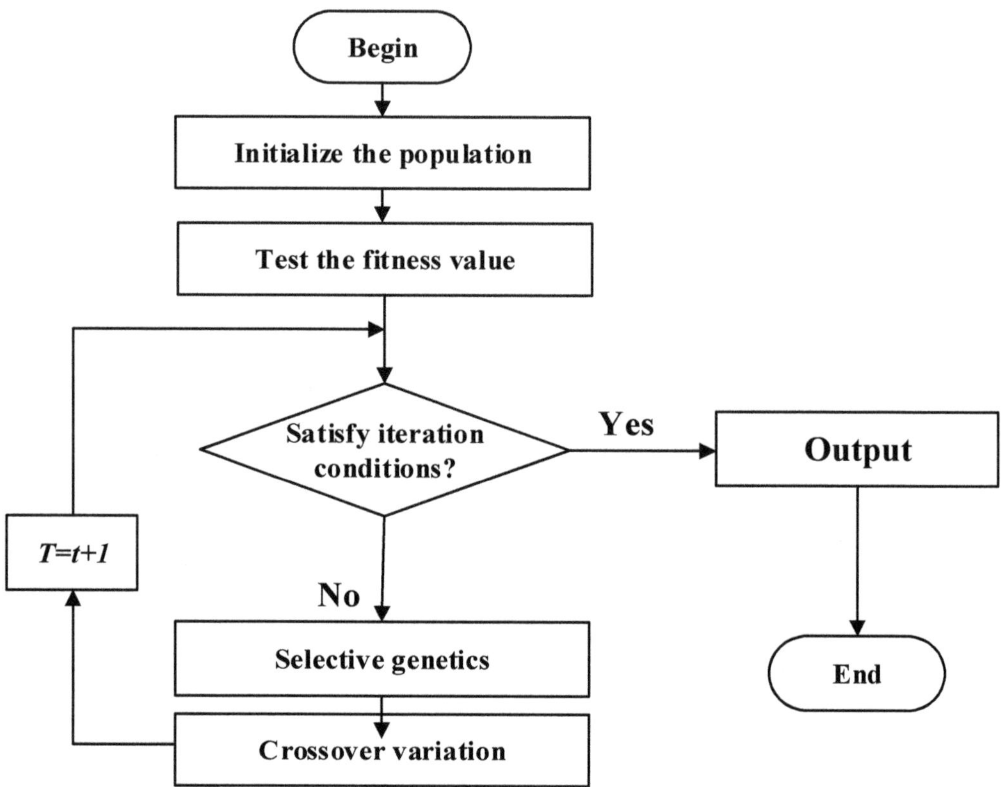

Figure 1. Genetic algorithm flowchart.

In order to ensure that the value of the fitness function of the algorithm evolves in the best direction, a special combinatorial operator is first devised, using which the population can be kept in an ascending state. This paper involves the use of three combinations of operators:

(1) Variation in the combinatorial operator. Two genes, a and b, are randomly selected from the parent. The genes are arranged in reverse order, and gene a is inserted in front of gene b. The fitness value is then observed to generate new offspring. If the fitness value becomes larger, the former reverses the order of the genes between a and b to generate offspring, and modifies the fitness value of S. The latter inserts a in front of b to generate offspring and modifies the fitness value of S. This combined arithmetic operation is completed, and if the fitness value of the offspring is not lower than that of the parent, an upward spiral is achieved. Otherwise, no change occurs;

(2) Crossover operator. Assuming that the parents are SA and SB, first, a gene is randomly selected from either parent as the starting point. Then, a cyclic crossover is applied, and finally the resulting children, SA' and SB', are used to replace the parents SA and SB;

(3) Variance operator. Two genes, c and d, in the parent are randomly selected, and then the positions of c and d are swapped and the fitness value of S is modified.

As regards populations, each individual has its own fitness, and the population also has an average fitness. The selection fitness function, adaptive crossover rate and variation rate formulae in this paper are shown below:

$$f(g_i) = \frac{N-I+1}{N}, i = 1,2,3,\ldots N \tag{21}$$

$$p_c = \begin{cases} p_{c1} - \dfrac{(p_{c1} - p_{c2}) \times (f_{\max} - f')}{f_{\max} - \bar{f}} & , f' \geq \bar{f} \\ p_{c1}, & , f' \prec \bar{f} \end{cases} \tag{22}$$

$$p_m = \begin{cases} p_{m1} - \dfrac{(p_{m1} - p_{m2}) \times (f_{\max} - f')}{f_{\max} - \bar{f}} & , f' \geq \bar{f} \\ p_{m1}, & , f' \prec \bar{f} \end{cases} \tag{23}$$

where N is the number of individuals in the population; f denotes the fitness value of the variant individual; f' denotes the relatively large fitness value of the two crossover individuals; $\bar{f}$ denotes the mean value of fitness in the population; $f_{\max}$ denotes the highest value of fitness in the population; p_{c1} and p_{c2} denote the maximum and minimum crossover probabilities, respectively, and p_{m1} and p_{m2} denote the maximum and minimum variance probabilities, respectively. p_{c1}, p_{c2}, p_{m1} and p_{m2} are parameters between 0 and 1, set herein as $p_{c1} = 0.85$, $p_{c2} = 0.65$, $p_{m1} = 0.15$, and $p_{m2} = 0.001$.

3.2. Algorithm Flow

In this paper, based on the concept of the traditional algorithm, the genetic algorithm is improved to make it more applicable to the distribution path optimization problem constructed in this paper. This algorithm combines a genetic algorithm with other advanced algorithms [52–55]. The improved algorithm flow is as follows:

Step 1—Chromosome coding. In this paper, a natural number code is used, with the distribution center represented by 0, and K and n representing the number of vehicles and the number of workstations, respectively, which results in a chromosome code of k + n + 1. For example, if the chromosome "0, 8, 0, 5, 9, 4, 7, 0, 23, 13, 25, 0, 15, 6, 0" is used, this means that there are 10 workstations served by four vehicles, and the route of the four vehicles represented by this chromosome is shown in Figure 2.

Vehicle path 1: $0 \longrightarrow 8 \longrightarrow 0$

Vehicle path 2: $0 \longrightarrow 5 \longrightarrow 9 \longrightarrow 4 \longrightarrow 7 \longrightarrow 0$

Vehicle path 3: $0 \longrightarrow 23 \longrightarrow 13 \longrightarrow 25 \longrightarrow 0$

Vehicle path 4: $0 \longrightarrow 15 \longrightarrow 6 \longrightarrow 0$

Figure 2. The route of the four vehicles in a hypothetical scenario.

Step 2—Generate the initial population. The purpose of population initialization is to generate the number of feasible solutions in the feasible domain, i.e., determine the size of the population. This is usually carried out in a random way. However, the rate of survival of the populations generated in this way is relatively low, and it is difficult for them to adapt to the environment. Therefore, this paper adopts a greedy algorithm for the local optimization of initial population generation based on the original approach. The chromosome generation process is repeated according to the aforementioned coding method until an initial population size of N ($N = 100$ is set in this study) is randomly generated for the initial population;

Step 3—Determine the fitness function. The path optimization problem studied in this paper uses the reciprocal of the minimum total cost of distribution as the fitness function, thereby ensuring that with a higher value of fitness, the cost will be lower. The fitness function is shown in Equation (24):

$$Fit_l = 1/Z_{1l} \quad (l = 1, 2, 3, \ldots, N) \tag{24}$$

where Fit_l denotes the fitness value corresponding to the l-th chromosome, Z_{1l} denotes the distribution cost corresponding to the l-th chromosome, and N denotes the population size;

Step 4—Selection. The most commonly used selection methods include the bidding tournament method, the roulette wheel method and the random traversal sampling method.

The roulette wheel method has good adaptability, so this paper uses it for the selection operation, i.e., the selection is made according to the proportion of individual fitness to the sum of individual fitnesses in the population. The selection probability formula is shown in Equation (25).

$$P_l = Fit_l \left/ \sum_{l=1}^{N} Fit_l \right. \tag{25}$$

Step 5—Crossover. The adaptive crossover rate function used in this paper is shown in Equation (22);

Step 6—Variability. The variability function used in this paper is shown in Equation (23);

Step 7—Evolutionary reversal operation;

Step 8—Update new populations;

Step 9—Determine if the stopping condition is satisfied. If yes, output the optimal solution; otherwise, go to step 3.

4. Case Study of a Discrete Assembly Manufacturing Company Workshop

4.1. Data Sources and Parameter Settings

This paper verifies the feasibility of the improved genetic algorithm via the optimization of the production material distribution vehicle path problem for two production lines in the smart gas meter and smart electric meter workshops of *K* Company. *K* Company is a large private enterprise producing energy-metering products in China, integrating R&D, production and sales, and a typical discrete assembly manufacturing enterprise. It mainly produces smart electricity meters, smart water meters, smart gas meters, ultrasonic heat meters and other metering instruments. The discrete manufacturing workshop has material distribution center, 32 stations to be distributed to, and a number of distribution trolleys. In this paper, the vehicle distribution path is optimally designed to minimize the total distribution cost, minimize the distribution route and maximize customer satisfaction (aiming at a service satisfaction of no less than 85%), under the condition that the maximum load constraint and time window requirements of each vehicle are met, and the problem is solved using an improved genetic algorithm. The simulation was carried out using MATLABR2020 in the Windows 10 environment. The relevant parameters of the model are shown in Table 2, and the table of distribution tasks for each workstation is shown in Table 3.

Table 2. Model parameters.

Parameter Symbols	Parameter Name	Parameter Values
Q	Maximum load capacity of material distribution vehicles	100 kg
V_o	Average travel rate of material distribution vehicles	50 m/min
F_k	Fixed cost per material distribution vehicle	RMB 100/Vehicle
C_p	Transport costs per unit distance traveled by vehicle	RMB 2/km
μ_1	Waiting costs for early arrival	RMB 20/h
μ_2	Delay costs for late arrivals	RMB 60/h
e_o	Carbon emissions per unit of fuel consumption	2.8 kg/L
λ	Carbon emissions per unit of cargo transported per unit of distance	0.0075 g/kg·km
ρ_o	Fuel consumption per unit distance when the vehicle is unladen	0.122 L/km
ρ^*	Fuel consumption per unit distance when the vehicle is fully loaded	0.388 L/km
p_e	Carbon tax	RMB 2/kg

Table 3. Distribution tasks by workstation.

Workstation	Coordinate (m)	Material Requirement (kg)	Delivery Time Window (min)	Workstation	Coordinate (m)	Material Requirement (kg)	Delivery Time Window (min)
0	(60,60)	0	0	17	(40,48)	13	[2,7]
1	(55,85)	26	[2,3]	18	(45,14)	7	[1,3]
2	(20,18)	19	[4,10]	19	(50,70)	6	[3,10]
3	(40,66)	33	[3,5]	20	(40,60)	12	[3,5]
4	(56,30)	21	[1,6]	21	(90,70)	4	[4,6]
5	(22,50)	22	[2,5]	22	(60,80)	6	[1.5,5]
6	(105,16)	22	[7,10]	23	(30,90)	10	[12,15]
7	(10,30)	29	[5,9]	24	(50,40)	18	[7,9]
8	(30,95)	20	[6,12]	25	(40,80)	22	[10,13]
9	(45,125)	25	[1,5]	26	(30,30)	8	[3,6]
10	(110,70)	24	[4,9]	27	(70,30)	9	[5,7]
11	(156,100)	31	[3,11]	28	(80,40)	16	[5,10]
12	(99,100)	24	[4,13]	29	(40,40)	17	[5,9]
13	(99,45)	26	[4,9]	30	(80,60)	15	[6,10]
14	(88,100)	25	[5,13]	31	(60,30)	11	[7,13]
15	(55,85)	26	[4,11]	32	(20,35)	13	[5,12]
16	(110,15)	25	[6,12]	-	-	-	-

4.2. MATLAB Software Solution

In this paper, a logistics distribution route optimization model for a discrete manufacturing enterprise workshop, focusing on carbon emissions, is established, with the lowest total cost of material distribution, the shortest route and the highest satisfaction as the model's optimization objectives. The MATLAB software is used to solve the model. The algorithm parameters are set as follows: the initial population size is 100 and the number of iterations is 3000.

4.2.1. Solving the Distribution Path Optimization Model without Considering Carbon Emissions

Among the Pareto solutions obtained without considering carbon emissions, the solution that best achieves customer satisfaction and minimizes cost is shown in Figure 3.

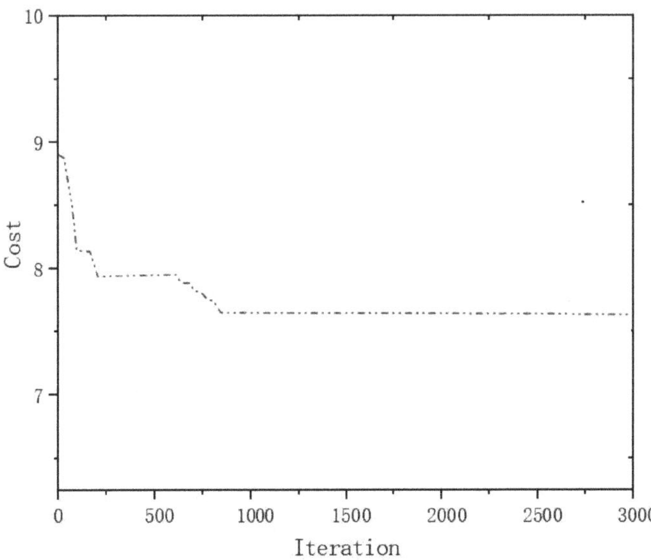

Figure 3. Cost iteration convergence graph without considering carbon emissions.

When not considering carbon emissions, the total cost composition only includes three parts: vehicle fixed cost, transportation cost and penalty cost. Here, under each constraint, the optimal distribution path is obtained, as shown in Table 4. Table 4 shows that, without considering carbon emissions, the distribution center needs to arrange seven material distribution vehicles to perform distribution for 32 discrete assembly workstations in the workshop. The total distance traveled during distribution is 26.5 km and the total cost of distribution is RMB 753, of which the fixed vehicle cost is RMB 700, the transportation cost is RMB 53 and the penalty cost is RMB 0.

Table 4. Vehicle route planning results without considering carbon emissions.

Distribution Vehicles	Distribution Path	Loading Rate	Delivery Distance (km)	Is It within the Time Window?
1	0-16-4-22-5-17-0	87%	2.8	Yes
2	0-26-1-13-2-0	79%	5.5	Yes
3	0-27-30-9-12-7-0	93%	2.4	Yes
4	0-15-6-10-21-29-0	93%	4.9	Yes
5	0-3-32-14-24-0	91%	5.2	Yes
6	0-25-8-11-18-31-0	89%	3.4	Yes
7	0-20-19-23-28-0	44%	2.3	Yes

4.2.2. Solving the Distribution Route Optimization Model Considering Carbon Emissions

Among the Pareto solutions obtained when considering carbon emissions, the solution that best achieves customer satisfaction and minimizes costs is shown in Figure 4.

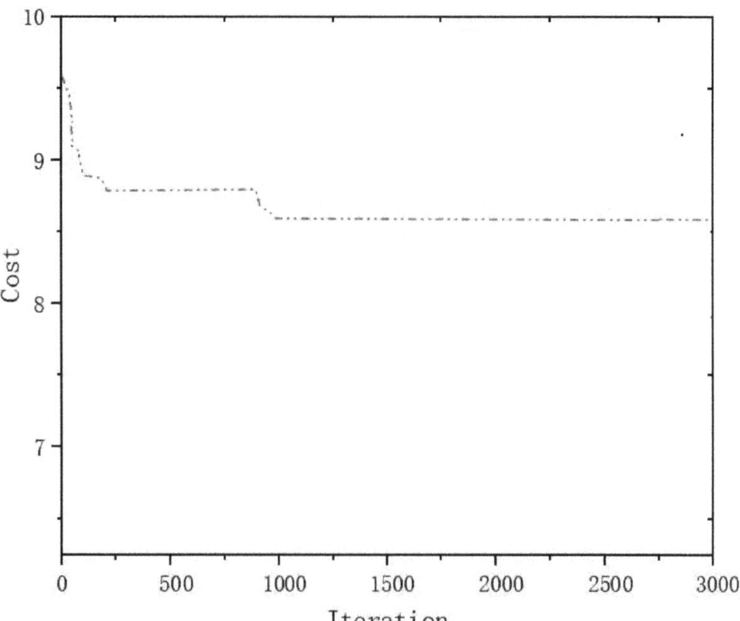

Figure 4. Cost iteration convergence graph considering carbon emissions.

When considering carbon emissions, the total cost comprises carbon emission costs, fixed vehicle costs, transportation costs and penalty costs. Here, the optimal distribution path is obtained under each constraint, as shown in Table 5. According to Table 5, the distribution center needs to arrange seven refrigerated vehicles to achieve distribution for 32 shops under the condition of carbon emission. The total distance traveled for distribution is 24.8 km and the total cost of distribution is RMB 965.08, of which the fixed vehicle cost is RMB 700, the transportation cost is RMB 49.6, the penalty cost is RMB 0 and the carbon emission cost is RMB 216.08.

Table 5. Results of vehicle route planning considering carbon emissions.

Distribution Vehicles	Distribution Path	Loading Rate	Delivery Distance (km)	Is It within the Time Window?
1	0-6-1-13-22-2-0	99%	2.8	Yes
2	0-29-5-17-26-30-0	93%	4.5	Yes
3	0-20-28-25-0	50%	2.9	Yes
4	0-27-3-12-7-0	95%	2.6	Yes
5	0-10-4-9-14-0	95%	4.5	Yes
6	0-19-16-21-11-31-0	77%	3.6	Yes
7	0-23-24-15-18-32-8-0	94%	3.9	Yes

4.3. Comparison of Data Results

A comparative analysis of the results obtained from multiple successive solutions for both cases, without and with carbon emissions, is shown in Table 6.

Table 6. Comparison of the solution results for the two cases.

Program	Distribution Vehicles	Delivery Distance	Carbon Emissions	Average Loading Rate	Fixed Cost	Transport Cost	Penalty Cost	Cost of Carbon Emissions	Total Cost
No consideration of carbon emissions	7	26.5	145.19	82.29%	700	53.0	0	290.38	1043.38
Considers carbon emissions	7	24.8	108.04	86.00%	700	49.6	0	216.08	965.08

5. Conclusions

In the process of planning the distribution route for materials on the shop floor, if the carbon emission factor is taken into account, the optimal distribution route within a discrete manufacturing enterprise can not only reduce distribution costs, but also reduce carbon emissions, achieving a win–win situation in terms of economic benefits and environmental protection. As seen in Table 6, the numbers of vehicles used, the distances traveled, the average loading rates, the carbon emissions, and each associated cost corresponding to the optimal solution paths for both cases are compared and analyzed. In terms of the number of vehicles used, the average loading rate and the penalty cost, the route without carbon emissions and the one with carbon emissions both require seven distribution vehicles, and the penalty cost is 0. The average vehicle load factor without carbon emissions is 82.29%, the total distance traveled is 26.5 km, and the transport cost is RMB 53; the transport cost is 53% and the carbon emission is 145.19 kg. The average vehicle load factor when considering carbon emissions is 86%, the total distance traveled is 24.8 km, the transportation cost is RMB 49.6 and the carbon emission is 108.04 kg. Therefore, the optimal distribution route considering carbon emissions entails a 7.5% lower total distribution cost, 6.4% lower total distance traveled and 25.6% lower total carbon emission than the optimal route without considering carbon emissions. It can be seen that the optimal distribution path considering carbon emissions has more obvious advantages than the optimal distribution path without consideration of carbon emissions. The optimal distribution path considering carbon emissions not only reduces the pollution emitted into the environment, but also helps to reduce the distribution costs of enterprises and improves their market competitiveness.

By comparing and analyzing the results, it is concluded that the path that considers carbon emissions is better. The distribution path planning process should consider not only transportation costs and penalty costs, but also carbon emissions, such that the planned distribution path can be more economical and environmentally friendly. The optimization results can be applied to the actual distribution process of workshop material distribution so as to achieve the objectives of reducing distribution costs, improving service satisfaction, reducing carbon emissions, improving the distribution efficiency of workshop logistics and making the distribution vehicle paths more economical and scientifically defendable.

In this paper, we have used the data obtained from research to solve the distribution path optimization model using MATLAB programming, taking the material distribution in the workshop stations of a discrete assembly manufacturing K company as an example. The optimal distribution path obtained for this company, taking into account carbon emissions, not only reduces distribution costs, but also controls carbon emissions, which has certain relevance to discrete manufacturing enterprises. In reality, there are many factors to be considered in the optimization of the distribution path of the shop floor logistics of a discrete manufacturing enterprise, and this paper makes use of a simplified model, so there are some shortcomings. For example, the distribution vehicles set in this paper are of the same model and have a uniform loading capacity, but in reality, due to the different demands related to distribution, many different models of distribution vehicles with different loading capacities will be used, resulting in changes in fixed costs and transportation costs, which can be increased in subsequent studies of multi-model distribution; at the same time, the driving speed of the vehicle set in this paper is fixed, and only one distribution vehicle is set on each road. In reality, the driving speeds of distribution vehicles are usually directly related to the road conditions, and will not remain unchanged, and the resulting fuel consumption will also vary due to changes in speed. The research process can be further extended by using not only genetic algorithms but also other algorithms, such as ant algorithms and bee colony algorithms. In addition, to compensate for the limitations of various heuristics, a combination of multiple heuristics can be used, for example, a genetic algorithm with 3-OPT local search as a variational operator and a road slope factor can be introduced into the carbon emission model to reduce carbon emissions. The Firefly algorithm can also used in combination with two local search and genetic operators to solve the VRP problem limited by vehicle volume, and fusing a genetic algorithm and fireworks algorithm can design an improved fireworks genetic algorithm, etc. Finally, to compare the effects, multiple heuristic algorithms can be used simultaneously to study the same problem, and then the optimal solution among them can be selected.

Author Contributions: Conceptualization, X.Z., L.J. and Y.X.; methodology, X.Z., L.J. and Y.X.; formal analysis, Y.X. and L.J.; investigation, X.Z. and L.J.; writing—original draft preparation, X.Z., L.J. and Y.X.; writing—review and editing, X.Z., L.J. and Y.X. All authors have read and agreed to the published version of the manuscript.

Funding: This research was funded by the Natural Science Foundation of Hunan Province, China (Project number: 2022JJ50244); Education Department of Hunan Province (Project number: 21B0695, 21A0475); Project of Hunan Social Science Achievement Evaluation Committee in 2022 (Project number: XSP22YBC081); Project of Shaoyang Social Science Achievement Evaluation Committee in 2022 (Project number: 22YBB10).

Data Availability Statement: Not applicable.

Conflicts of Interest: The authors declare no conflict of interest.

References

1.   Ting, Q.; Zhang, K.; Luo, H.; Wang, Z.; Jia, D.; Chen, X.; Huang, G.; Li, X. Dynamic linkage mechanism, system and case of "production-logistics" driven by Internet of Things. *J. Mech. Eng.* **2015**, *51*, 36–44. (In Chinese)
2.   Chen, T. Simulation of e-commerce logistics and distribution route planning during express delivery burst period. *Comput. Simul.* **2021**, *38*, 355–359. (In Chinese)
3.   Yin, C.; Deng, P.; Li, X. Intelligent Manufacturing Mode for Sophisticated Equipment Assembly Workshop. *J. Adv. Manuf. Syst.* **2018**, *17*, 533–549. [CrossRef]

4. Zhou, J. Intelligent manufacturing—The main direction of "Made in China 2025". *China Mech. Eng.* **2015**, *26*, 2273–2284. (In Chinese)

5. Dantzig, G.B.; Ramser, J.H. The truck dispatching problem. *Manag. Sci.* **1959**, *6*, 80–91. [CrossRef]

6. De Oliveira da Costa, P.R.; Mauceri, S.; Carroll, P.; Pallonetto, F. A genetic algorithm for a green vehicle routing problem. *Electron. Notes Discret. Math.* **2018**, *64*, 65–74. [CrossRef]

7. Zhou, L. Integrated optimization of low-carbon time-varying urban distribution vehicle path-dispatch scheduling. *Comput. Eng. Appl.* **2019**, *55*, 264–270. (In Chinese)

8. Bullnheimer, B.; Hartl, R.F.; Strauss, C. An improved ant system algorithm for the vehicle routing problem. *Ann. Oper. Res.* **1999**, *89*, 319–328. [CrossRef]

9. Lou, Z.K. Multi-objective optimization of distribution problems with fuzzy time windows. *Fuzzy Syst. Math.* **2017**, *31*, 183–190. (In Chinese)

10. Muller, J. Approximative solutions to the bicriterion vehicle routing problem with windows. *Eur. J. Oper. Res.* **2010**, *202*, 223–231. [CrossRef]

11. Murao, H.; Tommata, K.; Konishi, M. Pheromone based transportation scheduling system for the multivehicle routing problem. In Proceedings of the IEEE International Conference on Systems, Man and Cybernetis, Tokyo, Japan, 12–15 October 1999; Institute of Electrical and Electronics Engineers Inc.: Piscataway, NJ, USA, 1999.

12. Xia, Y.; Fu, Z. Improved tabu search algorithm for the open vehicle routing problem with soft time windows and satisfaction rate. *Clust. Comput.* **2018**, *22*, 8725–8733. [CrossRef]

13. Yan, Z.; Mei, F.; Ge, M.; Ling, L. A fuzzy soft time windows-based material flow path optimization method for shop floor. *Comput. Integr. Manuf. Syst.* **2015**, *21*, 2760–2767. (In Chinese)

14. Li, S.; Guo, Y.; Wang, Y.; Wu, Q.; Ge, Y. A real-time material distribution path planning method based on improved genetic algorithm. *J. Nanjing Univ. Aeronaut. Astronaut.* **2017**, *49*, 789–796. (In Chinese)

15. Zulvia, F.E.; Kuo, R.J.; Nugroho, D.Y. A many-objective gradient evolution algorithm for solving a green vehicle routing problem with time windows and time dependency for perishable products. *J. Clean. Prod.* **2020**, *242*, 118428.1–118428.14. [CrossRef]

16. Altabeeb, A.M.; Mohsen, A.M.; Ghallab, A. An improved hybrid firefly algorithm for capacitated vehicle routing problem. *Appl. Soft Comput.* **2019**, *84*, 105728. [CrossRef]

17. Zhong, X.; Jiang, S.; Song, H. ABCGA Algorithm for the Two Echelon Vehicle Routing Problem. In Proceedings of the IEEE International Conference on Computational Science & Engineering IEEE Computer Society, Guangzhou, China, 21–24 July 2017.

18. Liang, Q.B.; Chu, M.C. Research on lean production logistics of pharmaceutical enterprises in Liaoning based on Flexsim simulation. *J. Liaoning Univ. Technol.* **2022**, *24*, 5. (In Chinese)

19. Lu, Y. Simulation study of automotive painting logistics process based on Arena. *Logist. Technol.* **2015**, *34*, 203–207. (In Chinese)

20. Wang, Y. Simulation and optimization of production logistics system of an automobile manufacturing company. *Mod. Manuf. Eng.* **2017**, 59–62. (In Chinese)

21. Lee, C.-Y.; Lee, Z.-J.; Lin, S.-W.; Ying, K.-C. An enhanced ant colony optimization (EACO) applied to capacitated vehicle routing problem. *Appl. Intell.* **2010**, *32*, 88–95. [CrossRef]

22. Nishi, T.; Izuno, T. Column generation heuristics for ship routing and scheduling problems in crude oil transportation with split deliveries. *Comput. Chem. Eng.* **2014**, *60*, 329–338. [CrossRef]

23. Ahmed, A.; Sun, J. Bilayer Local Search Enhanced Particle Swarm Optimization for the Capacitated Vehicle Routing Problem. *Algorithms* **2018**, *11*, 31. [CrossRef]

24. Reihaneh, M.; Ghoniem, A. A branch-cut-and-price algorithm for the generalized vehicle routing problem. *J. Oper. Res. Soc.* **2018**, *69*, 307–318. [CrossRef]

25. Moura, A.; Pinto, T.; Alves, C.; Valério de Carvalho, J. A Matheuristic Approach to the Integration of Three-Dimensional Bin Packing Problem and Vehicle Routing Problem with Simultaneous Delivery and Pickup. *Mathematics* **2023**, *11*, 713. [CrossRef]

26. Smiti, N.; Dhiaf, M.M.; Jarboui, B.; Hanafi, S. Skewed general variable neighborhood search for the cumulative capacitated vehicle routing problem. *Int. Trans. Oper. Res.* **2020**, *27*, 651–664. [CrossRef]

27. Vidal, T.; Crainic, T.G.; Gendreau, M.; Prins, C. A hybrid genetic algorithm with adaptive diversity management for a large class of vehicle routing problems with time-windows. *Comput. Oper. Res.* **2013**, *40*, 475–489. [CrossRef]

28. Nalepa, J.; Blocho, M. Adaptive memetic algorithm for minimizing distance in the vehicle routing problem with time windows. *Soft Comput.* **2016**, *20*, 2309–2327. [CrossRef]

29. Molina, J.C.; Salmeron, J.L.; Eguia, I. An ACS-based memetic algorithm for the heterogeneous vehicle routing problem with time windows. *Expert Syst. Appl.* **2020**, *157*, 113379. [CrossRef]

30. Bogue, E.T.; Ferreira, H.S.; Noronha, T.F.; Prins, C. A column generation and a post optimization VNS heuristic for the vehicle routing problem with multiple time windows. *Optim. Lett.* **2020**, *16*, 79–95. [CrossRef]

31. Jalilvand, M.; Bashiri, M.; Nikzad, E. An effective Progressive Hedging algorithm for the two-layers time window assignment vehicle routing problem in a stochastic environment. *Expert Syst. Appl.* **2021**, *165*, 113877. [CrossRef]

32. Tilk, C.; Olkis, K.; Irnich, S. The last-mile vehicle routing problem with delivery options. *OR Spectr.* **2021**, *43*, 877–904. [CrossRef]

33. Hoogeboom, M.; Adulyasak, Y.; Dullaert, W.; Jaillet, P. The Robust Vehicle Routing Problem with Time Window Assignments. *Transp. Sci.* **2021**, *55*, 395–413. [CrossRef]

34. Pietrabissa, A. Distributed stochastic multi-vehicle routing in the Euclidean plane with no communications. *Int. J. Control* **2016**, *89*, 1664–1674. [CrossRef]
35. Avci, M.; Topaloglu, S. A hybrid metaheuristic algorithm for heterogeneous vehicle routing problem with simultaneous pickup and delivery. *Expert Syst. Appl.* **2016**, *53*, 160–171. [CrossRef]
36. Gholami-Zanjani, S.M.; Jafari-Marandi, R.; Pishvaee, M.S.; Klibi, W. Dynamic vehicle routing problem with cooperative strategy in disaster relief. *Int. J. Shipp. Transp. Logist.* **2019**, *11*, 455–475. [CrossRef]
37. Wang, J.; Yao, S.; Sheng, J.; Yang, H. Minimizing total carbon emissions in an integrated machine scheduling and vehicle routing problem. *J. Clean. Prod.* **2019**, *229*, 1004–1017. [CrossRef]
38. Behnke, M.; Kirschstein, T.; Bierwirth, C. A column generation approach for an emission-oriented vehicle routing problem on a multigraph. *Eur. J. Oper. Res.* **2021**, *288*, 794–809. [CrossRef]
39. Montoya, A.; Gueret, C.; Mendoza, J.E.; Villegas, J.G. A multi-space sampling heuristic for the green vehicle routing problem. *Transp. Res. Part C Emerg. Technol.* **2016**, *70*, 113–128. [CrossRef]
40. Koc, C.; Karaoglan, I. The green vehicle routing problem: A heuristic based exact solution approach. *Appl. Soft Comput.* **2016**, *39*, 154–164. [CrossRef]
41. Jabir, E.; Panicker, V.; Sridharan, R. Design and development of a hybrid ant colony-variable neighborhood search algorithm for a multi-depot green vehicle routing problem. *Transp. Res. Part D Transp. Environ.* **2017**, *57*, 422–457. [CrossRef]
42. Zhang, S.; Lee, C.K.M.; Choy, K.L.; Ho, W.; Ip, W.H. Design and development of a hybrid artificial bee colony algorithm for the environment vehicle routing problem. *Transp. Res. Part D Transp. Environ.* **2014**, *31*, 85–99. [CrossRef]
43. Niu, Y.; Yang, Z.; Chen, P.; Xiao, J. Optimizing the green open vehicle routing problem with time windows by minimizing comprehensive routing cost. *J. Clean. Prod.* **2018**, *171*, 962–971. [CrossRef]
44. Bektas, T.; Laporte, G. A comparative analysis of several vehicle emission models for road freight transportation. *Transp. Res. Part D Transp. Environ.* **2011**, *16*, 347–357.
45. Marcel, T. The accuracy of carbon emission and fuel consumption computations in green vehicle routing. *Eur. J. Oper. Res.* **2017**, *262*, 647–659.
46. Kwon, Y.; Choi, Y.; Lee, D. Heterogeneous fixed fleet vehicle routing considering carbon emission. *Transp. Res. Part D Transp. Environ.* **2013**, *23*, 81–89. [CrossRef]
47. Shen, X. Design and optimization of power communication early warning system based on neural network. *Autom. Instrum.* **2021**, *42*, 48–51+56. (In Chinese)
48. Yu, J.; Du, H.; Luo, T. Study on the optimization of just-in-time delivery path based on customer classification. *Transp. Syst. Eng. Inf.* **2020**, *20*, 202–208. (In Chinese)
49. Zhao, H. Hybrid quantum genetic algorithm-based optimal scheduling of logistics and distribution vehicles for foreign trade enterprises. *J. Qiqihar Univ.* **2021**, *37*, 26–30+35. (In Chinese)
50. Zhu, Q.; Wang, Z.; Huang, M. Fireworks algorithm with gravitational search operator method. *Control and Decision* **2016**, *31*, 1853–1859. (In Chinese)
51. Wang, Y.; Zheng, Y.; Xue, H.; Mi, Y. Enhanced fireworks algorithm-based mobile Based on enhanced fireworks algorithm for optimal scheduling of peak-shaving and valley-filling of mobile energy storage. *Power Syst. Autom.* **2021**, *45*, 48–56. (In Chinese)
52. Saxena, R.; Jain, M.; Kumar, A.; Jain, V.; Sadana, T.; Jaidka, S. An improved genetic algorithm-based solution to vehicle routing problem over open MP with load consideration. *Lect. Notes Electr. Eng.* **2019**, *537*, 285–296.
53. Lee, K.I.; Oh, H.S.; Jung, S.H.; Chung, Y.-S. Moving least square-based hybrid genetic algorithm square-based hybrid genetic algorithm for optimal design of W-band dual-reflector antenna. *IEEE Trans. Magn.* **2019**, *55*, 1–4.
54. Xiao, Y.Y.; Konak, A. A genetic algorithm with exact dynamic programming for the green vehicle routing & scheduling problem. *J. Clean. Prod.* **2017**, *167*, 1450–1463.
55. Zhang, B.; Zheng, Y.J.; Zhang, M.X.; Chen, S.-Y. Fireworks algorithm with enhanced fireworks interaction. *IEEE ACM Trans. Comput. Biol. Bioinform.* **2017**, *14*, 42–55. [CrossRef] [PubMed]

Article

A Novel Model Prediction and Migration Method for Multi-Mode Nonlinear Time-Delay Processes

Ping Yuan, Tianhong Zhou and Luping Zhao *

College of Information Science and Engineering, Northeastern University, Shenyang 110819, China;
yuanping1@ise.neu.edu.cn (P.Y.); zhoutianhongoff@163.com (T.Z.)
* Correspondence: zhaolp@ise.neu.edu.cn; Tel.: +86-18842552385

Abstract: Most industrial processes have nonlinear and time-delay characteristics leading to difficulty in prediction modeling. In addition, the working conditions of most industrial processes are complex, which results in multiple modes. Testing and modeling for each mode is a waste of time and resources. Therefore, it is urgent to complete model migration between different modes. In this work, a new prediction model, a nonlinear autoregressive model with exogenous inputs and back propagation neural network (NARX-BP), is proposed for the nonlinear and time-delay processes, where the input data order of the model is determined by the feedforward neural network (FNN) method, and the nonlinear relation is realized by the BP neural network. For the multi-mode characteristic, a new migration optimization algorithm, input–output slope/bias correction and differential evolution (IOSBC-DE), is provided for using a small amount of data under a new mode to correct the slope and bias of the relationship between the input and output variables through DE. The modeling and migration methods are applied to a wind tunnel system, and the simulation result shows the effectiveness of the proposed method.

Keywords: nonlinear; time-delay; migration model; NARX-BP; IOSBC-DE

1. Introduction

At present, industrial manufacturing is progressing towards high integration, deep automation, digitization, and intelligence. This is achieved through the utilization of various electronic sensing devices with high accuracy and advanced control systems, enabling real-time collection and analysis of production process data in factories. Industrial automation has become increasingly mature and is rapidly advancing toward intellectualization. In order to enhance competitiveness and achieve control objectives within the same industry, strict control over industrial production facilities is necessary to meet customer demands [1]. With the continuous proliferation and upgrading of data acquisition equipment, industrial process datasets have become more complex. These datasets exhibit highly nonlinear and time-delay characteristics, along with a variety of parameters, resulting in the emergence of multiple operational modes. In this context, the application of predictive modeling technology and the exploration of migration models have become inevitable trends.

Predictive modeling offers the ability to predict target variables by utilizing relevant and easily measurable process variables, thereby avoiding the inconvenience of direct measurement. This technology has found applications in certain industrial processes. However, the majority of controlled processes in the process industry exhibit nonlinear and time-delay characteristics. Traditional linear methods typically assume that the industrial process operates near a stable operating point, with variables exhibiting approximately linear correlations within a narrow range. However, nonlinearities are prevalent in industrial systems, and time delay is a common occurrence. Time delay can degrade the performance of closed-loop systems and even lead to their instability. Modeling nonlinear systems with time delay is thus a significant area of research within the field of automatic control.

Citation: Yuan, P.; Zhou, T.; Zhao, L. A Novel Model Prediction and Migration Method for Multi-Mode Nonlinear Time-Delay Processes. *Processes* **2023**, *11*, 1699. https://doi.org/10.3390/pr11061699

Academic Editor: Sergey Y. Yurish

Received: 30 April 2023
Revised: 30 May 2023
Accepted: 31 May 2023
Published: 2 June 2023

Model prediction in industrial processes involves predicting target variables based on input variables. The concept of model prediction was initially introduced by Brosilov and Tong [2]. However, this method assumes a linear relationship between output and input variables, which is often not the case in real industrial processes. Nonlinearity and time delay are common occurrences in industrial processes, making a simple linear model inadequate for meeting practical requirements. Consequently, numerous researchers have conducted extensive studies on the prediction of nonlinear and time-delay systems in industrial processes. For example, Xiang proposed a time-delay nonlinear backpropagation (BP) neural network model, which was utilized to predict the import and export trade development trends of an electronic product [3]. Xu, considering the time-delay nonlinear characteristics of rail transit passenger flow, employed support vector regression (SVR), non-autoregressive (NAR), and nonlinear autoregressive network with exogenous input (NARX) models to predict passenger flow, with the NARX model demonstrating higher accuracy [4]. Fu and Zhu developed a caving height prediction model for surrounding rock in caving mining at the Xiadian gold mine based on the time-delay nonlinear MGM model theory. However, the prediction accuracy of the model may be affected as disturbances continuously enter the system over time, resulting in an increased signal-to-noise ratio (SNR) [5]. In the study by Luo and Ding, they considered the nonlinear and time-delay characteristics of groundwater resource accumulation and proposed the GM model by combining grey system theory with the time-delay phenomenon [6]. These examples highlight the efforts of researchers to address the challenges posed by nonlinear and time-delay systems in industrial process prediction, utilizing various modeling approaches to improve accuracy and effectiveness.

Although significant progress has been made in the predictive modeling of nonlinear systems with time delays, there are still some shortcomings that need to be addressed. Firstly, the majority of existing prediction models primarily focus on minimizing training sample errors as their main criterion. However, this approach often results in poor generalization ability for the prediction models. Therefore, BP neural network with strong generalization ability rises with it. The BP neural network was proposed by Rumelhart in 1986 [7]. To address the limitations mentioned earlier, researchers have explored alternative approaches. One such approach is transforming the original forward modeling method into a self-supervised learning process that incorporates feedback. This is achieved through the concept of result forward propagation and error back propagation. Despite the emergence of various machine learning methods, the backpropagation (BP) neural network remains an important algorithm for tackling nonlinear problems [8]. Another aspect to consider is the dynamic and time-delay nature of complex industrial processes. In time series modeling, a widely utilized method is the Nonlinear Autoregressive model with eXogenous input (NARX) [9]. This method exhibits excellent learning and memory capabilities, as well as delayed feedback functions, enabling it to effectively describe the strong nonlinear dynamics often present in industrial processes. By incorporating nonlinear data-driven learning, NARX makes significant contributions to addressing time-delay problems within systems.

Moreover, due to the complex system setting parameters, the system characteristics change greatly under different working conditions, so the problem of multiple modes arises. With the development of various technologies, enterprises and society have higher and higher requirements for the quality of products, which means that the accuracy of establishing prediction models for industrial process data is higher and higher. For different modes, the process modeling based on data must be repeated, and a large number of experiments must be repeated to develop new models. Obviously, this is inefficient and uneconomical. Therefore, in order to save time and reduce costs, it is necessary to study the migration algorithm.

The model migration method serves to optimize the basic model established using known source domain data and derive a model suitable for the target domain based on data from new working conditions [10]. Researchers such as Luo, Yao, and Gao utilized Bayesian algorithms to extract the most statistically valuable information from data with similar properties. This information was then employed to estimate relevant data for the migration model [11]. In the context of battery operation modes, Tang et al. successfully migrated an accelerated aging model to describe normal speed aging behavior via the Bayesian Monte Carlo algorithm. They parameterized the new model within this framework and used it to predict the aging trajectory of the remaining batteries [12]. To address the issue of multiple working conditions in a wind tunnel control system, Ju, Wang, and Gao developed multiple sub-models. They implemented a scheduling mechanism that selects the closest local model based on the actual situation. However, this modeling method does not ensure the inclusion of all local models [13]. Yu proposed a cross-mode biomedical image multi-label classification algorithm. This method captured pattern characteristics from image content and explanatory text, introduced both homogeneous and heterogeneous data for transfer learning, and alleviated problems arising from small annotation data scales and imbalanced label distributions in the biomedical image field. However, in the case of incomplete data in industrial processes, ensuring accurate feature division becomes challenging [14]. Therefore, there is a need for a model migration method that has lower requirements regarding data characteristics and features a simpler structure. This would allow for effective model migration in situations where data may be incomplete or challenging to work with.

The input–output slope bias correction (IOSBC) method was first proposed by Gao for the injection molding process [15,16]. The mentioned method treats the new process as a shift and scaling of the historical process. With a small amount of data from the new process, the slope and bias parameters of the relationship between the input and output variables can be corrected. This correction process requires finding the optimal slope and bias parameters, which involves iterative operations. However, traditional iterative methods can sometimes get stuck in local minima, leading to a phenomenon known as a "dead cycle" that prevents further progress. To overcome this limitation, global optimization algorithms such as Genetic Algorithm (GA) and Particle Swarm Optimization (PSO) are commonly used [17]. However, compared to GA and PSO, Differential Evolution (DE) stands out as a more user-friendly, faster, and more powerful optimization technique [18]. DE can effectively search for the global optimum by maintaining a population of candidate solutions and iteratively improving them. Its simplicity and efficiency make it a preferred choice for optimizing the parameters in the context of model migration and adjustment.

In order to establish a prediction model for nonlinear and time-delayed industrial data and achieve model migration under different operating conditions to obtain accurate prediction models in all conditions, this article first employs the NARX model as the predictive model structure and the False Nearest Neighbor (FNN) algorithm to determine the order of input variables. The BP neural network is then used as the nonlinear fitting function of the NARX model to establish the predictive model under historical operating conditions. Based on the historical operating model and a small amount of new operating condition data, this article proposes a new model migration method, IOSBC-DE. This method uses the IOSBC structure as the mapping relationship between the historical model and the new model, combines a small amount of data from new operating conditions, and utilizes the DE algorithm with global optimization characteristics to optimize the bias slope in the new model and establish a prediction model for the new operating condition. Finally, the method is applied to real wind tunnel tests to verify the effectiveness of the strategy. The method proposed in this paper is easy to implement and provides a feasible technical approach for precise control of industrial systems in various modes.

2. Methodology

In order to achieve prediction of complex industrial data under harsh operating conditions, it is necessary to first establish a prediction model suitable for nonlinear and time-delayed industrial process scenarios. Furthermore, to accurately predict data under multiple operating conditions in industrial processes, transfer learning for the prediction model is required. In this section, the establishment of NARX-BP model and the implementation of IOSBC-DE migration algorithm are introduced.

2.1. Establishment of NARX-BP Model

The NARX model can provide accurate predictions for short time series, which has important applications in solving the modeling of systems with nonlinear and time-delay characteristics. The basic equation of NARX network for time series prediction can be expressed as follows:

$$\hat{y}(t) = \Psi\big(y(t-1),\ldots,y(t-n_y),u(t-1),\ldots,u(t-n_u)\big) \tag{1}$$

where $y(t-1),\ldots,y(t-n_y)$ and $u(t-1),\ldots,u(t-n_u)$, respectively, represent the measurement output and input sequence of the system; n_u, n_y represent the order of input and output variables at the past time, respectively; $\hat{y}(t)$ represents the prediction output; ψ represents the nonlinear function.

The structure of NARX model is shown in Figure 1.

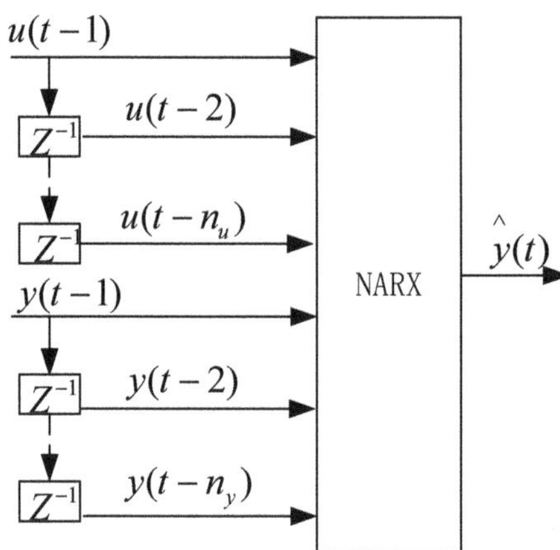

Figure 1. Structure of NARX model.

In the NARX model, the selection of variable order and nonlinear function significantly impacts the model's accuracy. In theory, a higher order can capture more information and better describe system characteristics. However, an excessively large order can lead to exponential data growth and increased noise. Hence, it is crucial to carefully select the appropriate order for modeling. Several commonly used methods for order determination include the correlation integral (CI) method [19], singular value decomposition (SVD) method [20], and the FNN method.

The CI method requires a substantial sample size and is susceptible to noise. The SVD method, being inherently linear, is not suitable for nonlinear systems. In this paper, the FNN method is employed to determine the variable order. The FNN method calculates the distance correlation before and after feature selection for each variable, evaluating the original feature's explanatory ability for the target variable. Hence, the FNN method is chosen as the order determination technique.

When selecting the nonlinear function, it is crucial to avoid overfitting and ensure the model possesses fault tolerance. The BP neural network exhibits a reasonable fault tolerance and is well-suited for capturing nonlinear mapping relationships.

Therefore, this paper employs NARX-BP neural network as the prediction model of nonlinear and time-delayed industrial conditions. The input data of the NARX model structure is determined by the FNN method, and the BP neural network is used as nonlinear function.

The design process of the NARX-BP model under a single mode is shown in Figure 2.

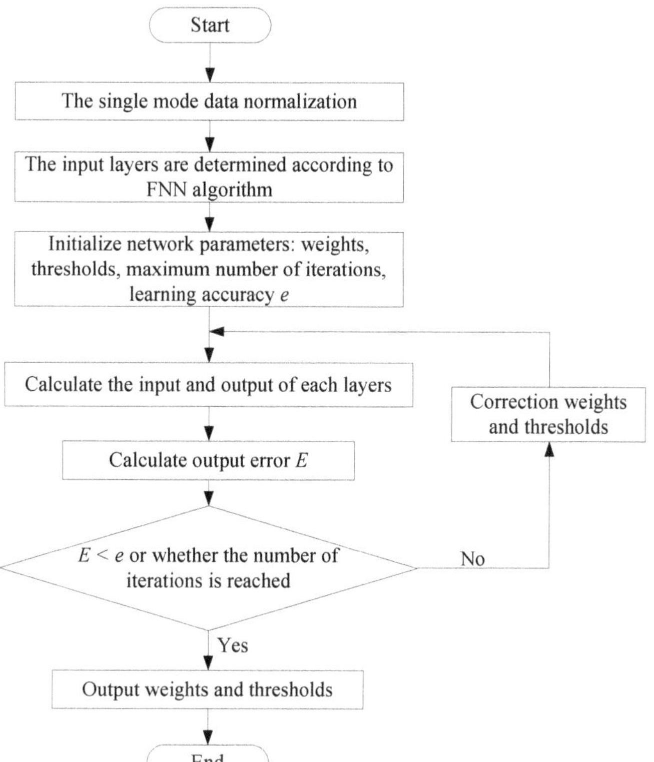

Figure 2. Flow chart of NARX-BP modeling.

Given the time series $\mathbf{X} = [x_{(1)}, \ldots, x_{(k)}]^{\mathrm{T}}$, suppose that one point $x_{(j)}$ in n dimensional space is the nearest point of another point $x_{(i)}$. The distance between them is $d_n = \left\| x_{(j)} - x_{(i)} \right\|^{(n)}$. When the dimension increases to $n + 1$, the distance between them becomes $d_{n+1} = \left\| x_{(j)} - x_{(i)} \right\|^{(n+1)}$. If d_{n+1} is much larger than d_n, it can be considered that this is caused by the fact that the two non-adjacent points in the high-dimensional attractor become adjacent points after being projected onto the low-dimensional line. Such points are false. Therefore, if Equation (2) holds, $x_{(j)}$ is the false nearest neighbor of $x_{(i)}$, where R is the threshold, usually between 10 and 50 [21].

$$\sqrt{\frac{d_{n+1}^2 - d_n^2}{d_{n+1}^2}} \geq R \tag{2}$$

Starting from $n = 1$, the ratio of false nearest neighbor points is calculated until the ratio of false nearest neighbor points is less than 5% or the false nearest neighbor points do not decrease with the increase of dimension. Then, the geometric structure of the attractor can be considered to be completely open, and n at this time is selected as variable order.

If there are n process variables $x_1, \ldots, x_n$, q output variables $y_1, \ldots, y_q$, the orders of variables determined by the FNN method $n_1, \ldots, n_n$ and $n_{o1}, \ldots, n_{oq}$, the input of the model $r(k)$ can be expressed as:

$$\begin{aligned}
\mathbf{r}(k) = \quad & (x_1(k-1),\ldots,x_1(k-n_1),\ldots,x_n(k-1),\ldots,x_n(k-n_n),\\
& y_1(k-1),\ldots,y_1(k-n_{o1}),\ldots,y_q(k-1),\ldots,y_q(k-n_{oq}))
\end{aligned} \tag{3}$$

According to the input of the model, the number of input layers can be determined as r, $r = n_1 + \ldots + n_n + n_{o1} + \ldots + n_{oq}$; the number of output layers can be determined as q, which is the number of output variables; the number of middle layers can be determined as h. Generally, the selection of the number of middle layers is calculated according to the following equation:

$$h = \sqrt{r+q} + a \tag{4}$$

where a is constant between 1 and 10, usually set around 5 [22].

Through the nonlinear mapping of the model, the output vector of the output layers at time k, $\mathbf{y}(k) = (y_1(k),\ldots,y_q(k))$ is calculated:

$$\mathbf{y}(k) = f_o\left(\mathbf{W}_{mo}^k\left(f_m\left(\mathbf{W}_{im}^k \mathbf{r}(k) - \mathbf{b}_m^k(k)\right) - \mathbf{b}_o^k\right)\right) \tag{5}$$

$$\mathbf{W}_{mo}^k(q \times h) = \begin{pmatrix} w_{11}'^k & \cdots & w_{1h}'^k \\ \vdots & \ddots & \vdots \\ w_{q1}'^k & \cdots & w_{qh}'^k \end{pmatrix} \tag{6}$$

$$\mathbf{W}_{im}^k(h \times c) = \begin{pmatrix} w_{11}^k & \cdots & w_{1r}^k \\ \vdots & \ddots & \vdots \\ w_{h1}^k & \cdots & w_{hr}^k \end{pmatrix} \tag{7}$$

$$\mathbf{b}_m^k(h \times 1) = \begin{pmatrix} b_1^k \\ \vdots \\ b_h^k \end{pmatrix} \tag{8}$$

$$\mathbf{b}_o^k(q \times 1) = \begin{pmatrix} b_1'^k \\ \vdots \\ b_q'^k \end{pmatrix} \tag{9}$$

where f_o is the mapping function selected from the middle layers to the output layers; $\mathbf{W}_{mo}^k$ is the link weight matrix between the middle layers and the output layers; f_m is the mapping function selected from the input layers to the middle layers; $\mathbf{W}_{im}^k$ is the link weight matrix between the input layers and the middle layers; $\mathbf{b}_m^k$ is the threshold matrix of each neuron in the middle layers; $\mathbf{b}_o^k$ is the threshold matrix of each neuron in the output layers.

Then, the gradient descent method is used to update the link weight matrix $\mathbf{W}_{mo}^k$, $\mathbf{W}_{im}^k$ and the threshold matrix $\mathbf{b}_m^k, \mathbf{b}_o^k$.

$$e(k) = \frac{1}{2q}\sum_{c=1}^q (\mathbf{y}_c(k) - \overline{\mathbf{y}_c}(k))^2 \tag{10}$$

$$\mathbf{W}_{mo}^{k+1} = \mathbf{W}_{mo}^k + \Delta\mathbf{W}_{mo}^k = \mathbf{W}_{mo}^k - \mu\frac{\partial e(k)}{\partial \mathbf{W}_{mo}^k} \tag{11}$$

$$\mathbf{W}_{im}^{k+1} = \mathbf{W}_{im}^k + \Delta\mathbf{W}_{im}^k = \mathbf{W}_{im}^k - \mu\frac{\partial e(k)}{\partial \mathbf{W}_{im}^k} \tag{12}$$

$$\mathbf{b}_m^{k+1} = \mathbf{b}_m^k + \Delta\mathbf{b}_m^k = \mathbf{b}_m^k - \mu\frac{\partial e(k)}{\partial \mathbf{b}_m^k} \tag{13}$$

$$\mathbf{b}_o^{k+1} = \mathbf{b}_o^k + \Delta\mathbf{b}_o^k = \mathbf{b}_o^k - \mu\frac{\partial e(k)}{\partial \mathbf{b}_o^k} \tag{14}$$

where $e(k)$ is the prediction error obtained from the k-th training. $\mathbf{y}_c(k)$ is the actual output; $\overline{\mathbf{y}_c}(k)$ is the model output; μ is the learning rate, usually between 0.01 and 0.8 [23].

Suppose the number of data samples is K. The first iteration is completed to obtain the weight matrix $\mathbf{W}_{mo}^K$, $\mathbf{W}_{im}^K$ and the threshold matrix $\mathbf{b}_m^K$, $\mathbf{b}_o^K$ of the model. Then, the global output $\overline{\mathbf{y}_c}(k)$ of the model is calculated as follows:

$$\overline{\mathbf{y}_c}(k) = f_o(\mathbf{W}_{mo}^K(f_m(\mathbf{W}_{im}^K\mathbf{r}(k) - \mathbf{b}_m^K)) - \mathbf{b}_o^K) \tag{15}$$

where $k = 1, \ldots, K$. The global error E is calculated as follows:

$$E = \frac{1}{2K}\sum_{k=1}^{K}\sum_{c=1}^{q}(\mathbf{y}_c(k) - \overline{\mathbf{y}_c}(k))^2 \tag{16}$$

If the global error E is greater than the set minimum error E_{min}, and the set number of iterations N has not been reached, the next iteration begins. Until the global error E is less than the set minimum error E_{min} or the number of iterations N has been reached, the calculation stops. The optimal weight matrix and threshold matrix are obtained.

$$\mathbf{W}_{mo} = \begin{pmatrix} w'_{11} & \cdots & w'_{1h} \\ \vdots & \ddots & \vdots \\ w'_{q1} & \cdots & w'_{qh} \end{pmatrix}, \; \mathbf{W}_{im} = \begin{pmatrix} w_{11} & \cdots & w_{1r} \\ \vdots & \ddots & \vdots \\ w_{h1} & \cdots & w_{hr} \end{pmatrix}, \; \mathbf{b}_m = \begin{pmatrix} b_1 \\ \vdots \\ b_h \end{pmatrix}, \; \mathbf{b}_o = \begin{pmatrix} b'_1 \\ \vdots \\ b'_q \end{pmatrix} \tag{17}$$

The output of the final model is expressed as follows:

$$\overline{\mathbf{y}}(k) = f_o(\mathbf{W}_{mo}(f_m(\mathbf{W}_{im}\mathbf{r}(k) - \mathbf{b}_m)) - \mathbf{b}_o) \tag{18}$$

To sum up, the model parameters and structure are determined as shown in Figure 3.

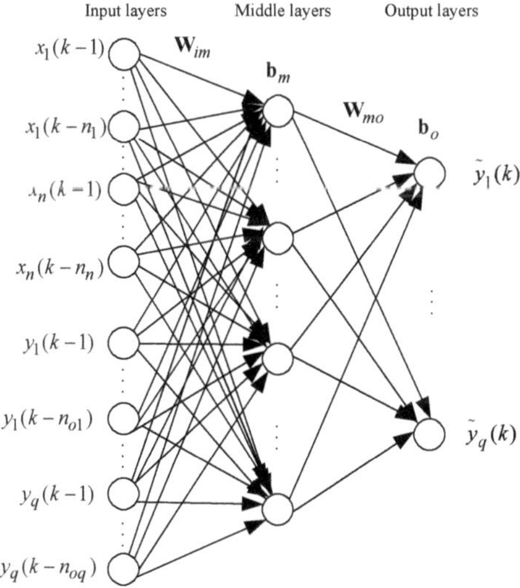

Figure 3. NARX-BP model.

2.2. Implementation of Model Migration

Based on the NARX-BP model discussed earlier, an accurate basic model for a single mode is established. However, when a new mode arises and limited operating data becomes available, the Input–Output Structure-Based Cooperation (IOSBC) method is employed to establish the parameter relationship between the basic model and the new model. Similar to Genetic Algorithm (GA), Differential Evolution (DE) exhibits the capability to be easily combined with other algorithms, making it suitable for optimizing parameters within the IOSBC method.

Therefore, this paper proposes a novel migration optimization algorithm called IOSBC-DE. This algorithm effectively migrates the model based on limited data, meeting the required level of accuracy. The model migration design process is illustrated in Figure 4.

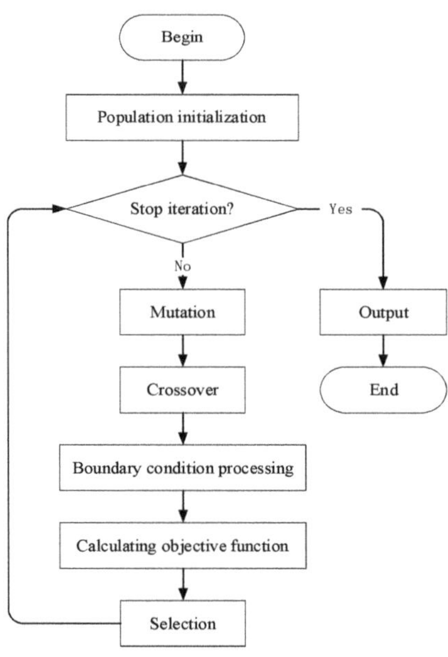

Figure 4. The flow chart of model migration.

The IOSBC method obtains the model under the new mode according to the basic model. The basic NARX-BP model established under a single mode is (13). The optimal weights and thresholds of the basic model are $\mathbf{W}_{mo}$, $\mathbf{W}_{im}$, $\mathbf{b}_m$, and $\mathbf{b}_o$.

The weight of the input layers to the middle layers under the new mode model is defined as $\mathbf{V}_{im}$ and the threshold is $\mathbf{c}_m$. The weight of the middle layers to the output layers is $\mathbf{V}_{mo}$ and the threshold is $\mathbf{c}_o$.

$$\mathbf{V}_{im} = \begin{pmatrix} s_{11} \times w_{11} + b_{11} & \cdots & s_{1r} \times w_{1r} + b_{1r} \\ \vdots & \ddots & \vdots \\ s_{h1} \times w_{h1} + b_{h1} & \cdots & s_{hr} \times w_{hr} + b_{hr} \end{pmatrix} \tag{19}$$

$$\mathbf{c}_m = \begin{pmatrix} s_{m1} \times b_1 + b_{m1} \\ \vdots \\ s_{mh} \times b_h + b_{mh} \end{pmatrix} \tag{20}$$

$$\mathbf{V}_{mo} = \begin{pmatrix} s'_{11} \times w'_{11} + b'_{11} & \cdots & s'_{1h} \times w'_{1h} + b'_{1h} \\ \vdots & \ddots & \vdots \\ s'_{q1} \times w'_{q1} + b'_{q1} & \cdots & s'_{qh} \times w'_{qh} + b'_{qh} \end{pmatrix} \tag{21}$$

$$\mathbf{c}_o = \begin{pmatrix} s_{o1} \times b'_1 + b_{o1} \\ \vdots \\ s_{oq} \times b'_q + b_{oq} \end{pmatrix} \tag{22}$$

where s_{ij}, b_{ij}, s_{mi}, and b_{mi} $(I = 1 \ldots h; j = 1 \ldots r)$ are the migration coefficients of the weights and thresholds of the input layers to the middle layers, and s'_{pi}, b'_{pi}, s_{op}, and b_{op} $(i = 1 \ldots h; p = 1 \ldots q)$ are the migration coefficients of the middle layers to the output layers.

For the calculation of the migration coefficients, DE is used to find the optimal solution via a small amount of data on the new mode.

First, the initial population is randomly generated, and the population size is denoted as P. The general value range is 20 to 100. Then, the most common binary encoding method is used to complete the encoding and decoding design. Then, the fitness error corresponding to the *pop*-th individual is calculated, *pop* = 1, ... , P. The fitness error function usually selects the average value of the square of the prediction error.

$$Fitness^{pop} = \frac{1}{K}\sum_{k}^{K}[\mathbf{y}_{new}(k) - \overline{\mathbf{y}}_{new}(k)]^2 \tag{23}$$

$$\overline{\mathbf{y}}_{new}^{pop}(k) = f_o(\mathbf{V}_{mo}^{pop}(f_h(\mathbf{W}_{im}^{pop}\mathbf{r}_{new}(k) - \mathbf{c}_{m}^{pop})) - \mathbf{c}_{o}^{pop}) \tag{24}$$

where $\mathbf{y}_{new}(k)$ is the actual measured value under the new mode at time k; $\overline{\mathbf{y}}_{new}^{pop}(k)$ is the output of the new mode calculated under the *pop*-th individual at time k; $\mathbf{r}_{new}(k)$ is the input under the new mode. $\mathbf{V}_{im}^{pop}$, $\mathbf{c}_{m}^{pop}$, $\mathbf{V}_{mo}^{pop}$, and $\mathbf{c}_{o}^{pop}$ are the weights and thresholds defined by the *pop*-th individual in the population.

During the model migration process, the initial population size Np of the differential evolution algorithm is set to 80, the initial relative scale factor F is set to 0.3, and the initial crossover factor Cr is set to 0.5. The number of iterations Gm is set to 80. In order to ensure a higher optimization success rate and faster convergence speed, an adaptive factor that varies with the number of iterations is selected to be used. The expressions are listed as follows:

$$F = F_0 \cdot 2^{exp(1-\frac{Gm}{1+Gm-G})} \tag{25}$$

$$Cr = \left(\frac{G-1}{Gm}\right)^2 \tag{26}$$

where F_0 is initial relative scale factor and G *is* current number of iterations.

In a wind tunnel system, the most important thing is the stability of the Mach number. Therefore, it is required to minimize the error between the output value and the predicted value and select the average of the error squares as the fitness function. In order to quickly find the migration model with the best fitness, select *DE/test/2* as the mutation strategy, which is defined as follows:

$$v_i^{(G)} = x_{best}^{(G)} + F\left(x_{R_1^i}^{(G)} - x_{R_2^i}^{(G)}\right) + F\left(x_{R_3^i}^{(G)} - x_{R_4^i}^{(G)}\right) \tag{27}$$

After initialization, DE creates a donor/mutant vector $v_i^{(G)}$ corresponding to each population member or target vector $x_i^{(G)}$ in the G-th iteration through mutation.

The indices R_1^i, R_2^i, R_3^i, and R_4^i are mutually exclusive integers randomly chosen from the range [1, Np], and all are different from the base index i. These indices are randomly generated anew for each donor vector. $x_{best}^{(G)}$ is the best individual vector with the best fitness in the population at G-th iteration [24].

Then, the fitness function and the number of iterations are judged until a certain set value is reached to complete the cycle. At this time, the optimal migration correction parameters of the weights and thresholds of the basic model are obtained, and then the optimal weights and thresholds of the new model are obtained, marked as $\mathbf{V}_{im}^{best}$, $\mathbf{c}_{m}^{best}$, $\mathbf{V}_{mo}^{best}$, and $\mathbf{c}_{o}^{best}$ At this time, the model of the new mode is expressed as follows:

$$\overline{\mathbf{y}}_{new}^{best}(k) = f_o(\mathbf{V}_{mo}^{best}(f_h(\mathbf{W}_{im}^{best}\mathbf{r}_{new}(k) - \mathbf{c}_{m}^{best})) - \mathbf{c}_{o}^{best}) \tag{28}$$

In conclusion, according to the determined NARX-BP model and the advantages of IOSBC-DE method, the migration optimization algorithm under multiple modes is designed. Utilizing a small amount of operating data obtained under the new mode, the optimal weights and thresholds of the new model are determined.

2.3. Model Evaluation Index

For a complex system, in order to judge whether it reaches the best level, the model evaluation index is needed. Generally, the root mean square error (*RMSE*) is selected as the evaluation index. The smaller the *RMSE* is, the more accurate the model is:

$$RMSE = \sqrt{\frac{1}{K}\sum_{k=1}^{K}(\mathbf{y}(k) - \overline{\mathbf{y}}(k))^2} \tag{29}$$

where $\mathbf{y}(k)$ indicates the actual value, $\overline{\mathbf{y}}(k)$ represents the predicted value, and K represents the number of samples.

On the other hand, the purpose of model prediction is to control. Usually, there is a set value for the output variable, which is recorded as $\mathbf{y}_{set}$. When putting the model into the controller, the control effect should be evaluated through the accuracy of the model via the indexes accuracy (*A*) and maximum deviation (*MD*). *A* is the deviation between the average predicted value and the set value over time. *MD* is the maximum predicted deviation between the actual value and the predicted value during the whole prediction process.

$$A = \left| \mathbf{y}_{set} - \frac{1}{K}\sum_{k=1}^{K}\overline{\mathbf{y}}(k) \right| \tag{30}$$

$$MD = \max|\mathbf{y}(k) - \overline{\mathbf{y}}(k)| \tag{31}$$

3. Process Description of Wind Tunnel System

The wind tunnel is an annular tubular experimental device used to artificially generate and control airflow, simulating the airflow around aircraft or objects. It allows for the measurement of the airflow's impact on the object and the observation of physical phenomena. The wind tunnel is a typical nonlinear time-delay system with multiple working modes due to various set parameters. Its importance lies in the study of aerodynamics, the analysis and design of aircraft structures, appearance, and material usage. The continuous wind tunnel is powered by compressors, and the airflow within is controlled by the compressor speed. Key process parameters include total pressure, Mach number, and attack angle.

To meet the desired conditions, the wind tunnel is first operated to achieve a stable Mach number, after which the desired change in attack angle is set, and data is collected during this change.

The continuous wind tunnel represents a complex system with multiple variables. The Mach number serves as an important parameter for analyzing the wind tunnel's flow field performance and acts as a primary control index. During the blowing process in the wind tunnel, the total pressure in the stable section of the flow field is influenced by factors such as gas temperature, gas constant, and gas density. While the gas temperature remains uncontrolled during experiments, the total pressure is primarily adjusted by modifying the gas density. Directly obtaining the Mach number using a mechanistic model in the complex wind tunnel system proves challenging. However, leveraging the knowledge of aerodynamic theory allows for calculating a more accurate Mach number by considering the total pressure and static pressure.

Currently, the total pressure in the continuous wind tunnel is controlled independently, and changes in total pressure significantly impact the Mach number. The compressor speed, in turn, affects the static pressure. Additionally, the attack angle refers to the angle between the wing and the direction of airflow when an aircraft model is positioned within the test section. Precise control of the attack angle is crucial for aircraft model startup testing. The main influential variables are summarized in Table 1.

Table 1. Key variables of wind tunnel process.

No.	Variable Description	Unit
1	Mach number (Ma)	/
2	Total pressure (P_0)	kPa
3	Speed (S_p)	r/s
4	Attack angle (A_{tt})	°

4. Results and Discussion

In the wind tunnel flow field, the flow field characteristics change greatly under different working modes. Due to the special needs of the experimental model, the set value of the Mach number is different, which has caused great difficulties in the establishment of the model.

In this paper, the experimental transonic data with Mach number operating parameters ranging from 0.6 to 1.0 are studied, with the injection gap set at 24 mm. The opening to closing ratio is set to 1.5%. The attack angle ranges from $-4°$ to $10°$. According to different Mach number set values, the selection of test conditions and the number of samples are shown in Table 2.

Table 2. Different modes.

Mode Number	Ma Set Value	Number of Samples
1	0.6	1261
2	0.7	2717
3	0.75	1989
4	0.8	985
5	0.85	2113
6	0.9	1451
7	0.95	1455
8	1.0	965

4.1. Ma Prediction Results Based on NARX-BP Model

According to the actual operation process of the wind tunnel and expert experience, the total pressure, speed, and attack angle are the input variables, and the Mach number is the output variable of the model. According to the structure of the NARX-BP model, the selection of the order of model variables directly affects the accuracy of the model prediction. The order of model variables is determined via FNN. Taking the Mach number and attack angle order analysis under Mode 1 as an example, the order of the variables is from 1 to 15, and the analysis results are shown in Figure 5. As seen from Figure 5, for Mode 1, the optimal order of the Mach number is 6, and the optimal order of the attack angle is 6.

Similarly, the FNN method is used to analyze all variables under all modes. The results are shown in Table 3.

As seen from the above table, the order of each variable is slightly different under different modes, but the order of the same variable can change by up to three orders under different modes. In order to better reflect the applicability of the selected order, the mode of the orders of each variable under different modes is selected as the final order of the model, and the optimal order of the variables is shown in Table 4.

According to the optimal order of the variables, the number of input layers of the NARX-BP neural network is determined to be 20, the number of output layers is 1, and the number of middle layers is 10 calculated via Equation (4).

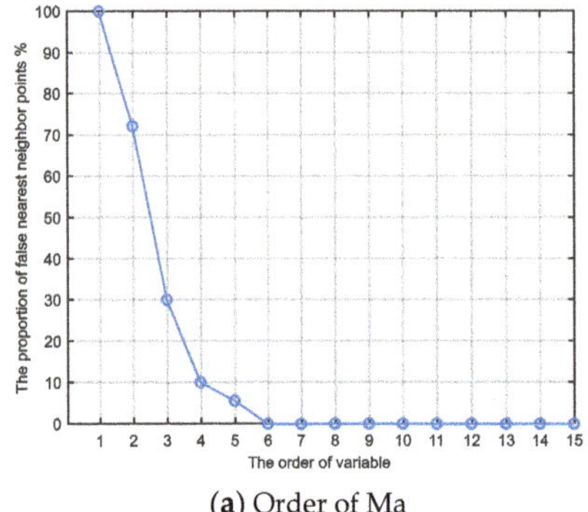

(**a**) Order of Ma

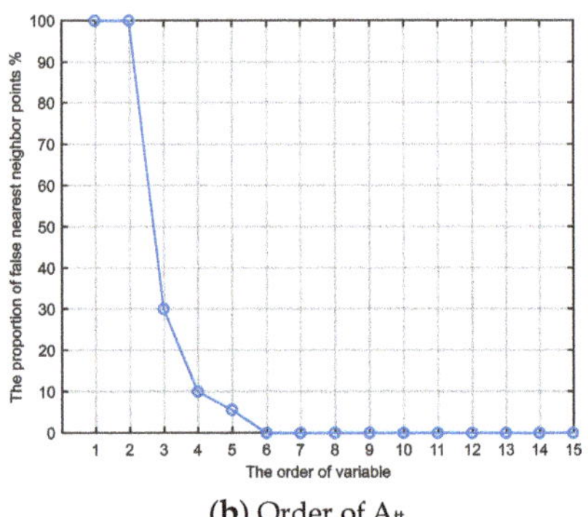

(**b**) Order of A$_{tt}$

Figure 5. Order analysis of model variables.

Table 3. Order analysis of model variables.

Test Number	Ma (n_{ma})	P$_0$ (n_o)	S$_p$ (n_{sp})	A$_{tt}$ (n_a)
1	6	3	6	6
2	5	4	8	4
3	5	5	6	5
4	4	5	6	4
5	4	4	5	5
6	3	3	6	4
7	5	4	7	5
8	3	4	7	5

Table 4. Optimal orders of variables.

Variable	Ma (n_{ma})	P$_0$ (n_o)	S$_p$ (n_{sp})	A$_{tt}$ (n_a)
Optimal order	5	4	6	5

Taking Mode 5 as an example, 70% of the data is selected for training, and 30% of the data is tested. The prediction results of the proposed model are compared with the prediction results of the BP neural network and PLS regression prediction adopted in previous related work [25], as shown in Figure 6. As seen from Figure 6a, the changing trend of the predicted Mach number is closer to the actual Mach number via the proposed method. In contrast, the changing trend of the predicted Mach number via the BP network method and the PLS method is quite different from the actual Mach number. It can be seen from Figure 6b that the error between the predicted Mach number and the actual Mach number calculated via the proposed method is less than that predicted via the BP network method and the PLS method.

Similarly, the Mach number prediction of each mode is carried out, respectively, and the prediction model evaluation indexes of each mode are shown in Table 5.

According to the above table, the *RMSE* values of the proposed NARX-BP model are less than 0.001 under all modes. Under each mode, the *RMSE* value predicted via the NARX-BP model is less than that predicted via the BP network model and the PLS model, which shows that the predicted Mach number of the NARX-BP model is closer to the actual Mach number and the prediction is more accurate. Under a single mode, the *A* value of the Mach number is analyzed. Except for Mode 3 and Mode 7, the *A* values of the NARX-BP model are lower than that of the BP network model and the PLS model. In addition, under each mode, the *MD* between the predicted Mach number and the actual Mach number obtained via the NARX-BP model is less than that predicted via the BP model and the PLS model. In conclusion, the proposed NARX-BP model has more advantages than the BP network model and the PLS model. The NARX-BP model can further improve the ability of Mach number tracking and prediction.

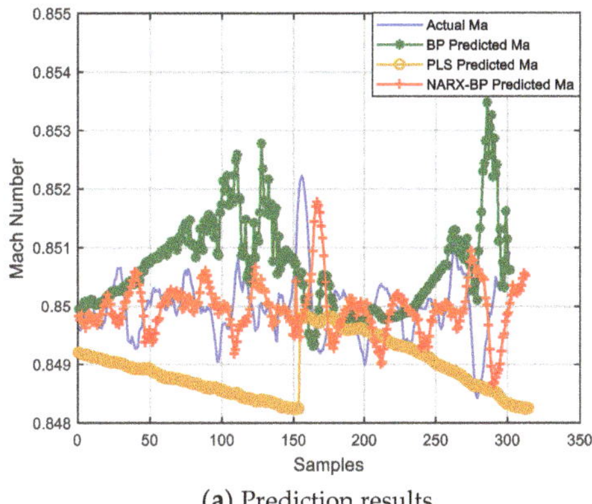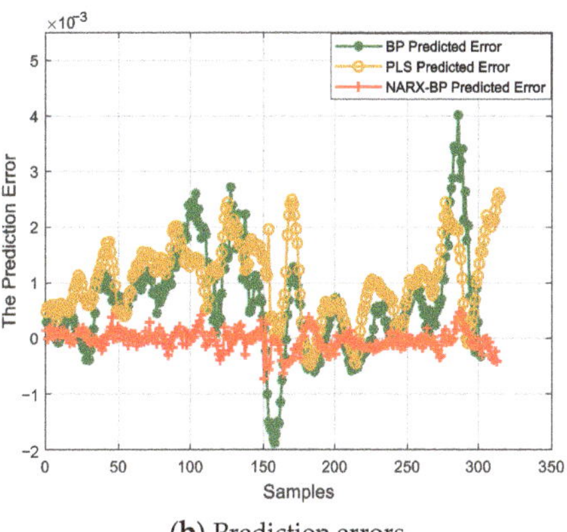

(**a**) Prediction results   (**b**) Prediction errors

Figure 6. Mach number prediction results and errors of mode 5.

Table 5. Evaluation indexes of prediction models.

Test Number	BP			PLS			NARX-BP		
	RMSE	*A*	*MD*	*RMSE*	*A*	*MD*	*RMSE*	*A*	*MD*
1	0.00030	0.00015	0.00081	0.00052	0.00018	0.00139	0.00015	0.00007	0.00063
2	0.00069	0.00044	0.00190	0.00097	0.00071	0.00220	0.00015	0.00008	0.00088
3	0.00019	0.00062	0.00051	0.00036	0.00048	0.00102	0.00011	0.00058	0.00050
4	0.00044	0.00014	0.00132	0.00061	0.00014	0.00156	0.00019	0.00005	0.00064
5	0.00115	0.00067	0.00408	0.00125	0.00107	0.00261	0.00018	0.00004	0.00072
6	0.00068	0.00034	0.00234	0.00150	0.00087	0.00354	0.00018	0.00009	0.00103
7	0.00073	0.00005	0.00309	0.00166	0.00066	0.00431	0.00019	0.00023	0.00182
8	0.00045	0.00031	0.00177	0.00105	0.00047	0.00382	0.00013	0.00017	0.00036

4.2. Model Migration Results of Mach Number Prediction Model

In the wind tunnel, the core equipment is expensive and consumes a significant amount of energy. Conducting a complete test for each working mode and establishing corresponding models would require substantial manpower and material resources. Therefore, it is crucial for researchers to obtain accurate models while minimizing the number of tests. To achieve this goal and reduce costs and energy consumption, the migration algorithm is applied to the Mach number prediction model of the wind tunnel flow field in this study.

Based on the selected modes, Mode 5 is chosen as the historical mode to establish the NARX-BP basic model, while Mode 4 and Mode 6 are selected as the new modes. Only the first 3% of data from each new mode is used for migrating and modifying the basic model to obtain the new prediction model. The remaining data under the new mode is then predicted using the newly obtained model.

To demonstrate the superiority of the DE model migration, separate NARX-BP models are established for Mode 4 and Mode 6 as new modes, using only 3% of the available data. The prediction results of the DE migration model are compared with the non-migration model established for each mode separately. Additionally, to provide an objective evaluation of the DE algorithm's performance, the prediction model optimized via DE is compared with prediction models obtained through optimization and screening via GA and PSO algorithms. The prediction results for Mode 4 are presented in Figure 7a, and the corresponding prediction errors are shown in Figure 7b. The prediction results for Mode 6 are depicted in Figure 8a, while the prediction errors are displayed in Figure 8b.

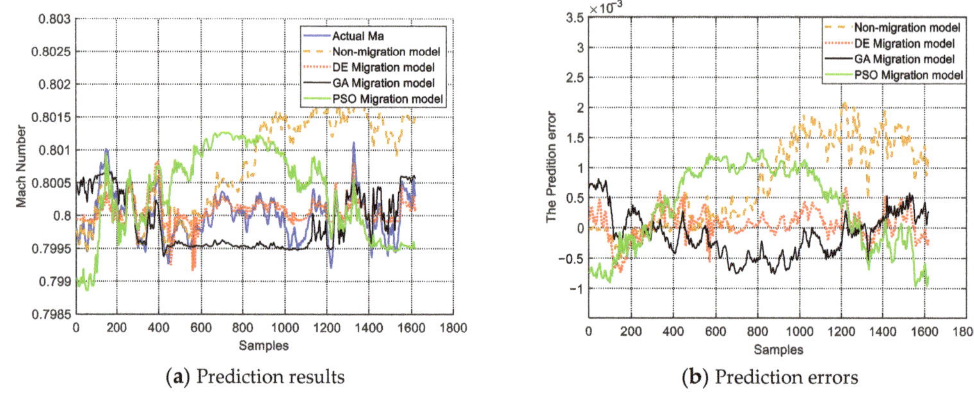

(a) Prediction results
(b) Prediction errors

Figure 7. Mach number prediction results and errors of Mode 4.

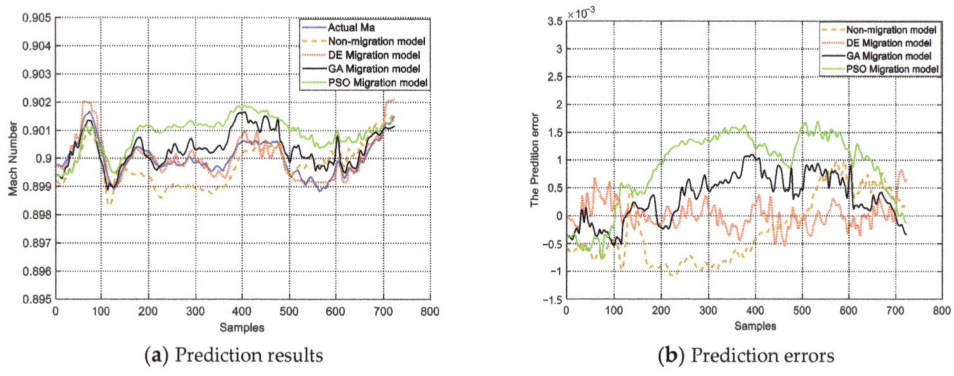

(a) Prediction results
(b) Prediction errors

Figure 8. Mach number prediction results and errors of Mode 6.

From Figures 7a and 8a, it is evident that when using only a small amount of data from the new modes, both the migration models and non-migration models struggle to accurately reflect the changing trend of the Mach number. However, the predicted Mach number of the migration model is closer to the actual Mach number. Figures 7b and 8b further illustrate that the error in the predicted Mach number via the DE migration model is lower compared to the non-migration model.

The prediction results and evaluation indexes of the above migration model and non-migration model are shown in Table 6. The prediction results and evaluation indexes of migration models with different optimization algorithms are shown in Table 7.

When only the first 3% of data are used for the non-migration model, the RMSE values exceed 0.001 under Mode 4 and Mode 6. For the DE migration model, the RMSE can ensure that the prediction result is within 0.001 under Mode 4. Although the RMSE under Mode 6 using the migration model exceeds 0.001, this value is also less than the RMSE of the non-migration model. At the same time, under these two modes, the MD values of the Mach number after model migration are less than the non-migration model. Therefore, the results show that when there is only a small amount of data on the new mode, the migration model method has a better prediction effect.

When comparing DE with GA and PSO optimization methods, it was found that under Mode 4, DE outperformed GA and PSO in three types of evaluation indicators, indicating that the performance of DE in finding globally optimal solutions in the wind tunnel migration model is higher than GA and PSO's. Through this method, the running time of the test can be greatly reduced, and the blowing cost of the wind tunnel can be saved, which is of great significance to the study of wind tunnel tests.

Table 6. Evaluation indexes of prediction models.

Test Number	Non-Migration Model			DE Migration Model		
	RMSE	*A*	*MD*	*RMSE*	*A*	*MD*
4	0.00141	0.00020	0.00358	0.00025	0.00010	0.00074
6	0.00065	0.00263	0.00114	0.00024	0.00005	0.00082

Table 7. Evaluation indexes of prediction models optimized via different algorithms.

Test Number	DE Migration Model			GA Migration Model			PSO Migration Model		
	RMSE	*A*	*MD*	*RMSE*	*A*	*MD*	*RMSE*	*A*	*MD*
4	0.00025	0.00010	0.00074	0.00076	0.00057	0.00194	0.00075	0.00034	0.00130
6	0.00024	0.00005	0.00082	0.00053	0.00035	0.00110	0.00110	0.00090	0.00170

5. Conclusions

This article presents a novel prediction modeling and migrating method designed to address the nonlinear and time-delay characteristics commonly observed in industrial processes. To tackle multimodal problems, an effective model transfer strategy called IOSBC-DE is proposed. The strategy involves determining the input data order of the NARX model using the FNN method and utilizing a BP neural network as the mapping function. Moreover, the IOSBC method establishes the relationship between historical patterns and new patterns, employing DE optimization to fine-tune slope and bias parameters during the model migration process. Experimental data from wind tunnel processes demonstrate that NARX-BP achieves more accurate predictions for nonlinear and time-delay industrial processes, improving accuracy by 50% (RMSE from 0.00030 to 0.00015) compared to traditional BP neural networks. The migrated model through IOSBC-DE exhibits an 82.3% (RMSE from 0.00141 to 0.00025) increase in prediction accuracy compared to the non-migrated model, with DE outperforming other optimization algorithms in terms of prediction performance. The feasibility of the proposed method is effectively validated.

Author Contributions: Conceptualization, P.Y. and L.Z.; methodology, L.Z. and T.Z.; software, T.Z.; validation, L.Z.; formal analysis, L.Z. and T.Z.; investigation, L.Z.; resources, P.Y. and L.Z.; data curation, T.Z.; writing—original draft preparation, L.Z. and T.Z.; writing—review and editing, L.Z. and T.Z.; visualization, T.Z.; supervision, P.Y. and L.Z.; project administration, P.Y. and L.Z.; funding acquisition, P.Y. and L.Z. All authors have read and agreed to the published version of the manuscript.

Funding: This research was funded by the National Natural Science Foundation of China (No. 61503069) and the Fundamental Research Funds for the Central Universities (N150404020).

Data Availability Statement: Not applicable.

Conflicts of Interest: The authors declare no conflict of interest.

References

1. Chen, Y.S. Intelligent control in industrial application. *Wirel. Internet Technol.* **2021**, *18*, 88–89.
2. Brosilow, C.; Tong, M. Inferential control of processes: Part II. The structure and dynamics of inferential control systems. *Aiche J.* **1978**, *24*, 492–500. [CrossRef]
3. Xiang, J.W. A new nonlinear time series Prediction method based on neural networks. *Mod. Elec. Tech.* **2007**, *4*, 118–122.
4. Xu, R.J. Research on Passenger Flow Prediction of Rail Transit Based on Time-Delay Nonlinear Autoregressive Neural Network. Master's Thesis, Chongqing University, Chongqing, China, 2017.
5. Fu, J.X.; Zhu, P.R.; Song, W.D.; Tan, Y.Y. Caving height prediction of caving method based on time-delay nonlinear MGM model. *J. Min. Saf. Eng.* **2020**, *37*, 741–749.

6. Luo, D.; Ding, H.H. Study on the prediction of groundwater resources based on time delay GM (1,1) model. *J. North China Univ. Water Resour. Electr. Power (Nat. Sci. Ed.)* **2018**, *41*, 19–24.

7. Rumelhart, D.E.; Hinton, G.E.; Williams, R.J. Learning Representations by Back Propagating Errors. *Nature* **1986**, *323*, 533–536 [CrossRef]

8. Zhou, Z. Review on the development of BP neural network. *Shanxi Electron. Technol.* **2008**, *2*, 90–92.

9. Blanchard, T.; Samanta, B. Wind speed forecasting using neural networks. *Wind. Eng.* **2020**, *44*, 33–38. [CrossRef]

10. Pan, S.J.; Yang, Q. A Survey on Transfer Learning. *IEEE Trans. Knowl. Data Eng.* **2010**, *22*, 1345–1359. [CrossRef]

11. Luo, L.K.; Yao, Y.; Gao, F.R. Iterative improvement of parameter estimation for model migration by means of sequential experiments. *Comput. Chem. Eng.* **2015**, *73*, 128–140. [CrossRef]

12. Tang, X.P. Aging trajectory prediction for lithium-ion batteries via model migration and Bayesian Monte Carlo method. *Appl. Energy* **2019**, *254*, 113591. [CrossRef]

13. Ju, X.F.; Wang, L.X.; Gao, Y. Wind Tunnel Flow Field Model Predictive Control Based on Multi-modes. *Control Eng.* **2018**, *25*, 1830–1835.

14. Yu, Y.H.; Lin, H.F.; Meng, J.N.; Guo, H.; Zhao, Z.H. Classification modeling and recognition for cross modal and multi-label biomedical image. *J. Image Graph.* **2018**, *23*, 917–927.

15. Lu, J.D.; Gao, F.R. Process Modeling Based on Process Similarity. *Ind. Eng. Chem. Res.* **2008**, *47*, 1967–1974. [CrossRef]

16. Lu, J.D.; Yao, Y.; Gao, F.R. Model Migration for Development of a New Process Model. *Ind. Eng. Chem. Res.* **2009**, *48*, 9603–9610 [CrossRef]

17. Dai, X.J.; Li, M.Q.; Kou, J.S. Review of genetic algorithm theory. *Control Decis.* **2000**, *3*, 263–273.

18. Wang, L.; Ma, P. Overview of system identification methods. *Power Inf.* **2001**, *4*, 63–66.

19. Broomhead, D.S.; King, P.G. Extracting qualitative dynamics from experimental data. *Phys. D Nonlinear Phenom.* **1986**, *20*, 217–236 [CrossRef]

20. Kennel, M.B.; Brown, R. Determining embedding dimension for phase-space reconstruction using a geometrical construction *Physic Rev. A* **1992**, *45*, 3303–3411. [CrossRef]

21. Gu, X.H.; Li, T.F.; Yang, L.P.; Yi, J.; Zhou, W. Original feature selection based on false nearest neighbor criterion in supervised local preserving subspace. *Opt. Precis. Eng.* **2014**, *22*, 1921–1928.

22. Han, W.; Nan, L.; Su, M.; Chen, Y.; Li, R.; Zhang, X. Research on the Prediction Method of Centrifugal Pump Performance Based on a Double Hidden Layer BP Neural Network. *Energies* **2019**, *12*, 2709. [CrossRef]

23. Kandil, N.; Khorasani, K.; Patel, R.V.; Sood, V.K. Optimum learning rate for backpropagation neural networks. In Proceedings of the Canadian Conference on Electrical and Computer Engineering, Vancouver, BC, Canada, 14–17 September 1993; Volume 1, pp 465–468.

24. Das, S.; Mullick, S.; Suganthan, P. Recent advances in differential evolution—An updated survey. *Swarm Evol. Comput.* **2016**, *27*, 1–30. [CrossRef]

25. Guo, J.; Zhang, R.; Cui, X.C.; Huang, X.; Zhao, L.P. Mach Number Prediction and Analysis of Multi-Mode Wind Tunnel System In Proceedings of the 2020 Chinese Automation Congress, Shanghai, China, 6–8 November 2020; pp. 6895–6900.

 *processes*

Article

Numerical Experiments on Performance Comparisons of Conical Type Direct-Acting Relief Valve—With or without Conical Angle in Valve Element and Valve Seat

Huiyong Liu [1],* and Qing Zhao [2]

[1] Department of Mechanical Design, School of Mechanical Engineering, Guizhou University, Guiyang 550025, China
[2] Department of Water Resources and Hydropower Engineering, College of Civil Engineering, Guizhou University, Guiyang 550025, China
* Correspondence: heartext@163.com

Citation: Liu, H.; Zhao, Q. Numerical Experiments on Performance Comparisons of Conical Type Direct-Acting Relief Valve—With or without Conical Angle in Valve Element and Valve Seat. *Processes* 2023, 11, 1792. https://doi.org/10.3390/pr11061792

Academic Editors: Sergey Y. Yurish and Jiaqiang E

Received: 11 February 2023
Revised: 30 March 2023
Accepted: 4 May 2023
Published: 12 June 2023

Abstract: This paper conducts numerical experiments on performance comparisons of CTDARV—with or without conical angle in the valve element and valve seat. The working principles of three kinds of CTDARV are introduced. The simulation models of three kinds of CTDARV are established by utilizing AMESIM. Numerical experiments on CTDARV, with or without a conical angle in the valve element and the valve seat, are conducted and the performance comparisons of three kinds of CTDARV are obtained. The results show that: (1) When all parameters of VED, VSD, VEM, SS, CAVE&CAVS, and OD have the same value, respectively, CA-VE has the highest stable pressure, CA-VE&VS has the highest stable displacement, CA-VS has the lowest stable pressure, and CA-VE has the lowest stable displacement. The stable pressure of CA-VE is significantly higher than that of CA-VS and CA-VE&VS. The stable displacement of CA-VE&VS is significantly higher than that of CA-VE and CA-VS, and the stable displacement of CA-VE and CA-VS has little difference. (2) With the increase of VED from 13 mm to 16 mm, the stable pressure of CA-VE remains constant, while that of CA-VS and CA-VE&VS both decreases. As the VSD increases from 3 mm to 6 mm, the stable pressure of CA-VE and CA-VE&VS decreases, and that of CA-VE decreases significantly. With the increase of VEM from 0.01 kg to 0.04 kg, the stable pressure of CA-VE, CA-VS, and CA-VE&VS remains unchanged. With the increase of SS from 5 N/mm to 20 N/mm, the stable pressure of CA-VE, CA-VS and CA-VE&VS increases. With the increase of CAVE&CAVS from 15 degrees to 60 degrees, the stable pressure of CA-VE, CA-VS, and CA-VE&VS decreases. With the OD increase from 0.8 mm to 1.4 mm, the stable pressure of CA-VE, CA-VS and CA-VE&VS remains unchanged. (3) With the increase of VED from 13 mm to 16 mm, the stable displacement of CA-VE will not change, while that of CA-VS and CA-VE&VS will increase. As the VSD increases from 3 mm to 6 mm, the stable displacement of CA-VE increases, while that of CA-VE&VS decreases. When VSD is 4 mm–6 mm, the stable displacement of CA-VS remains unchanged. With the increase of VEM from 0.01 kg to 0.04 kg, the stable displacement of CA-VE, CA-VS and CA-VE&VS remains unchanged. As SS increases from 5 N/mm to 20 N/mm, the stable displacement of CA-VE, CA-VS, and CA-VE&VS decreases. As CAVE&CAVS increases from 15 degrees to 60 degrees, the stable displacement of CA-VE, CA-VS, and CA-VE&VS decreases. With the OD increasing from 0.8 mm to 1.4 mm, the stable displacement of CA-VE, CA-VS, and CA-VE&VS remains unchanged. (4) With the increase of VED from 13 mm to 16 mm, the velocity of CA-VE remains unchanged, while that of CA-VS and CA-VE&VS increases. As the VSD increases from 4 mm to 6 mm, the velocity of CA-VS remains unchanged, while that of CA-VE and CA-VE&VS decreases. With the increase of VEM from 0.01 kg to 0.04 kg, the velocity oscillation of CA-VE gradually increases, and the velocity of CA-VS and CA-VE&VS has little change. As SS increases from 5 N/mm to 20 N/mm, the velocity of CA-VE increases, while that of CA-VS and CA-VE&VS decreases. When CAVE&CAVS is 15 degrees and 30 degrees, the velocity of CA-VE is lower than that of CA-VS and CA-VE&VS. With the OD increasing from 0.8 mm to 1.4 mm, the velocity oscillation of CA-VE increases gradually, and the velocity of CA-VS and CA-VE&VS changes little.

Keywords: CTDARV; performance comparisons; AMESIM

1. Introduction

The cone valve type direct-acting relief valve (CTDARV) is widely used in low-pressure and small-flow systems because of its advantages of good sealing performance, fast response, strong anti-pollution ability, etc. Generally, the CTDARV can be divided into three categories according to whether the valve element or the valve seat has a conical angle: with a conical angle in the valve element but without a conical angle in the valve seat (CA-VE), without a conical angle in the valve element but with a conical angle in the valve seat (CA-VS), with a conical angle in both the valve core and the valve seat (CA-VE&VS). In the hydraulic systems of industrial equipment, these three types of CTDARVs have been widely used. However, under the same working medium, geometric parameters and external working conditions, what are the response characteristics of the three types of CTDARVs? How is the performance comparison of these three types of CTDARVs? Can these three types of CTDARVs be directly replaced with each other? These problems have always plagued designers and on-site staff for a long time. Therefore, it is of great significance to carry out numerical experiments on performance comparisons of these three types of CTDARVs.

During recent decades, there have been lots of papers about the CTDARV. Yuan et al. [1] conducted a numerical study on the cavitating flow phenomenon inside poppet valves with two valve seat structures aiming to examine the flow mechanisms underlying varying cavitation phenomena at different openness. Min et al. [2] measured the discharge coefficient of different pilot-stage poppet valves and used Fluid-Structure Interaction to analyze the influence of the orifice-submerged jet on the discharge coefficient. Hao et al. [3] established a prediction model of flow force on the cone and cavitation for the poppet valve by using CFD combined with the Zwart–Gerber–Belamri cavitation model and investigated the effects of three poppet valve configurations and their parameters on the flow force and cavitation intensity in valves. Yuan, et al. [4] performed a three-dimensional simulation considering the compressibility of each constituent phase to clarify the governing mechanisms under the cavitating flow inside two water poppet valves. Fornaciari et al. [5] deal with experimental tests and numerical simulations (3D and 0D fluid-dynamic modeling) of a conical poppet pressure relief valve with flow force compensation. Min et al. [6] studied the unstable vibration of the poppet and cavitation using visual experiments. Filo et al. [7] used CFD to analyze an innovative directional control valve consisting of four poppet seat valves and two electromagnets enclosed inside a single body. Wang et al. [8] established the overflow rate of a poppet relief valve and the dynamics characteristics of a poppet relief valve and derived the stability conditions of a poppet relief valve system using the Routh–Hurmitz method. Gomez et al. [9] developed a computational model based on the finite volume method to characterize the flow at the interior of the valve while it is moving. Chiavola, et al. [10] experimentally characterized the behavior of a conical poppet valve and explored the effect of the feeding conditions on the flow characteristics of the valve. Lei et al. [11] established the flow model of the poppet valve orifice with a novel function of flow discharge coefficient and established the dynamic model including the aforementioned flow model of the poppet valve by considering the fluid forces caused by the valve body motion and the flowrate variation. Liu et al. [12] presented the regulation methods of flow forces acting on the main poppet in a large flow load control valve. Ji et al. [13] studied the dynamics of axial vibration and lateral vibration coexisting in a poppet valve. Han et al. [14] presented a numerical investigation into the flow force and cavitation characteristics inside water hydraulic poppet valves. Bo [15] presented the fluid-solid coupling model for the optimization of a poppet valve, and analyzed the actual flow performance of the poppet valve. Wang et al. [16] proposed a hydraulic poppet valve dynamic numerical model based on the CFD and dynamic mesh technique for obtaining the transient characteristics of hydraulic poppet valve during opening process. Yi et al. [17] investigated experimentally the interactions between the poppet vibration characteristics and cavitation property in relief valves with the unconfined poppet. Min et al. [18] presented a fluid-structure interaction modeling of poppet fluid, and simulated the response of poppet under the action of step and periodic excitation signal, and analyzed dynamic characteristics of viscous force and hydraulic force on poppet surface. Zheng and Quan [19] optimized the structure of the poppet valve based on the internal flow, and studied the flow-force on poppet valve in the case of the converging flow using CFD. Shi and Chen [20] simulated the inside flow field of the hydraulic poppet valve by the dynamic mesh technology based on the physical numerical modeling of the hydraulic poppet valve. Shi [21] established and meshed the 3-D model of the poppet valve, and simulated the inner flow field of the poppet valve at different cone diameter and different opening positions. Yi et al. [22] obtained the frequency spectrum of the squeal noise by analyzing the sampling data from the accelerometer mounted on the valve body. Chen and Bao [23] developed a simulation model to study the static characteristic of a regulator valve with large flow capacity. Luan et al. [24] established a numerical model of the hydraulic poppet valve using the CFD software in order to obtain the transient distribution of flow and valve core stress of the hydraulic poppet valve during opening process. Rundo and Altare [25] analyzed different methods for the evaluation of the flow forces in conical poppet valves, and analyzed three different poppet angles and two flow directions. Guo et al. [26] developed a 6 DOF model of a three-dimensional poppet valve to stimulate the transient flow field in the opening process by means of computational fluid dynamics software-Fluent.

However, the existing literatures have not yet compared the performance of CTDARV--with or without conical angle in valve element and valve seat. This paper carries out numerical experiments on three kinds of CTDARV, hoping to obtain the performance comparisons of three kinds of CTDARV, thus providing theoretical basis for design, manufacture and use of three kinds of CTDARV.

The rest of this paper is organized as follows. In Section 2, the working principles of three kinds of CTDARV are introduced. In Section 3, the simulation models of three kinds of CTDARV are established by utilizing AMESIM. The numerical experiments on three kinds of CTDARV are conducted and the performance comparisons of three kinds of CTDARV are obtained in Section 4. Finally, some conclusions are drawn in Section 5.

2. Working Principles of Three Kinds of CTDARV

Figure 1 is the structural diagram of three kinds of CTDARV. The typical feature of Figure 1a is that there is a conical angle in valve element and no conical angle in the valve seat (CA-VE). The typical feature of Figure 1b is that there is a conical angle in the valve seat and no conical angle in the valve element (CA-VS). The typical feature of Figure 1c is that there is a conical angle in the valve element and valve seat (CA-VE&VS). The working principle of the CTDARV can be described in two steps.

(1)   Before movement of the valve element, as shown in Figure 1a–c.

At this step, the valve element is subject to the joint action of the following forces: the upward hydraulic force on the valve element F_1, the downward hydraulic force on the valve element F_2, the mass force G, the spring force F_s, and the friction force f.

For CA-VE, CA-VS, CA-VE&VS, at this step, the equations of F_1, F_2, G, F_s, f are (1), (2), (3), respectively.

$$\begin{cases} F_1 = \frac{\pi}{4}d^2 p_1 + \frac{\pi}{4}(D^2 - d^2)p_2 \\ F_2 = \frac{\pi}{4}D^2 p_2 \\ G = mg \\ F_s = k(x_0 + x) \\ f = \mu A \frac{v}{h} \end{cases} \tag{1}$$

$$\begin{cases} F_1 = \frac{\pi}{4}d^2 p_1 \\ F_2 = \frac{\pi}{4}D^2 p_2 \\ G = mg \\ F_s = k(x_0 + x) \\ f = \mu A \frac{v}{h} \end{cases} \tag{2}$$

$$\begin{cases} F_1 = \frac{\pi}{4}d^2 p_1 + \frac{\pi}{4}(D^2 - (d + l\sin\alpha)^2)p_2 \\ F_2 = \frac{\pi}{4}D^2 p_2 \\ G = mg \\ F_s = k(x_0 + x) \\ f = \mu A \frac{v}{h} \end{cases} \tag{3}$$

where, D, d are the valve element diameter and the damping hole diameter; p_1, p_2 are the pressure in the front chamber of the valve element and in the spring chamber; m is the mass of the valve element; g is the gravity acceleration; k is the spring stiffness; x_0, x are the pre-compression and deformation of the spring, respectively; μ is the dynamic viscosity; A is the surface area between the valve element and the hydraulic oil; v is the velocity difference between the valve element and the valve body; and h is the distance between the valve element and the valve body.

It can be seen from Figure 1b that p_2 is connected to the outlet, so $p_2 = 0$, and $F_2 = 0$. When p_1 is low, thus $F_1 \leq F_s + f + G$, the valve element does not move, and the valve core and valve seat contact.

(2)　After movement of valve element, as shown in Figure 1d–f

At this step, the valve element is subject to the joint action of the following forces: the upward hydraulic force on the valve element F_1, the downward hydraulic force on the valve element F_2, the mass force G, the spring force F_s, the friction force f, and the flow force F_f. The equations of F_1, F_2, G, F_s, f, F_f are as follows:

For CA-VE, CA-VS, CA-VE&VS, at this step, the equations of F_1, F_2, G, F_s, f, F_f are (4), (5), (6), respectively.

$$\begin{cases} F_1 = \frac{\pi}{4}d^2 p_1 + \frac{\pi}{4}(D^2 - d^2)p_2 \\ F_2 = \frac{\pi}{4}D^2 p_2 \\ G = mg \\ F_s = k(x_0 + x) \\ f = \mu A \frac{v}{h} \\ F_f = \rho q(\beta_2 v_2 \cos(\alpha) - \beta_1 v_1) \end{cases} \tag{4}$$

$$\begin{cases} F_1 = \frac{\pi}{4}d^2 p_1 \\ F_2 = \frac{\pi}{4}D^2 p_2 \\ G = mg \\ F_s = k(x_0 + x) \\ f = \mu A \frac{v}{h} \\ F_f = \rho q(\beta_2 v_2 \cos(\alpha) - \beta_1 v_1) \end{cases} \tag{5}$$

$$\begin{cases} F_1 = \frac{\pi}{4}d^2 p_1 + \frac{\pi}{4}(D^2 - (d + l\sin\alpha)^2)p_2 \\ F_2 = \frac{\pi}{4}D^2 p_2 \\ G = mg \\ F_s = k(x_0 + x) \\ f = \mu A \frac{v}{h} \\ F_f = \rho q(\beta_2 v_2 \cos(\alpha) - \beta_1 v_1) \end{cases} \tag{6}$$

where, ρ is the density of the hydraulic oil; q is the flowrate of oil flowing through the valve element and the valve seat; β_1, β_2 are the momentum correction coefficient at the inlet and outlet of the valve element; v_1, v_2 are the velocity at the inlet and outlet of the valve element; and α is the half angle of the valve element. The description of other symbols is the same as mentioned above.

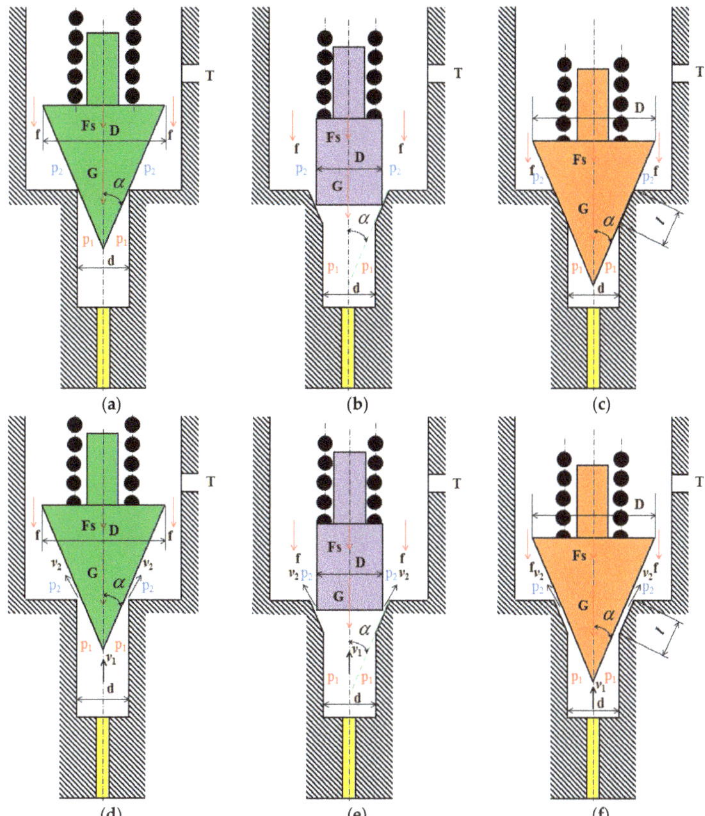

Figure 1. The structural diagram of three kinds of CTDARV. (**a**) CA-VE: Stage 1; (**b**) CA-VS: Stage 1; (**c**) CA-VE&VS: Stage 1; (**d**) CA-VE: Stage 2; (**e**) CA-VS: Stage 2; (**f**) CA-VE&VS: Stage 2.

It can be seen from Figure 1b that p_2 is connected to the outlet, so $p_2 = 0$, and $F_2 = 0$. When p_1 is high, thus $F_1 > F_s + f + G$, the valve element moves upwards, and the valve core and valve seat are separated, as shown in Figure 1c. When $F_1 = F_s + f + F_f + G$, the valve element moves upward for a certain distance and then maintains the stable state, and the high-pressure oil overflows to the tank through the outlet.

3. AMESIM Simulation Models of Three Kinds of CTDARV

In recent years, some famous software including AMESim, MATLAB, EASY5, HOP-SAN have been widely employed in hydraulic system simulation. The biggest advantage of AMESim, compared with other software, i.e., MATLAB, is that it can free users from time-consuming programming and tedious mathematical modeling.

AMESim is an advanced modeling and simulation environment built upon the basis of a bond graph, composed of many libraries such as Hydraulic, Mechanical, Control, etc. Users can build complex system models just in AMESim to conduct simulation and analysis. HCD is the library for designing hydraulic components in AMESim, which can help users to establish almost all hydraulic components to study the performance of hydraulic components.

Based on the working principle of the three kinds of CTDARV, the AMESIM simulation models of the three kinds of CTDARV with/without orifice are built by utilizing the Hydraulic library, the HCD library, and the Mechanical library, as shown in Figure 2. In these models, the thick solid line represents the oil pressure action surface, and the arrow represents the pressure action direction. The following basic assumptions were made when building the AMESIM simulation models:

(1) The operating temperature and ambient temperature remain constant;
(2) The physical and chemical properties of the working medium remain constant;
(3) No geometric shape error and assembly error in all parts;
(4) No internal and external leakage;
(5) No deformation in all parts during operation.

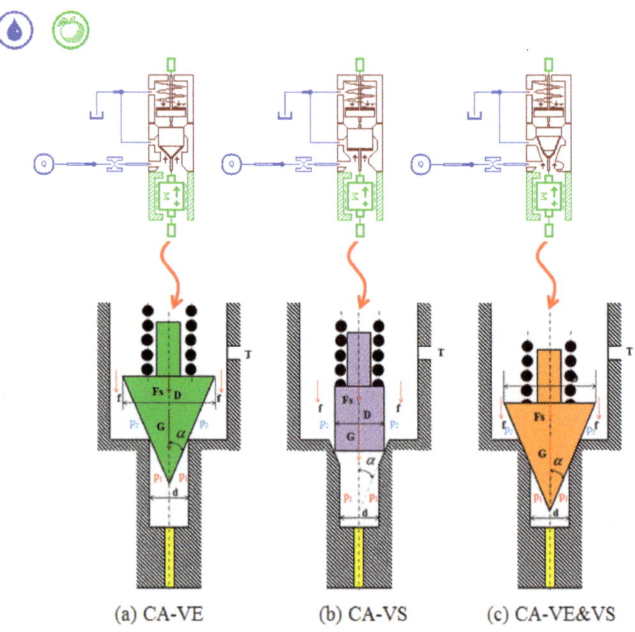

(a) CA-VE   (b) CA-VS   (c) CA-VE&VS

Figure 2. AMESIM simulation models of the three kinds of CTDARV.

4. Results and Discussion

The basic parameters are: the working temperature $T = 40\,^\circ\text{C}$, the density of hydraulic oil $\rho = 850\ \text{kg}/\text{m}^3$, the bulk modulus of hydraulic oil $K = 17000$ bar, the absolute viscosity of hydraulic oil $\nu = 51$ cP, and the flowrate of hydraulic oil $Q = 15\ \text{L}/\text{min}$. The simulation parameters are shown in Table 1, including the valve element diameter (VED), valve seat diameter (VSD), valve element mass (VEM), spring stiffness (SS), conical angle of valve element and valve seat (CAVE&CAVS), and the orifice diameter (OD). By setting the parameters in the AMESIM simulation models of the three kinds of CTDARV, performance comparisons can be obtained.

Table 1. Simulation parameters.

VED	VSD	VEM	SS	CAVE & CAVS	OD
(mm)	(mm)	(kg)	(N/mm)	(°)	(mm)
13, 14, 15, 16	5	0.01	10	45	1
15	3, 4, 5, 6	0.01	10	45	1
15	5	0.01, 0.02, 0.03, 0.04	10	45	1
15	5	0.01	5, 10, 15, 20	45	1
15	5	0.01	10	15, 30, 45, 60	1
15	5	0.01	10	45	0.8, 1.0, 1.2, 1.4

4.1. Pressure Performance Comparisons of the Three Kinds of CTDARV

4.1.1. Effect of VED on Pressure Response

Figure 3 shows the effect of VED (13 mm–16 mm) on the pressure response of the three kinds of CTDARV. It can be seen from Figure 3 that when the VED is 13 mm–16 mm, the three kinds of CTDARV will eventually reach their respective stable pressure, of which CA-VE has the highest stable pressure, and CA-VS has the lowest stable pressure. The stable pressure of CA-VE is significantly higher than that of CA-VS and CA-VE&VS. The pressure of CA-VE fluctuates, but the number of fluctuations is small, while the pressure of CA-VS and CA-VE&VS does not fluctuate. The pressure of CA-VE reaches its stable pressure at 0.02 s, while CA-VS and CA-VE&VS reach their respective stable pressure at 0.001 s. With the increase of VED from 13 mm to 16 mm, the stable pressure of CA-VE will not change, while the stable pressure of CA-VS and CA-VE&VS will decrease. Specifically, when the VED is 13 mm, 14 mm, 15 mm, and 16 mm, the stable pressure of CA-VE is 68.92 bar; the stable pressure of CA-VS is 8.24 bar, 7.07 bar, 6.14 bar, and 5.28 bar; and the stable pressure of CA-VE&VS is 12.51 bar, 11.21 bar, 10.08 bar, and 9.11 bar, respectively.

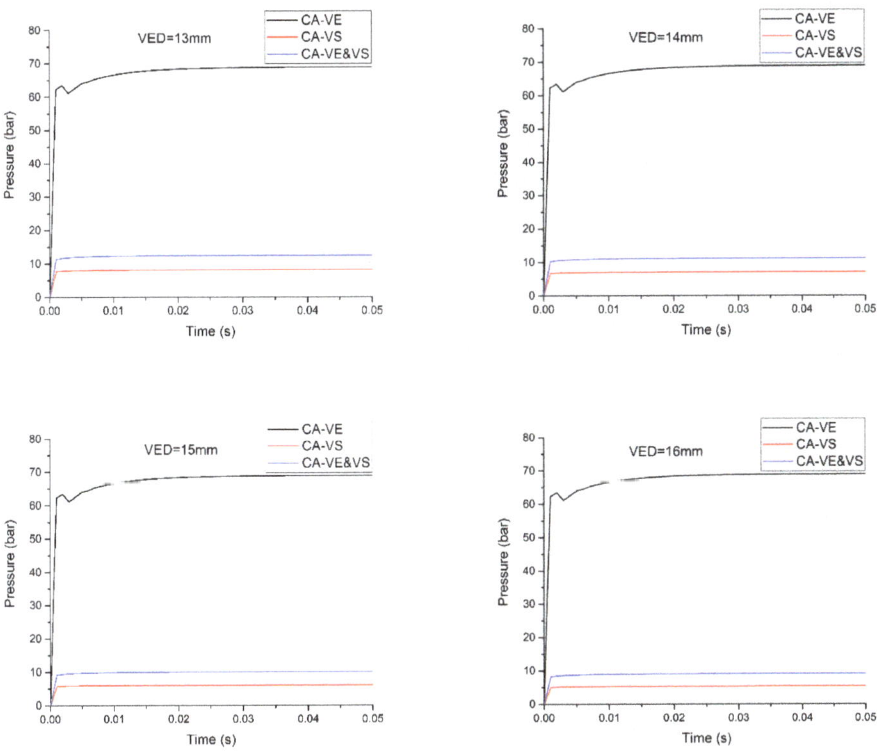

Figure 3. The pressure response of the three kinds of CTDARV: VED = 13 mm–16 mm.

4.1.2. Effect of VSD on Pressure Response

Figure 4 shows the effect of VSD (3 mm–6 mm) on the pressure response of the three kinds of CTDARV. It can be seen from Figure 4 that when VSD is 3 mm–6 mm, respectively, the three kinds of CTDARV will eventually reach their respective stable pressure, of which CA-VE has the highest stable pressure, and CA-VS has the lowest stable pressure. The stable pressure of CA-VE is significantly higher than that of CA-VS and CA-VE&VS. The smaller the VSD, the longer the pressure fluctuation times of CA-VE, and the longer the time required to reach the stable pressure. When VSD is 6 mm, the pressure of CA-VE does not fluctuate. When VSD is 3 mm–6 mm, the pressure of CA-VS and CA-VE&VS do not fluctuate. When the VSD is 3 mm, the pressure of CA-VS reaches the stable pressure of 0.02 s, while when the VSD is 4 mm–6 mm, the pressure of CA-VS reaches the stable pressure at 0.001 s. When VSD is 3 mm–6 mm, the pressure of CA-VE&VS reaches stable pressure at 0.001 s. With the increase of VSD from 3 mm to 6 mm, the stable pressure of CA-VE and CA-VE&VS decreases, and the stable pressure of CA-VE decreases significantly. The stable pressure of CA-VS is larger when the VSD is 3 mm, while it keeps the same small value when the VSD is 4 mm–6 mm. Specifically, when VSD is 3 mm, 4 mm, 5 mm, and 6 mm, the stable pressure of CA-VE is 229.08 bar, 115.24 bar, 68.92 bar, 45.73 bar, respectively; that of CA-VS is 10.83 bar, 6.14 bar, 6.14 bar, and 6.14 bar, respectively; and that of the CA-VE&VS is 10.84 bar, 10.28 bar, 10.08 bar, 9.96 bar, respectively.

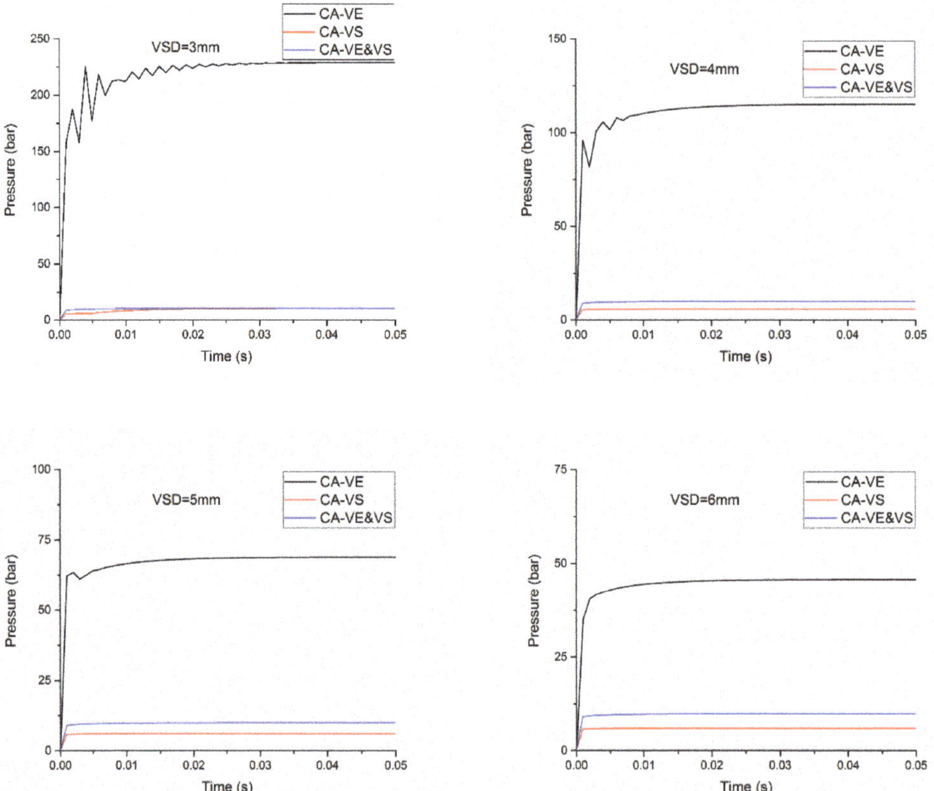

Figure 4. The pressure response of the three kinds of CTDARV: VSD = 3 mm–6 mm.

4.1.3. Effect of VEM on Pressure Response

Figure 5 shows the effect of VEM (0.01 kg–0.04 kg) on the pressure response of the three kinds of CTDARV. It can be seen from Figure 5 that when the VEM is 0.01 kg–0.04 kg, the three kinds of CTDARV will eventually reach their respective stable pressure, of which CA-VE has the highest stable pressure, and CA-VS has the lowest stable pressure. The stable pressure of CA-VE is significantly higher than that of CA-VS and CA-VE&VS. The greater the VEM, the longer the pressure fluctuation times of CA-VE, and the longer the time required to reach the stable pressure. When the VEM is 0.01 kg and 0.02 kg, the pressure of CA-VE&VS reaches the stable pressure in 0.001 s. When VEM is 0.03 kg and 0.04 kg, the pressure of CA-VE&VS reaches stable pressure at about 0.003 s and 0.005 s respectively. When VEM is 0.01 kg–0.04 kg, the pressure of CA-VS reaches stable pressure at 0.001 s. With the increase of VEM from 0.01 kg to 0.04 kg, the stable pressures of CA-VE

CA-VS and CA-VE&VS remain unchanged. Specifically, when the VEM is 0.01 kg, 0.02 kg, 0.03 kg, and 0.04 kg, the stable pressure of CA-VE, CA-VS, and CA-VE&VS is 68.92 bar, 6.14 bar, and 10.08 bar, respectively.

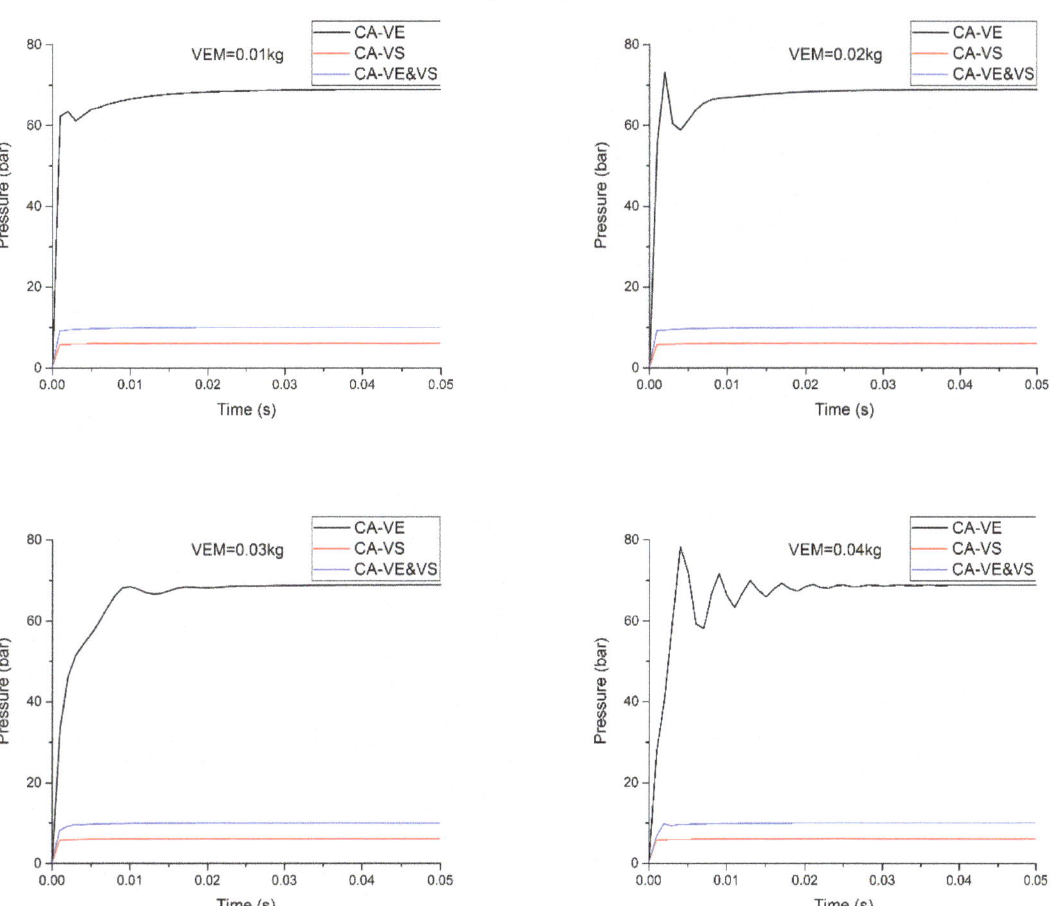

Figure 5. The pressure response of the three kinds of CTDARV: VEM = 0.01 kg–0.04 kg.

4.1.4. Effect of SS on Pressure Response

Figure 6 shows the effect of SS (5 N/mm–20 N/mm) on the pressure response of the three kinds of CTDARV. It can be seen from Figure 6 that when SS is 5 N/mm–20 N/mm, the three kinds of CTDARV will eventually reach their respective stable pressure, of which CA-VE has the highest stable pressure, and CA-VS has the lowest stable pressure. The stable pressure of CA-VE is significantly higher than that of CA-VS and CA-VE&VS. When SS is 5 N/mm–20 N/mm, the pressure of CA-VE fluctuates less, and the pressure of CA-VS and CA-VE&VS does not fluctuate. When SS is 5 N/mm–20 N/mm, respectively, the pressure of CA-VS and CA-VE&VS reaches stable pressure at 0.001 s. With the increase of SS from 5 N/mm to 20 N/mm, the stable pressure of CA-VE, CA-VS and CA-VE&VS increases, but the increase is very small. Specifically, when SS is 5 N/mm, 10 N/mm, 15 N/mm, and 20 N/mm, the stable pressure of CA-VE is 68.17 bar, 68.92 bar, 69.65 bar and 70.38 bar, respectively; and that of CA-VS is 6.06 bar, 6.14 bar, 6.22 bar, and 6.30 bar, respectively; and that of CA-VE&VS is 9.76 bar, 10.08 bar, 10.40 bar, and 10.70 bar, respectively.

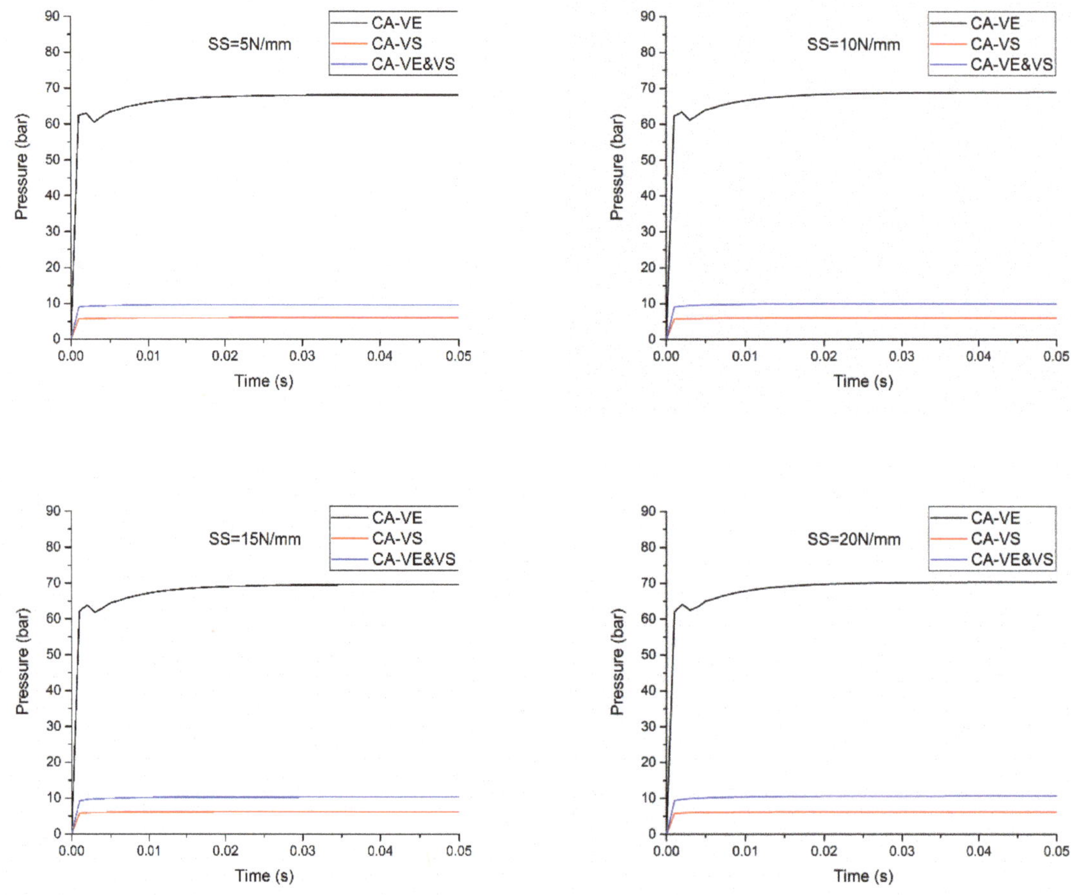

Figure 6. The pressure response of the three kinds of CTDARV: SS = 5 N/mm–20 N/mm.

4.1.5. Effect of CAVE&CAVS on Pressure Response

Figure 7 shows the effect of CAVE&CAVS (15 degrees–60 degrees) on the pressure response of the three kinds of CTDARV. It can be clearly seen from Figure 7 that when CAVE&CAVS are 15 degrees 60 degrees, respectively, the three kinds of CTDARV will eventually reach their respective stable pressure, of which CA-VE has the highest stable pressure and CA-VS has the lowest stable pressure. The stable pressure of CA-VE is significantly higher than that of CA-VS and CA-VE&VS. When CAVE&CAVS is 15 degree–60 degrees, the pressure of CA-VS and CA-VE&VS does not fluctuate. When CAVE&CAVS is 30 degrees and 45 degrees, respectively, the pressure of CA-VE does not fluctuate. When CAVE&CAVS is 15 degrees and 60 degrees, respectively, the pressure of CA-VE fluctuates, but the number of fluctuations is small. When CAVE&CAVS is 15 degrees 60 degrees, respectively, the pressure of CA-VS and CA-VE&VS reaches stable pressure at 0.001 s. With the increase of CAVE&CAVS from 15 degrees to 60 degrees, the stable pressure of CA-VE, CA-VS, and CA-VE&VS decreases, but the decrease is not significant. Specifically, when CAVE&CAVS are 15 degrees, 30 degrees, 45 degrees, and 60 degrees, the stable pressure of CA-VE is 78.65 bar, 73.86 bar, 68.92 bar, and 63.35 bar, respectively; the stable pressure of CA-VS is 6.53 bar, 6.28 bar, 6.14 bar, and 6.02 bar, respectively; and the stable pressure of CA-VE&VS is 11.40 bar, 10.50 bar, 10.08 bar, and 9.77 bar, respectively.

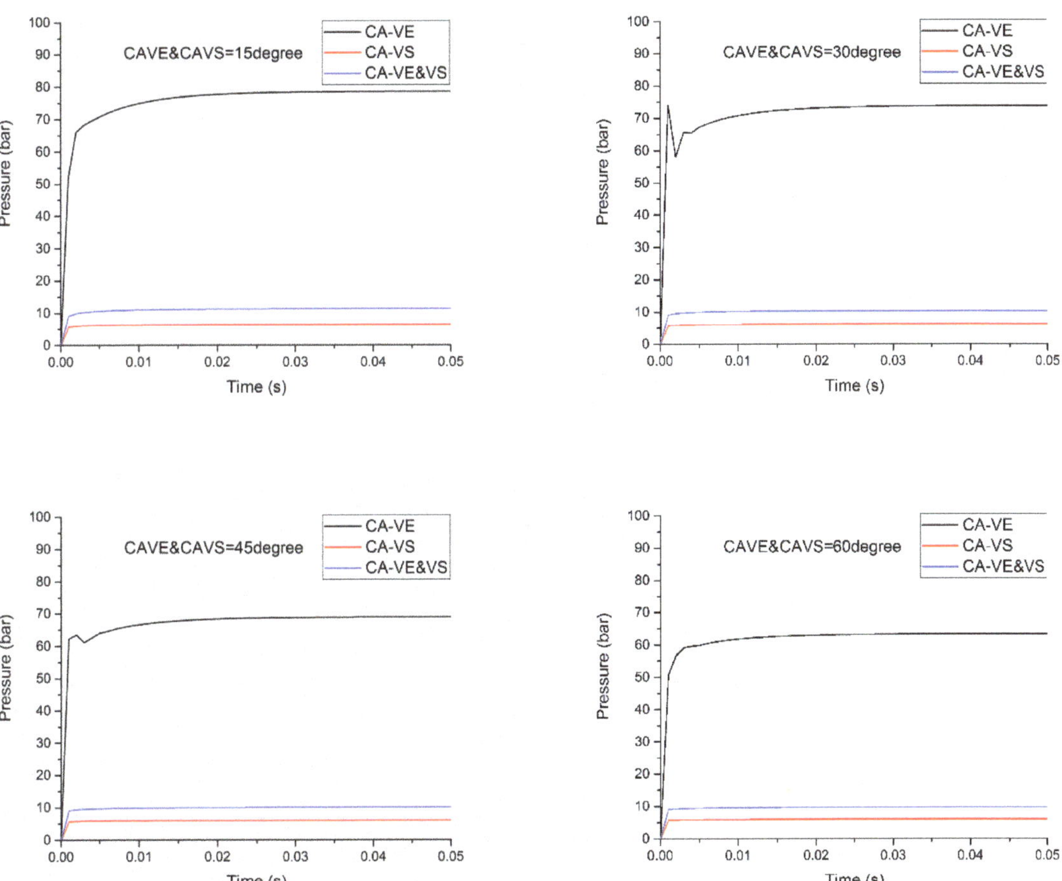

Figure 7. The pressure response of the three kinds of CTDARV: CAVE&CAVS = 15 degrees–60 degrees.

4.1.6. Effect of OD on Pressure Response

Figure 8 shows the effect of OD (0.8 mm–1.4 mm) on the pressure response of the three kinds of CTDARV. It can be clearly seen from Figure 8 that when the OD is 0.8 mm–1.4 mm, respectively, the three kinds of CTDARV will eventually reach their respective stable pressure, of which CA-VE has the highest stable pressure and CA-VS has the lowest stable pressure. The stable pressure of CA-VE is significantly higher than that of CA-VS and CA-VE&VS. When OD is 0.8 mm–1.4 mm, the pressure of CA-VS and CA-VE&VS does not fluctuate. When OD is 0.8 mm, the pressure of CA-VE does not fluctuate. When the OD is 1.0 mm–1.4 mm, respectively, the pressure of CA-VE fluctuates, but the number of fluctuations is small. The greater the OD, the greater the pressure fluctuation of CA-VE. When the OD is 0.8 mm–1.4 mm, respectively, the pressure of CA-VS and CA-VE&VS reaches stable pressure at 0.001 s. With the increase of OD from 0.8 mm to 1.4 mm, the stable pressures of CA-VE, CA-VS, and CA-VE&VS remain unchanged. Specifically, when the OD is 0.8 mm, 1.0 mm, 1.2 mm and 1.4 mm, the stable pressure of CA-VE is 68.92 bar, the stable pressure of CA-VS is 6.14 bar, and the stable pressure of CA-VE&VS is 10.08 bar, respectively.

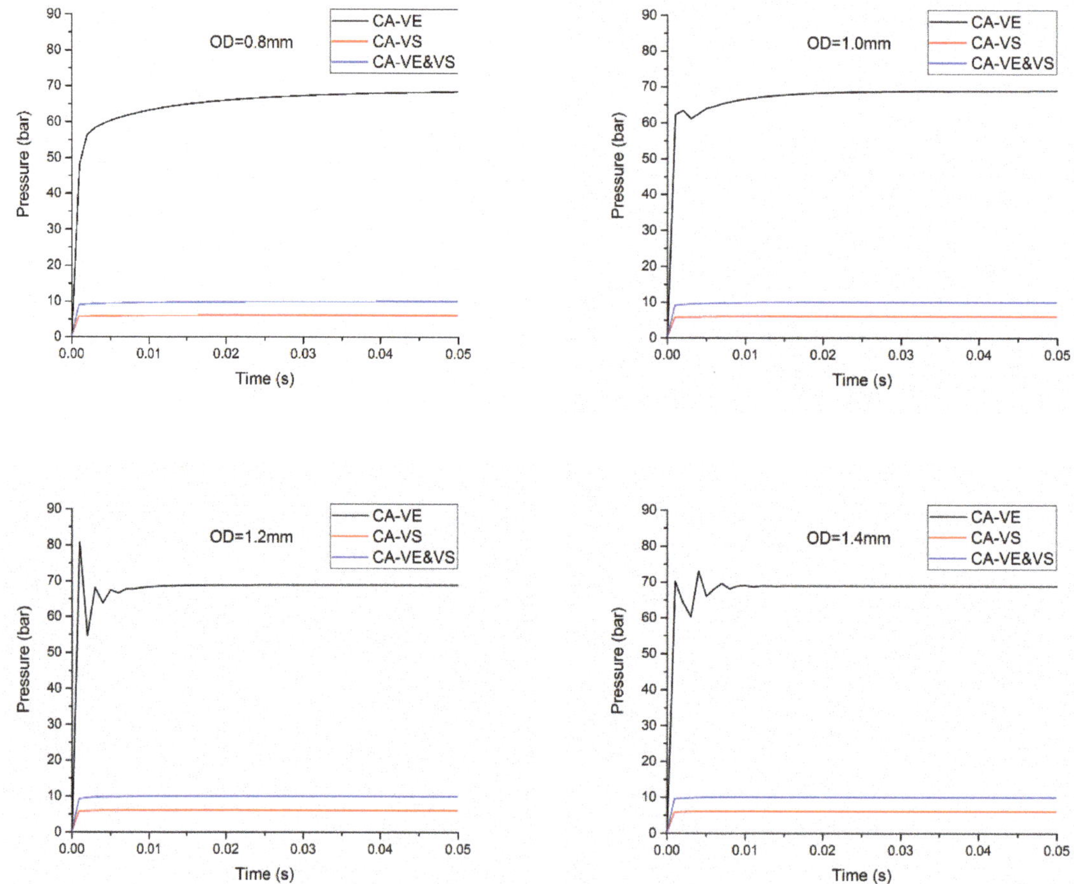

Figure 8. The pressure response of the three kinds of CTDARV: OD = 0.8 mm–1.4 mm.

4.2. Flowrate Performance Comparisons of the Three Kinds of CTDARV

4.2.1. Effect of VED on Flowrate Response

Figure 9 shows the effect of VED (13 mm–16 mm) on the flowrate response of the three kinds of CTDARV. It can be clearly seen from Figure 9 that when the VED is 13 mm–16 mm, the three kinds of CTDARV will eventually reach a stable flowrate of 15 L/min, and all have similar flowrate curves. It can be seen from the partial enlarged diagram that the CA-VS has relatively large flowrate and the CA-VE has relatively small flowrate in the initial stage before 0.001 s. At 0.001 s, the flowrate of CA-VE, CA-VS and CA-VE&VS decreased with the increase of VED. Specifically, when the VED is 13 mm, 14 mm, 15 mm, and 16 mm, the flowrate of CA-VE is 3.5102 L/min, 3.2074 L/min, 2.8822 L/min, and 2.5346 L/min, respectively; the flowrate of CA-VS is 5.7207 L/min, 5.65757 L/min, 5.5783 L/min, and 5.4804 L/min, respectively; and the flowrate of CA-VE&VS is 5.0839 L/min, 4.8622 L/min, 4.5983 L/min, and 4.2873 L/min, respectively.

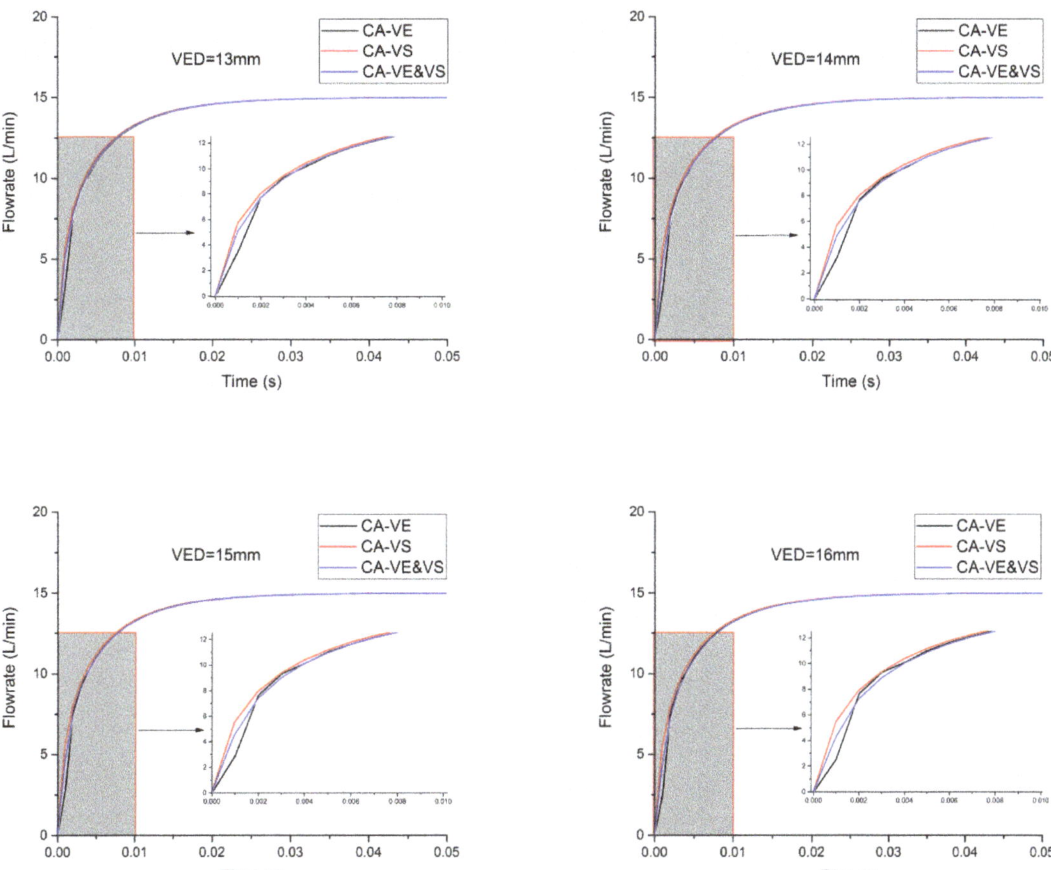

Figure 9. The flowrate response of the three kinds of CTDARV: VED = 13 mm–16 mm.

4.2.2. Effect of VSD on Flowrate Response

Figure 10 shows the effect of VSD (3 mm–6 mm) on the flowrate response of the three kinds of CTDARV. It can be clearly seen from Figure 10 that when the VED is 13 mm–16 mm, the three kinds of CTDARV will eventually reach a stable flowrate of 15 L/min. It is worth noting that when the VSD is 3 mm, the flowrate of CA-VS oscillates violently before 0.015 s. When the VSD is 4 mm–6 mm, the flowrate of CA-VS does not fluctuate. When VSD increases from 3 mm to 6 mm, the flowrate fluctuation of CA-VE decreases gradually, while the flowrate of CA-VE&VS does not fluctuate. When VSD is 5 mm and 6 mm, it can be seen from the local enlarged view that CA-VS has relatively large flowrate and CA-VE has relatively small flowrate in the initial stage before 0.001 s. At 0.001 s, with the VSD increasing from 5 mm to 6 mm, the flowrate of CA-VE and CA-VE&VS increased, and the flowrate of CA-VS remained unchanged. Specifically, when VSD is 5 mm and 6 mm, the flowrate of CA-VE is 2.8822 L/min and 4.6088 L/min, respectively; the flowrate of CA-VS is 5.5783 L/min, and 5.5783 L/min, respectively; and the flowrate of CA-VE&VS is 4.5983 L/min, and 4.8690 L/min, respectively.

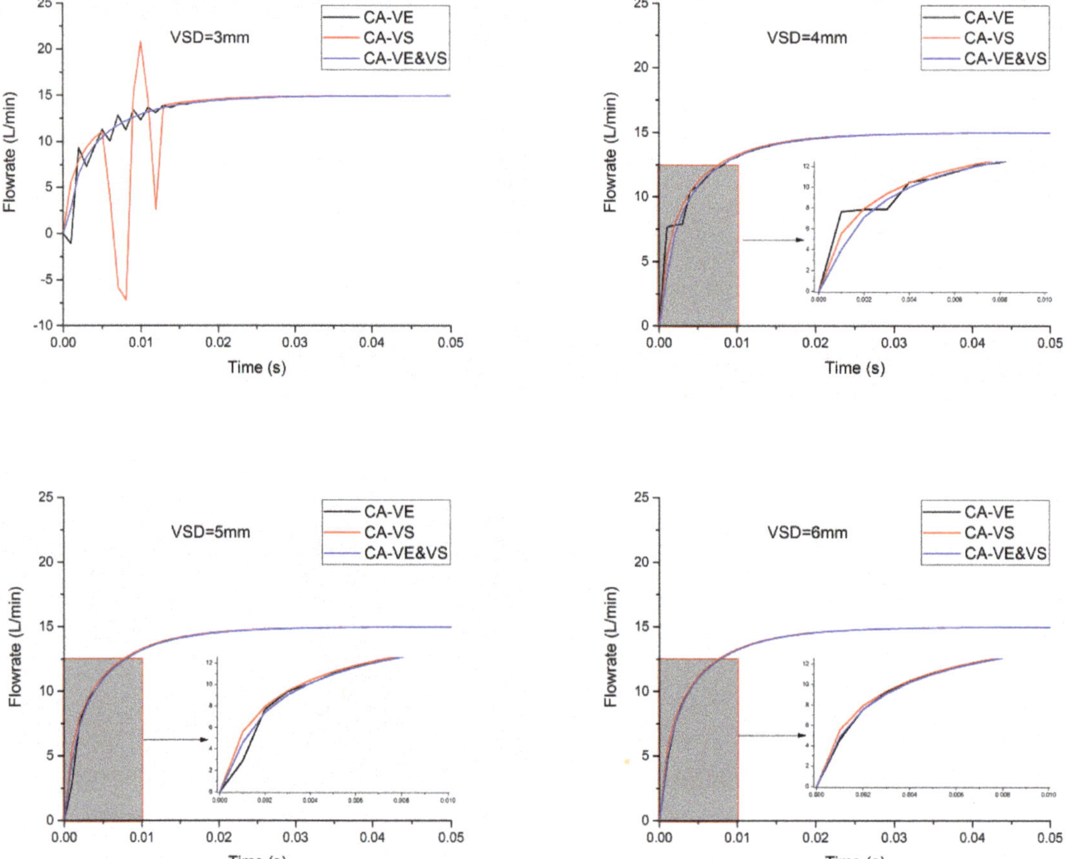

Figure 10. The flowrate response of the three kinds of CTDARV: VSD = 3 mm–6 mm.

4.2.3. Effect of VEM on Flowrate Response

Figure 11 shows the effect of VEM (0.01 kg–0.04 kg) on the flowrate response of the three kinds of CTDARV. It can be clearly seen from Figure 11 that when the VEM is 0.01 kg–0.04 kg, the three kinds of CTDARV will eventually reach a stable flow of 15 L/min. When VEM is 0.01 kg, the flow of CA-VE does not fluctuate. When VEM increases from 0.02 kg to 0.04 kg, the flow fluctuation of CA-VE increases gradually. When VEM increased from 0.01 kg to 0.04 kg, the flow of CA-VS and CA-VE&VS did not fluctuate. When VEM is 0.01 kg, it can be seen from the local enlarged view that CA-VS has relatively large flow rate and CA-VE has relatively small flow rate in the initial stage before 0.001 s. At 0.001 s, the flow of CA-VE is 2.8822 L/min, the flow of CA-VS is 5.5783 L/min, and the flow of CA-VE&VS is 4.5983 L/min.

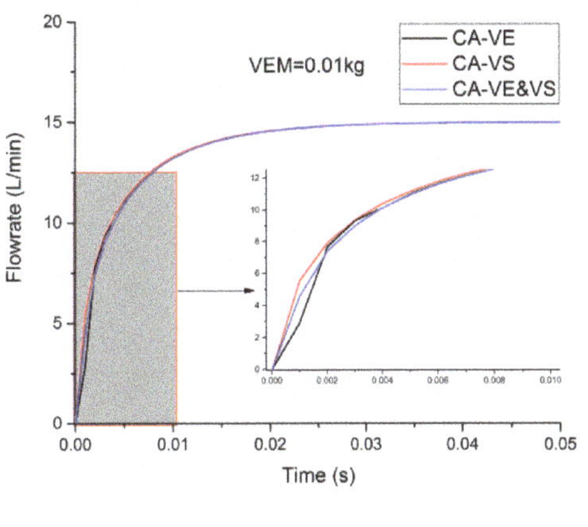
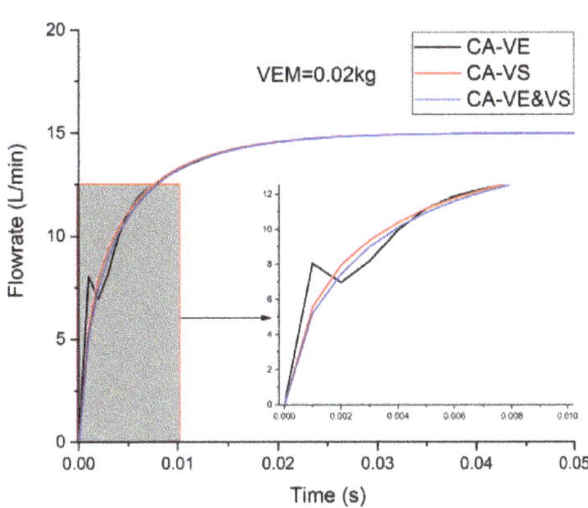
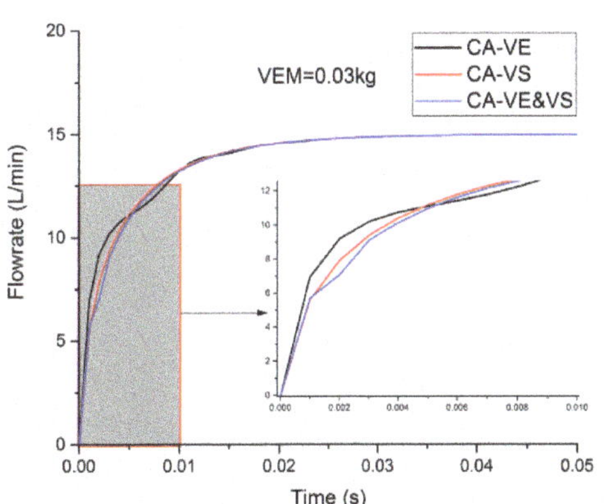
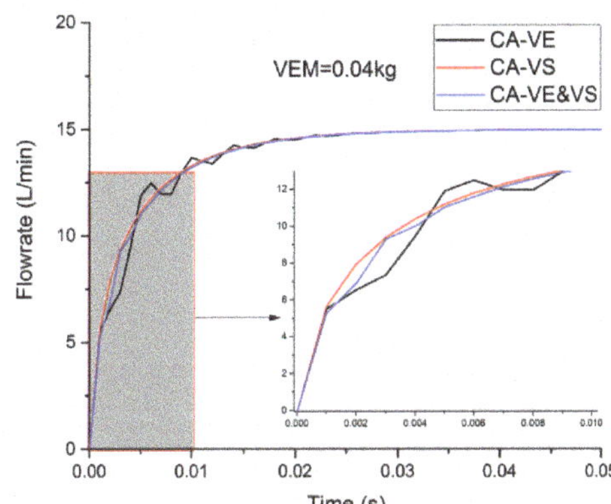

Figure 11. The flowrate response of the three kinds of CTDARV: VEM = 0.01 kg–0.04 kg.

4.2.4. Effect of SS on Flowrate Response

Figure 12 shows the effect of SS (5 N/mm–20 N/mm) on the flowrate response of the three kinds of CTDARV. It can be clearly seen from Figure 12 that when SS is 5 N/mm–20 N/mm, the three kinds of CTDARV will eventually reach a stable flowrate of 15 L/min, and all have similar flowrate curves. It can be seen from the partial enlarged diagram that the CA-VS has relatively large flowrate and the CA-VE has relatively small flowrate in the initial stage before 0.001 s. At 0.001 s, with SS increasing from 5 N/mm to 20 N/mm, the flowrate of CA-VE decreases, while the flowrate of CA-VS and CA-VE&VS increases, but the decrease and increase are not significant. Specifically, when SS is 5 N/mm, 10 N/mm, 15 N/mm and 20 N/mm, the flowrate of CA-VE is 2.9047 L/min, 2.8822 L/min, 2.8620 L/min, and 2.8440 L/min, respectively; the flowrate of CA-VS is 5.5754 L/min, 5.5783 L/min, 5.5813 L/min, and 5.5842 L/min, respectively; and the flowrate of CA-VE&VS is 4.5796 L/min, 4.5983 L/min, 4.6159 L/min, and 4.6327 L/min, respectively.

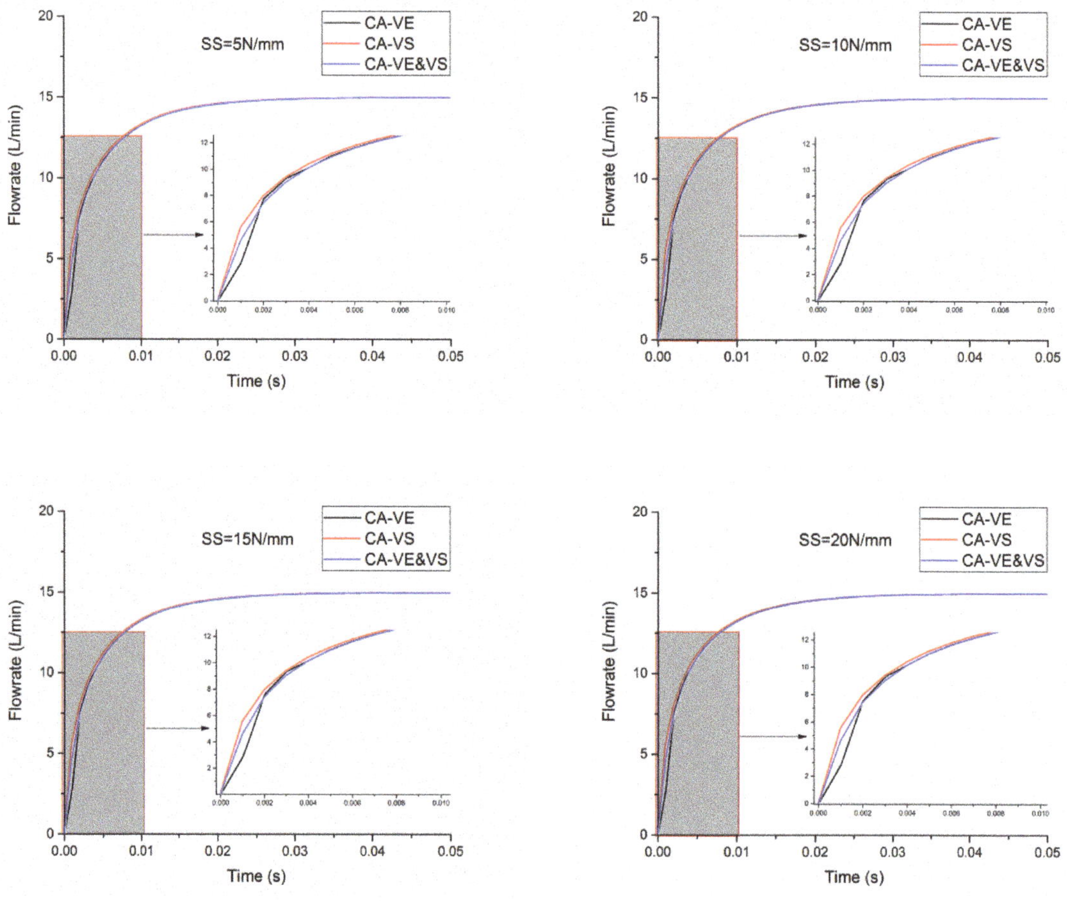

Figure 12. The flowrate response of the three kinds of CTDARV: SS = 5 N/mm–20 N/mm.

4.2.5. Effect of CAVE&CAVS on Flowrate Response

Figure 13 shows the effect of CAVE&CAVS (15 degrees–60 degrees) on the flowrate response of the three kinds of CTDARV. It can be clearly seen from Figure 13 that when CAVE&CAVS are 15 degrees–60 degrees, respectively, the three kinds of CTDARV will eventually reach a stable flowrate of 15 L/min. It can be seen from the partial enlarged diagram that when CAVE&CAVS is 15 degrees, CA-VE has relatively large flowrate and CA-VE&VS has relatively small flowrate in the initial stage before 0.001 s. When CAVE&CAVS is 15 degrees, at 0.001 s, the flowrate of CA-VE is 7.8151 L/min, the flowrate of CA-VS is 3.9823 L/min, and the flowrate of CA-VE&VS is 0.5041 L/min. At 0.001 s, with the increase of CAVE&CAVS from 15 degrees to 60 degrees, the flowrate of CA-VS and CA-VE&VS increases. At 0.001 s, the flowrate of CA-VE increases as CAVE&CAVS increases from 45 degrees to 60 degrees. Specifically, when CAVE&CAVS is 15 degrees, 30 degrees, 45 degrees, and 60 degrees; the flowrate of CA-VE is 7.8151 L/min, 5.2183 L/min, 2.8822 L/min, and 3.2828 L/min; the flowrate of CA-VS is 3.9823 L/min, 5.2281 L/min, 5.5783 L/min, and 5.7075 L/min; and the flowrate of CA-VE&VS is 0.5041 L/min, 3.7486 L/min, 4.5983 L/min, and 4.8908 L/min, respectively.

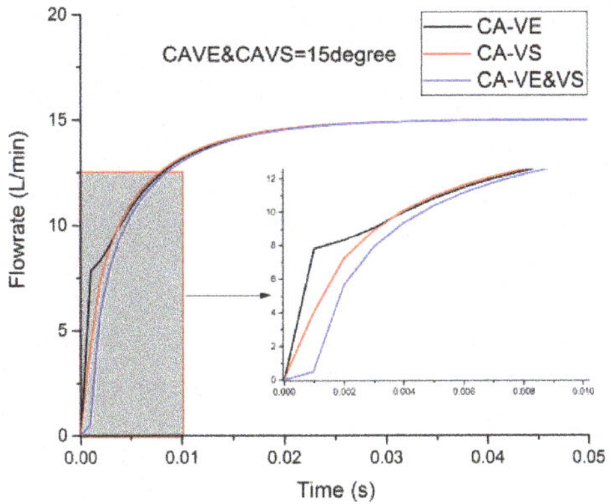
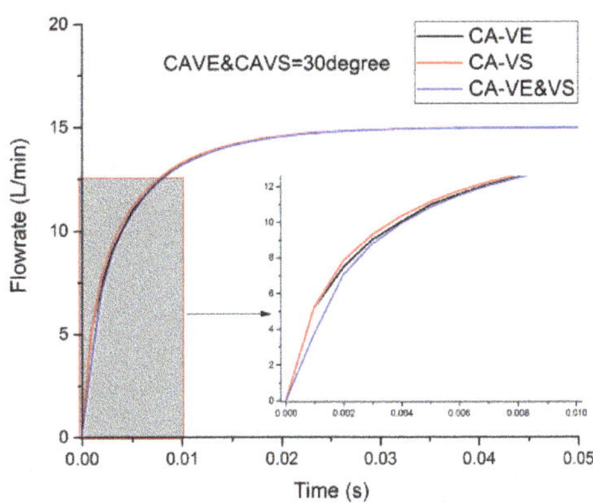
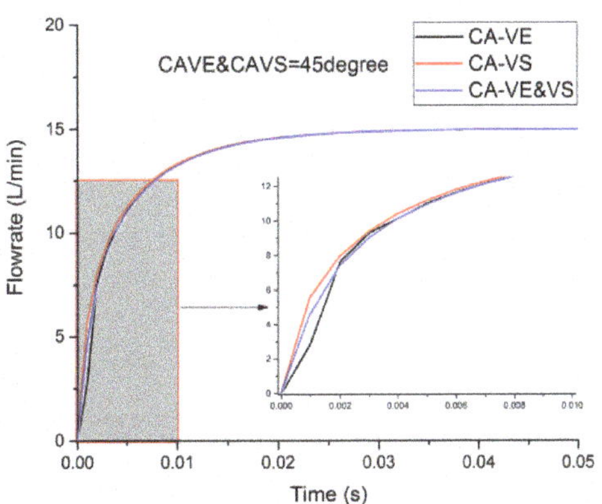
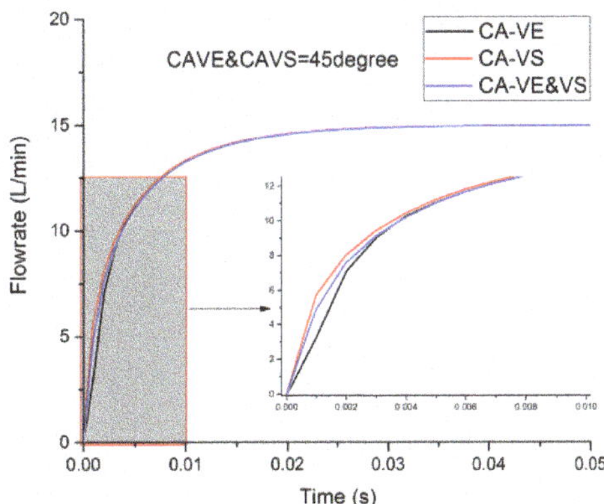

Figure 13. The flowrate response of the three kinds of CTDARV: CAVE&CAVS = 15 degrees–60 degrees.

4.2.6. Effect of OD on Flowrate Response

Figure 14 shows the effect of OD (0.8 mm–1.4 mm) on the flowrate response of the three kinds of CTDARV. It can be clearly seen from Figure 14 that when CAVE&CAVS are 15 degrees–60 degrees, respectively, the three kinds of CTDARV will eventually reach a stable flowrate of 15 L/min. It can be seen from the partial enlarged diagram that when CAVE&CAVS is 15 degrees, CA-VE has relatively large flowrate and CA-VE&VS has relatively small flowrate in the initial stage before 0.001 s. When CAVE&CAVS is 15 degrees, at 0.001 s, the flowrate of CA-VE is 7.8151 L/min, the flowrate of CA-VS is 3.9823 L/min, and the flowrate of CA-VE&VS is 0.5041 L/min At 0.001 s, with the increase of CAVE&CAVS from 15 degrees to 60 degrees, the flowrate of CA-VS and CA-VE&VS increases At 0.001 s, the flowrate of CA-VE increases as CAVE&CAVS increases from 45 degrees to 60 degrees. Specifically, when CAVE&CAVS is 15 degrees, 30 degrees, 45 degrees, and 60 degrees; the flowrate of CA-VE is 7.8151 L/min, 5.2183 L/min, 2.8822 L/min, and 3.2828 L/min; the flowrate of CA-VS is 3.9823 L/min, 5.2281 L/min, 5.5783 L/min, and 5.7075 L/min; and the flowrate of CA-VE&VS is 0.5041 L/min, 3.7486 L/min, 4.5983 L/min, and 4.8908 L/min, respectively.

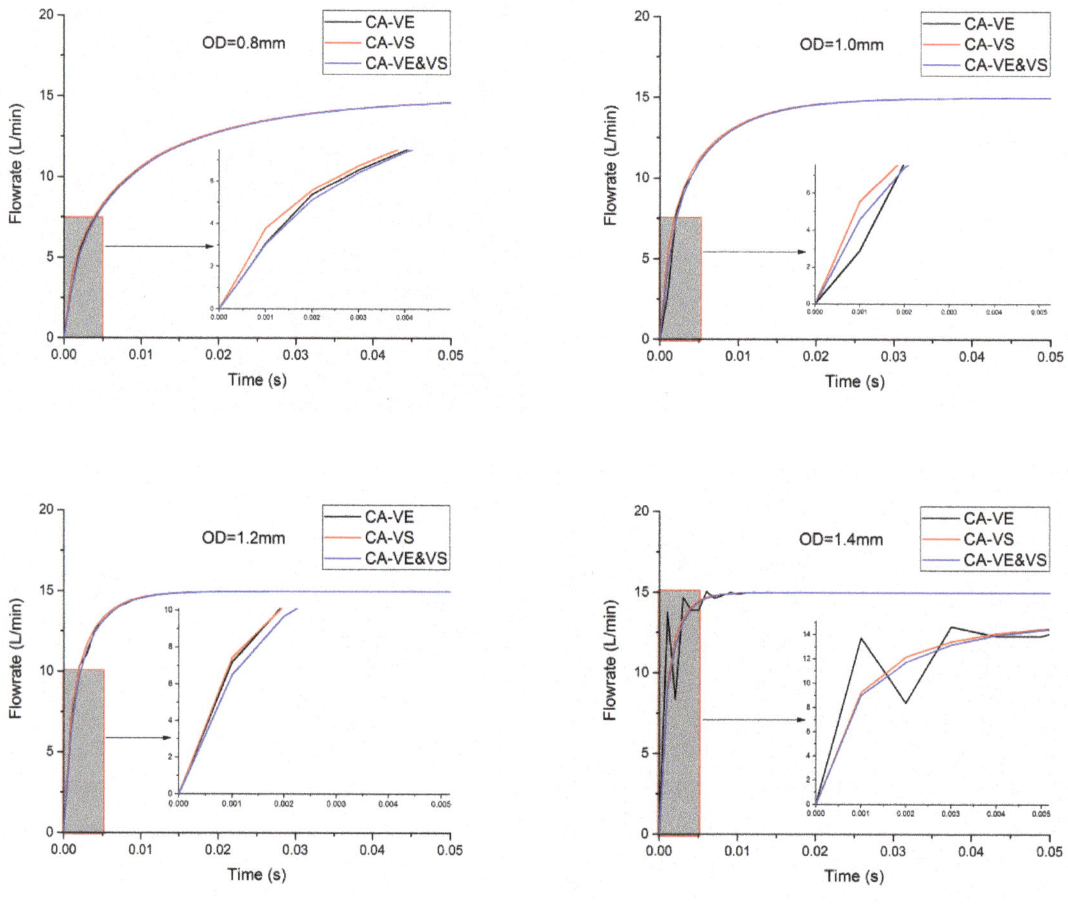

Figure 14. The flowrate response of the three kinds of CTDARV: OD = 0.8 mm–1.4 mm.

4.3. Displacement Performance Comparisons of the Three Kinds of CTDARV

4.3.1. Effect of VED on Displacement Response

Figure 15 shows the effect of VED (13 mm–16 mm) on the displacement response of the three kinds of CTDARV. It can be seen from Figure 15 that when VED is 13 mm–16 mm, the three kinds of CTDARV will eventually reach their respective stable displacement, of which CA-VE&VS has the highest stable displacement, and CA-VE has the lowest stable displacement. The stable displacement of CA-VE&VS is significantly higher than that of CA-VE and CA-VS. The stable displacement of CA-VE and CA-VS is similar. With the increase of VED from 13 mm to 16 mm, the stable displacement of CA-VE will not change, but the stable displacement of CA-VS and CA-VE&VS will increase, of which the stable displacement of CA-VS will increase slightly, while the stable displacement of CA-VE&VS will increase relatively large. Specifically, when the VED is 13 mm, 14 mm, 15 mm, and 16 mm, respectively, the stable displacement of CA-VE is 0.2587 mm; the stable displacement of CA-VS is 0.2776 mm, 0.2784 mm, 0.2791 mm, and 0.2798 mm, respectively, and the stable pressure displacement of CA-VE&VS is 0.6334 mm, 0.6723 mm, 0.7121 mm, and 0.7527 mm, respectively.

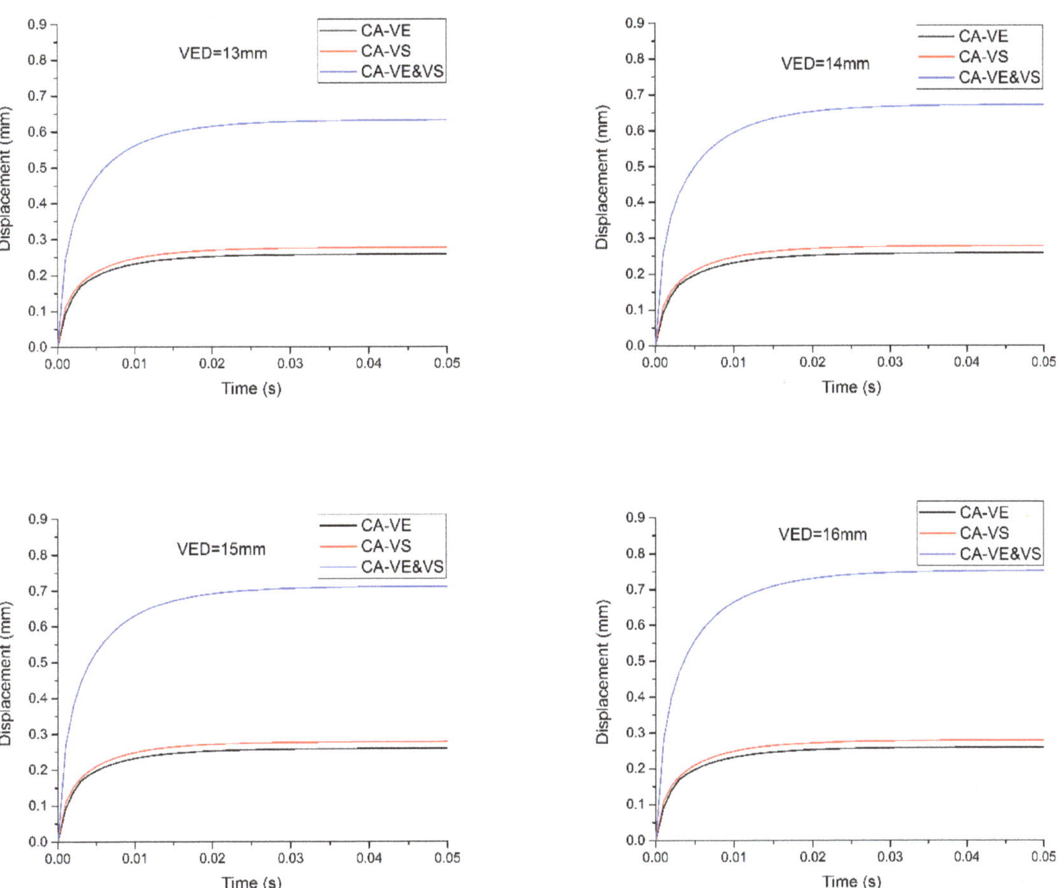

Figure 15. The displacement response of the three kinds of CTDARV: VED = 13 mm–16 mm.

4.3.2. Effect of VSD on Displacement Response

Figure 16 shows the effect of VSD (3 mm–6 mm) on the displacement response of the three kinds of CTDARV. It can be clearly seen from Figure 16 that when VSD is 3 mm–6 mm, respectively, the three kinds of CTDARV will eventually reach their respective stable displacement. It is particularly noteworthy that when the VSD is 3 mm, the stable displacement of CA-VS is the highest, reaching 5 mm, and the stable displacement of CA-VE is the lowest, and the stable displacement of CA-VE, CA-VS, and CA-VE&VS differ greatly. When VSD is 4 mm–6 mm, CA-VE&VS has the highest stable displacement and CA-VE has the lowest stable displacement. The stable displacement of CA-VE&VS is significantly higher than that of CA-VE and CA-VS, and the stable displacement of CA-VE and CA-VS has little difference. With the increase of VSD from 3 mm to 6 mm, the stable displacement of CA-VE increases, but the increase is small, while the stable displacement of CA-VE&VS decreases. The stable displacement of CA-VS is 5 mm when VSD is 3 mm. When VSD is 4 mm–6 mm, the stable displacement of CA-VS remains unchanged. Specifically, when the VSD is 3 mm, 4 mm, 5 mm, and 6 mm, respectively, the stable displacement of CA-VE is 0.2393 mm, 0.2513 mm, 0.2587 mm, and 0.2636 mm, respectively; the stable displacement of CA-VS is 5 mm, 0.2791 mm, 0.2791 mm, and 0.2791 mm, respectively; and the stable pressure displacement of CA-VE&VS is 1.5468 mm, 0.9321 mm, 0.7121 mm, and 0.5817 mm, respectively.

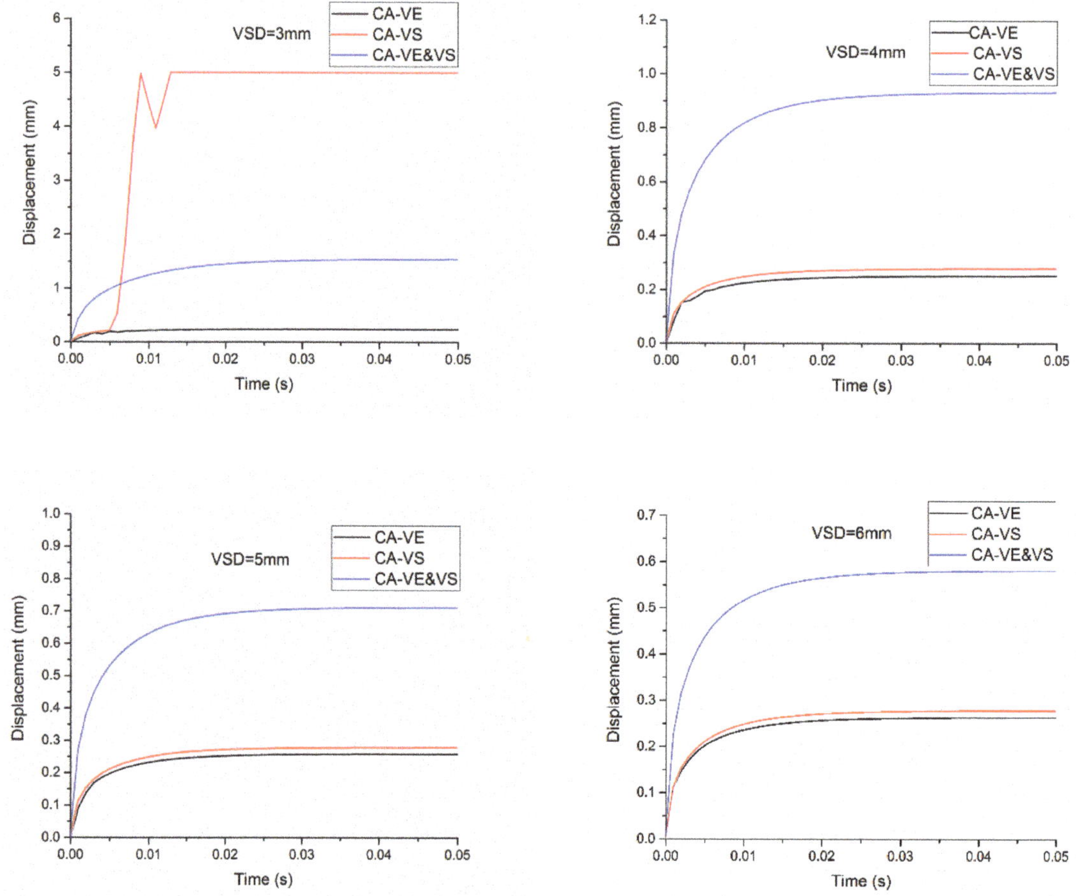

Figure 16. The displacement response of the three kinds of CTDARV: VSD = 3 mm–6 mm.

4.3.3. Effect of VEM on Displacement Response

Figure 17 shows the effect of VEM (0.01 kg–0.04 kg) on the displacement response of the three kinds of CTDARV. It can be clearly seen from Figure 17 that when the VEM is 0.01 kg–0.04 kg, the three kinds of CTDARV will eventually reach their respective stable displacement. The stable displacement of CA-VE&VS is the highest, and that of CA-VE is the lowest. The stable displacement of CA-VE&VS is significantly higher than that of CA-VE and CA-VS. The difference between the stable displacement of CA-VE and CA-VS is not very large. With the increase of VEM from 0.01 kg to 0.04 kg, the initial displacement of CA-VE oscillates gradually, and the number of oscillations increases gradually, while the displacement of CA-VS and CA-VE&VS does not oscillate. With the increase of VEM, the stable displacement of CA-VE, CA-VS, and CA-VE&VS remain unchanged. Specifically, when the VEM is 0.01 kg, 0.02 kg, 0.03 kg, and 0.04 kg, respectively, the stable displacement of CA-VE is 0.2587 mm, the stable displacement of CA-VS is 0.2791 mm, and the stable pressure displacement of CA-VE&VS is 0.7121 mm.

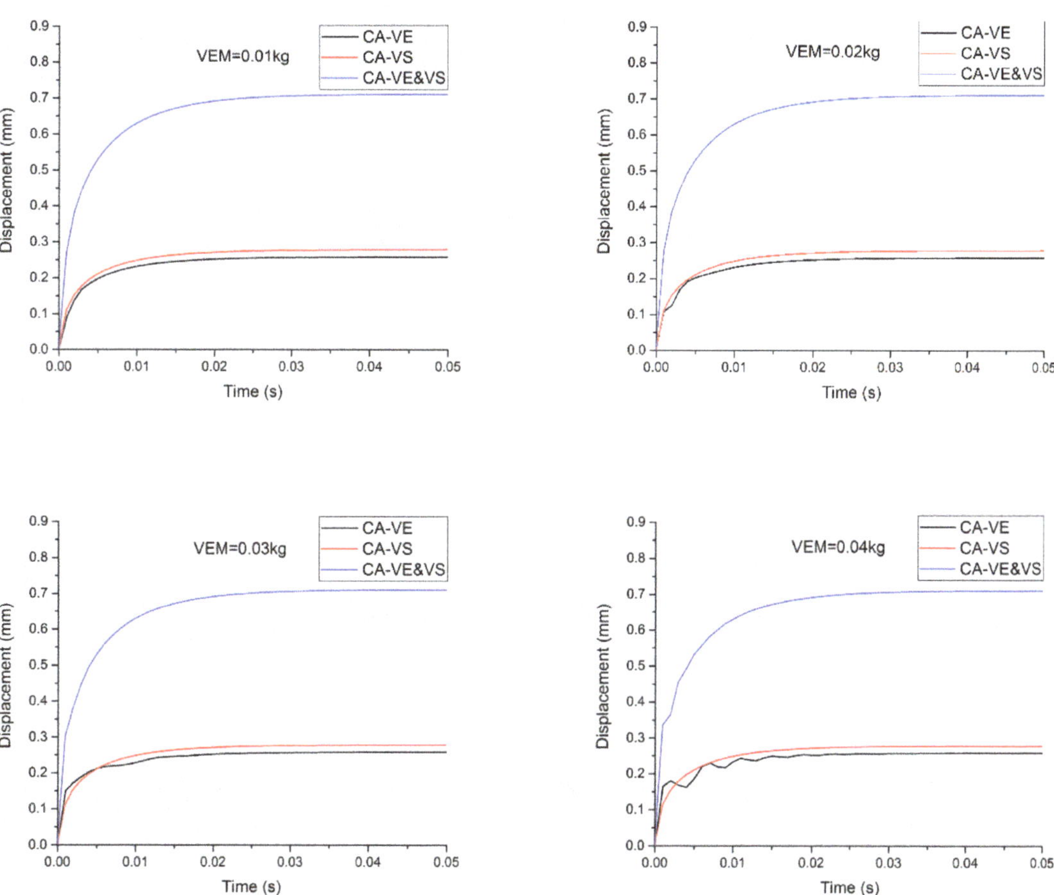

Figure 17. The displacement response of the three kinds of CTDARV: VEM = 0.01 kg–0.04 kg.

4.3.4. Effect of SS on Displacement Response

Figure 18 shows the effect of SS (5 N/mm–20 N/mm) on the displacement response of the three kinds of CTDARV. It can be clearly seen from Figure 18 that when SS is 5 N/mm–20 N/mm, the three kinds of CTDARV will eventually reach their respective stable displacement. The stable displacement of CA-VE&VS is the highest, and that of CA-VE is the lowest. The stable displacement of CA-VE&VS is significantly higher than that of CA-VE and CA-VS. The difference between the stable displacement of CA-VE and CA-VS is not very large. With the increase of SS from 5 N/mm to 20 N/mm, the stable displacement of CA-VE, CA-VS and CA-VE&VS decreased, but the decrease was not significant. Specifically, when SS is 5 N/mm, 10 N/mm, 15 N/mm, and 20 N/mm, respectively, the stable displacement of CA-VE is 0.2601 mm, 0.2587 mm, 0.2572 mm, and 0.2559 mm, respectively; the stable displacement of CA-VS is 0.2810 mm, 0.2791 mm, 0.2774 mm, and 0.2756 mm, respectively; and the stable pressure displacement of CA-VE&VS is 0.7246 mm, 0.7121 mm, 0.7005 mm, and 0.6897 mm, respectively.

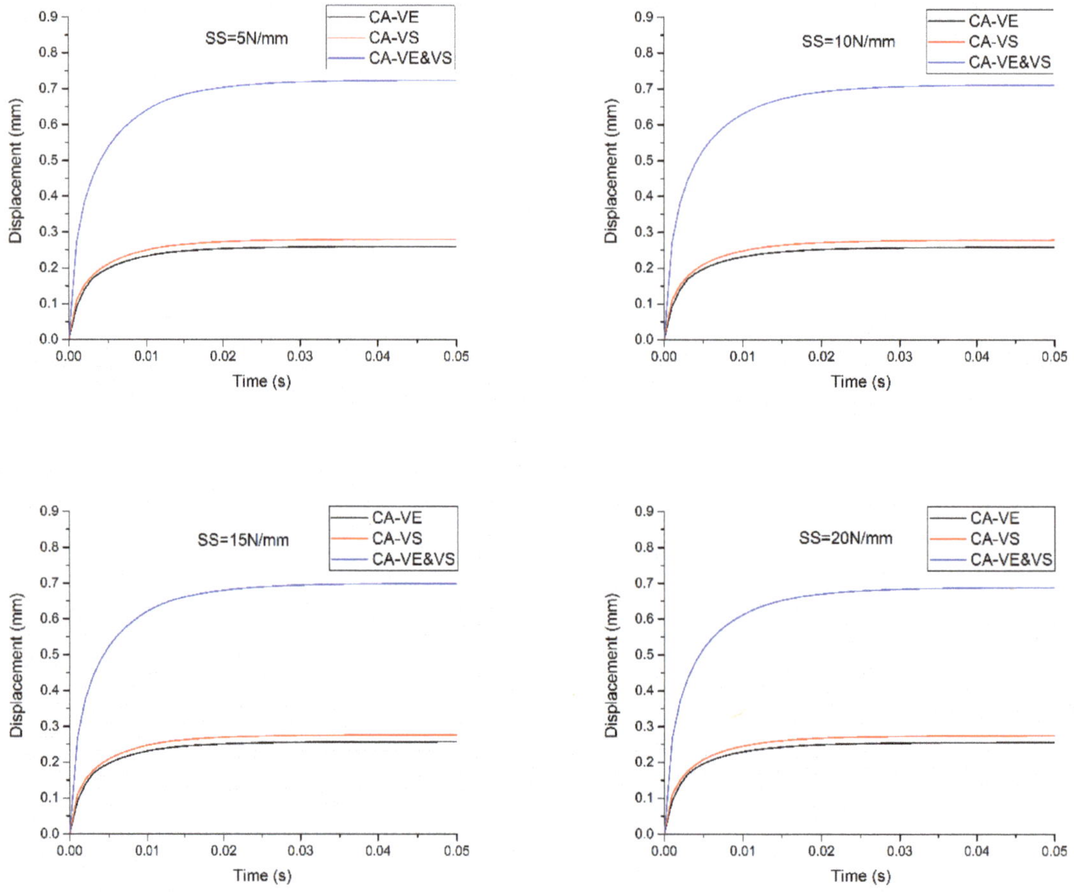

Figure 18. The displacement response of the three kinds of CTDARV: SS = 5 N/mm–20 N/mm.

4.3.5. Effect of CAVE&CAVS on Displacement Response

Figure 19 shows the effect of CAVE&CAVS (15 degrees–60 degrees) on the displacement response of the three kinds of CTDARV. It can be clearly seen from Figure 19 that when the CAVE&CAVS are 15 degrees–60 degrees, respectively, the three kinds of CTDARV will eventually reach their respective stable displacement. The stable displacement of CA-VE&VS is the highest, and that of CA-VE is the lowest. The stable displacement of CA-VE&VS is significantly higher than that of CA-VE and CA-VS. The difference between the stable displacement of CA-VE and CA-VS is not very large. With the increase of CAVE&CAVS from 15 degrees to 60 degrees, the stable displacement of CA-VE, CA-VS, and CA-VE&VS decreased, but the decrease was not significant. Specifically, when CAVE&CAVS are 15 degrees, 30 degrees, 45 degrees, and 60 degrees, respectively, the stable displacement of CA-VE is 0.6665 mm, 0.3551 mm, 0.2587 mm, and 0.2188 mm, respectively; the stable displacement of CA-VS is 0.7372 mm, 0.3896 mm, 0.2791 mm, and 0.2309 mm, respectively; and the stable compressive displacement of CA-VE&VS is 1.8701 mm, 1.0022 mm, 0.7121 mm, and 0.5782 mm, respectively.

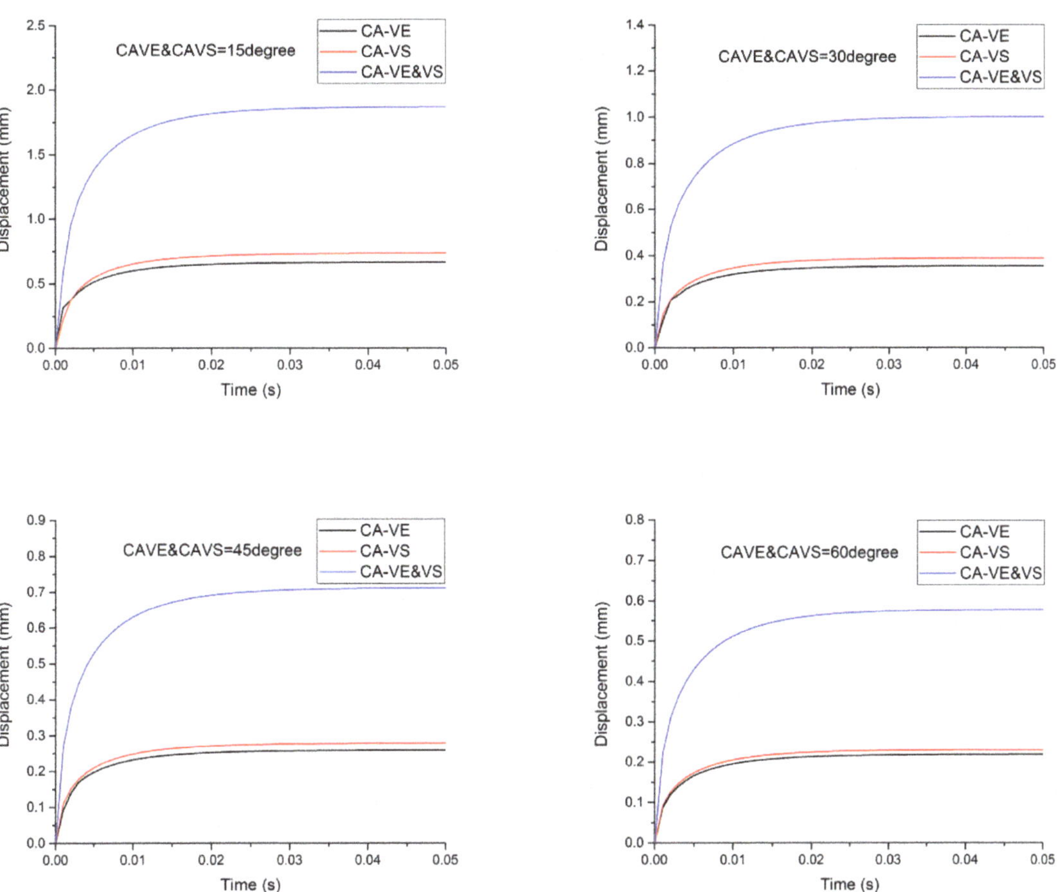

Figure 19. The displacement response of the three kinds of CTDARV: CAVE&CAVS = 15 degrees–60 degrees.

4.3.6. Effect of OD on Displacement Response

Figure 20 shows the effect of OD (0.8 mm–1.4 mm) on the displacement response of the three kinds of CTDARV. It can be clearly seen from Figure 20 that when the OD is 0.8 mm–1.4 mm, the three kinds of CTDARV will eventually reach their respective stable displacement. The stable displacement of CA-VE&VS is the highest, and that of CA-VE is the lowest. The stable displacement of CA-VE&VS is significantly higher than that of CA-VE and CA-VS. The difference between the stable displacement of CA-VE and CA-VS is not very large. As the OD increases from 0.8 mm to 1.4 mm, the initial displacement of CA-VE oscillates gradually, and the number of oscillations increases gradually, while the displacement of CA-VS and CA-VE&VS does not oscillate. With the OD increasing from 0.8 mm to 1.4 mm, the stable displacement of CA-VE, CA-VS, and CA-VE&VS remained unchanged. Specifically, when the OD is 0.8 mm, 1.0 mm, 1.2 mm, and 1.4 mm, respectively, the stable displacement of CA-VE is 0.2587 mm; the stable displacement of CA-VS is 0.2791 mm; and the stable pressure displacement of CA-VE&VS is 0.7121 mm.

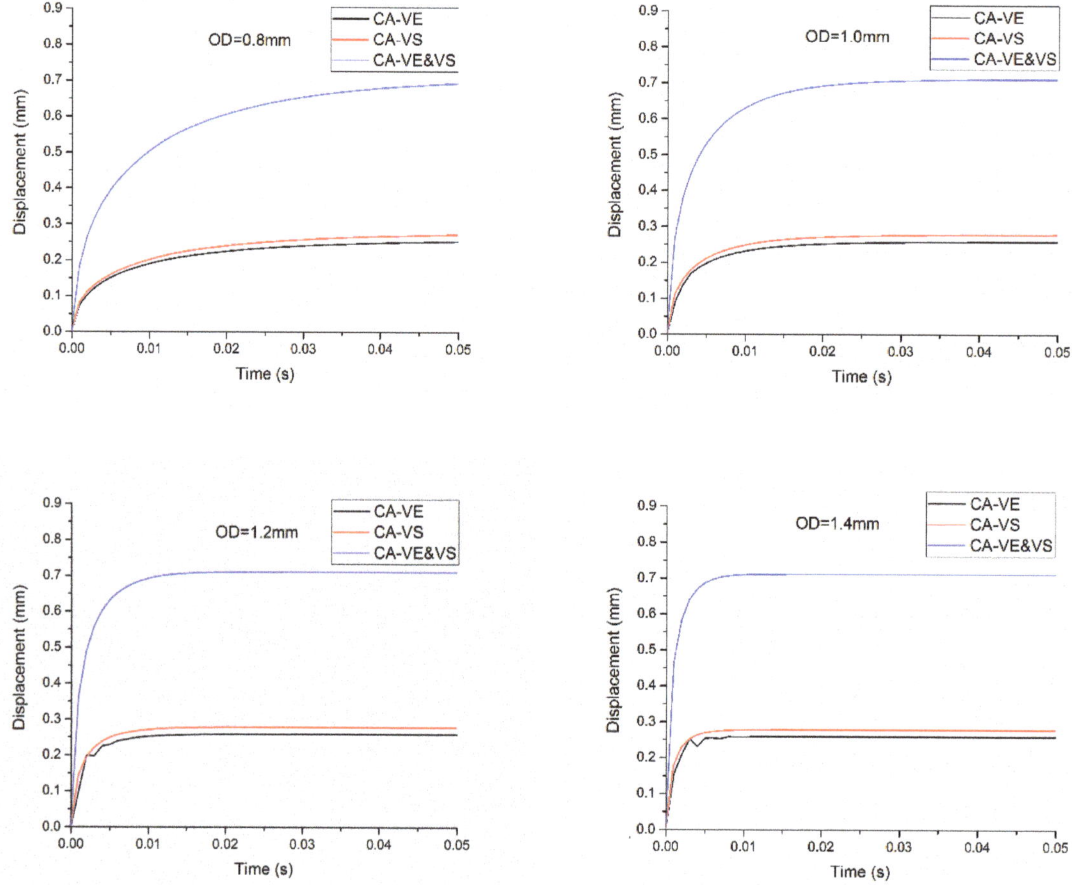

Figure 20. The displacement response of the three kinds of CTDARV: OD = 0.8 mm–1.4 mm.

4.4. Velocity Performance Comparisons of the Three Kinds of CTDARV

4.4.1. Effect of VED on Velocity Response

Figure 21 shows the effect of VED (13 mm–16 mm) on the velocity response of the three kinds of CTDARV. It can be clearly seen from Figure 21 that when the VED is 13 mm–16 mm, the three kinds of CTDARV will eventually reach the speed of 0 m/s. With the increase of VED from 13 mm to 16 mm, the speed of CA-VE remains unchanged, while the speed of CA-VS and CA-VE&VS increases, but the increase is not significant. The speed of CA-VE, CA-VS, and CA-VE&VS reaches the maximum value in 0.001 s. Specifically, when the VED is 13 mm, 14 mm, 15 mm, and 16 mm, respectively, the speed of CA-VE is 0.2380 m/s; the speed of CA-VS is 0.0555 m/s, 0.0561 m/s, 0.0572 m/s and 0.0591 m/s, respectively; and the speed of CA-VE&VS is 0.1331 m/s, 0.1405 m/s, 0.1487 m/s, and 0.1574 m/s, respectively.

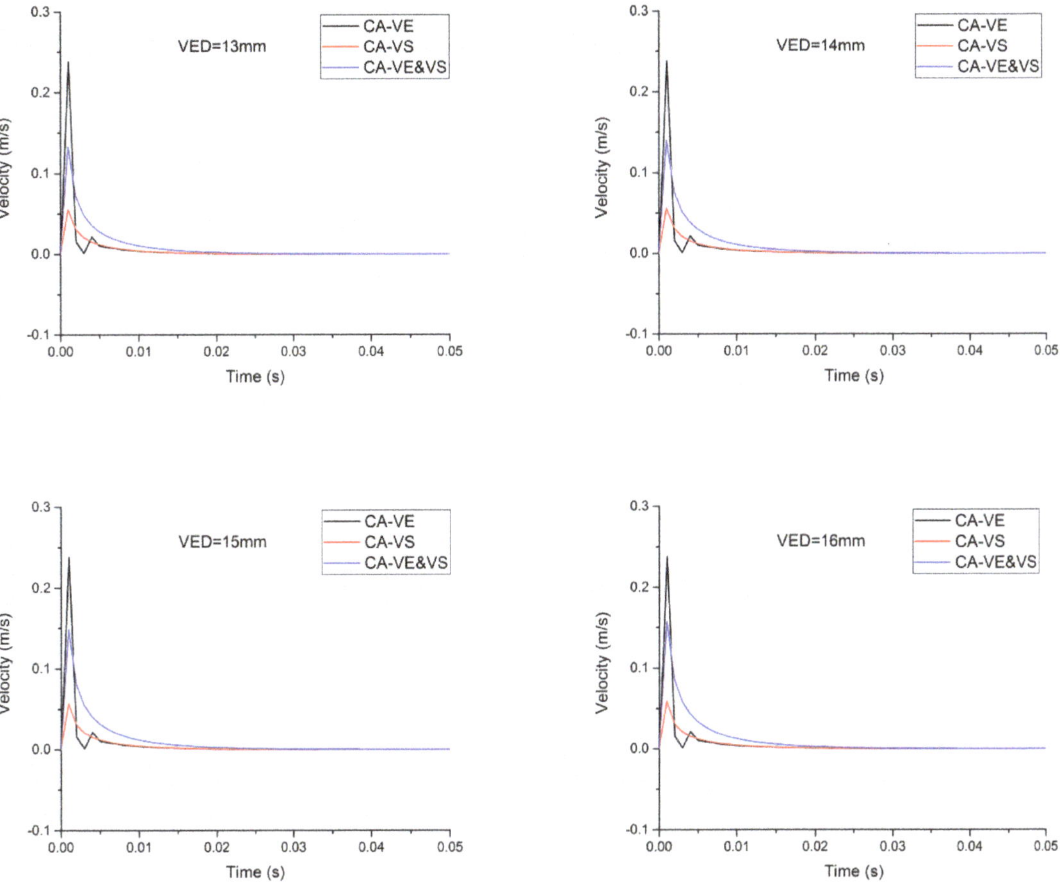

Figure 21. The displacement response of the three kinds of CTDARV without orifice. VED = 13 mm–16 mm.

4.4.2. Effect of VSD on Velocity Response

Figure 22 shows the effect of VSD (3 mm–6 mm) on the velocity response of the three kinds of CTDARV. It can be clearly seen from Figure 22 that when the VSD is 3 mm–6 mm, the three CTDARVs will eventually reach the speed of 0 m/s. When the VSD is 3 mm, the speed of CA-VS changes sharply, and the maximum speed reaches 1.8760 m/s. As VSD increases from 4 mm to 6 mm, the speed of CA-VS remains unchanged. As VSD increases from 4 mm to 6 mm, the speed of CA-VE and CA-VE&VS decreases. As the VSD increases from 4 mm to 6 mm, the speeds of CA-VE, CA-VS and CA-VE&VS all reach the maximum value in 0.001 s. Specifically, when the VSD are 4 mm, 5 mm, and 6 mm, respectively, in 0.001 s, the speeds of CA-VE are 0.2684 m/s, 0.2380 m/s, and 0.1180 m/s; CA-VS are 0.0572 m/s; and CA-VE&VS are 0.2013 m/s, 0.1487 m/s, and 0.1231 m/s, respectively.

4.4.3. Effect of VEM on Velocity Response

Figure 23 shows the effect of VEM (0.01 kg–0.04 kg) on the velocity response of the three kinds of CTDARV. It can be clearly seen from Figure 23 that when the VEM is 0.01 kg–0.04 kg, the three kinds of CTDARV will eventually reach the speed of 0 m/s. With the increase of VEM from 0.01 kg to 0.04 kg, the speed oscillation of CA-VE becomes more and more intense, and the speed of CA-VS and CA-VE&VS changes little.

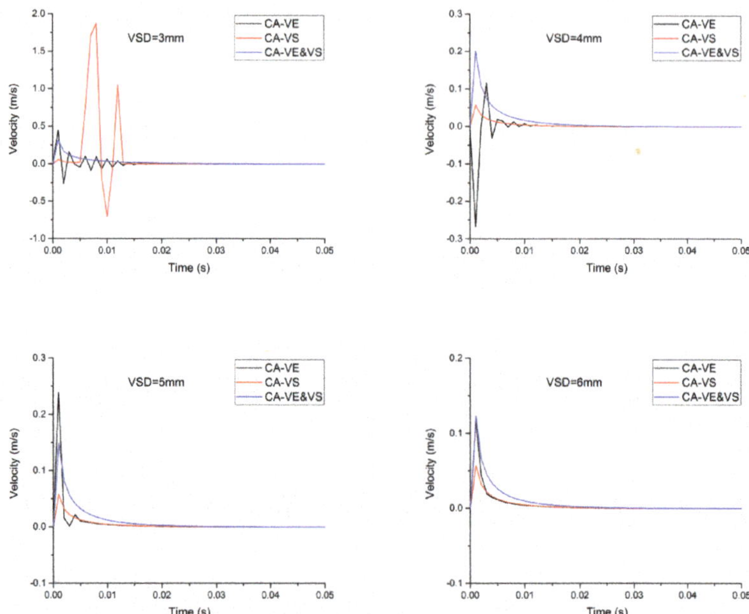

Figure 22. The velocity response of the three kinds of CTDARV: VSD = 3 mm–6 mm.

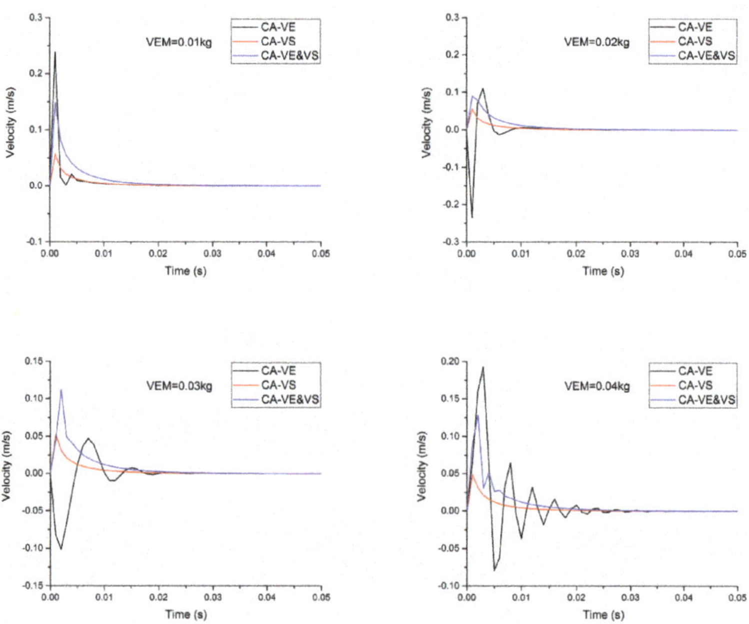

Figure 23. The velocity response of the three kinds of CTDARV: VEM = 0.01 kg–0.04 kg.

4.4.4. Effect of SS on Velocity Response

Figure 24 shows the effect of SS (5 N/mm–20 N/mm) on the velocity response of the three kinds of CTDARV. It can be clearly seen from Figure 24 that when SS is 5 N/mm–20 N/mm, the three CTDARVs will eventually reach the speed of 0 m/s and have similar speed curves. As SS increases from 5 N/mm to 20 N/mm, the speed of CA-VE increases, while the speed of CA-VS and CA-VE&VS decreases, but the increase and decrease are not significant. Specifically, when SS is 5 N/mm, 10 N/mm, 15 N/mm, and 20 N/mm, the speed of CA-VE is 0.2354 m/s, 0.2380 m/s, 0.2403 m/s, and 0.2425 m/s, respectively; the speed of CA-VS is 0.0576 m/s, 0.0572 m/s, 0.0569 m/s, and 0.0566 m/s, respectively; and the speed of CA-VE&VS is 0.1506 m/s, 0.1487 m/s, 0.1468 m/s, and 0.1450 m/s.

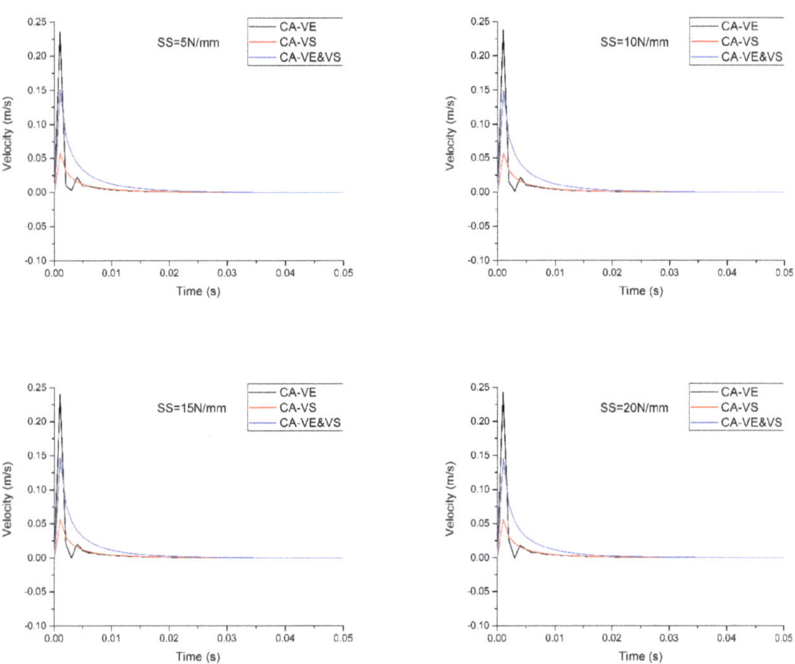

Figure 24. The velocity response of the three kinds of CTDARV: SS = 5 N/mm–20 N/mm.

4.4.5. Effect of CAVE&CAVS on Velocity Response

Figure 25 shows the effect of CAVE&CAVS (15 degrees–60 degrees) on the velocity response of the three kinds of CTDARV. It can be clearly seen from Figure 25 that when the CAVE&CAVS are 15 degrees–60 degrees, the three kinds of CTDARV will eventually reach a speed of 0 m/s. At 0.001 s, when CAVE&CAVS is 15 degrees and 30 degrees, the speed of CA-VE is lower than that of CA-VS and CA-VE&VS. At 0.001 s, when CAVE&CAVS is 45 degrees and 60 degrees, the speed of CA-VE is higher than that of CA-VS and CA-VE&VS. Specifically, when CAVE&CAVS is 15 degrees, 30 degrees, 45 degrees, and 60 degrees, the maximum speed of CA-VE is 0.2049 m/s, 0.0410 m/s, 0.2380 m/s, and 0.2242 m/s, and the maximum speed of CA-VS is 0.2077 m/s, 0.0902 m/s, 0.0572 m/s, and 0.0451 m/s, and the maximum speed of CA-VE&VS is 0.5350 m/s, 0.2289 m/s, 0.1487 m/s, and 0.1212 m/s.

4.4.6. Effect of OD on Velocity Response

Figure 26 shows the effect of OD (0.8 mm–1.4 mm) on the velocity response of the three kinds of CTDARV. It can be clearly seen from Figure 26 that when the OD is 0.8 mm–1.4 mm, the three kinds of CTDARV will eventually reach the speed of 0 m/s. With the increase of OD from 0.8 mm to 1.4 mm, the velocity oscillation of CA-VE becomes more and more intense, and the velocity of CA-VS and CA-VE&VS changes little.

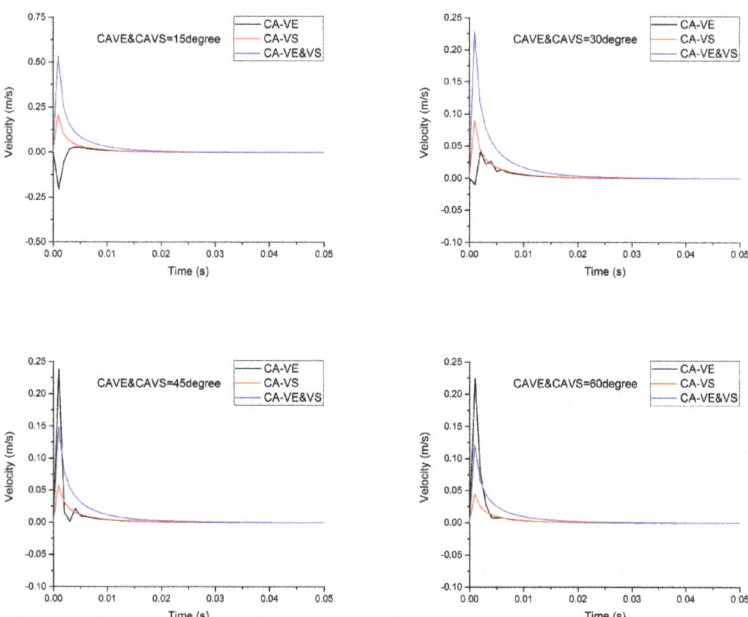

Figure 25. The velocity response of the three kinds of CTDARV: CAVE&CAVS = 15 degrees–60 degrees.

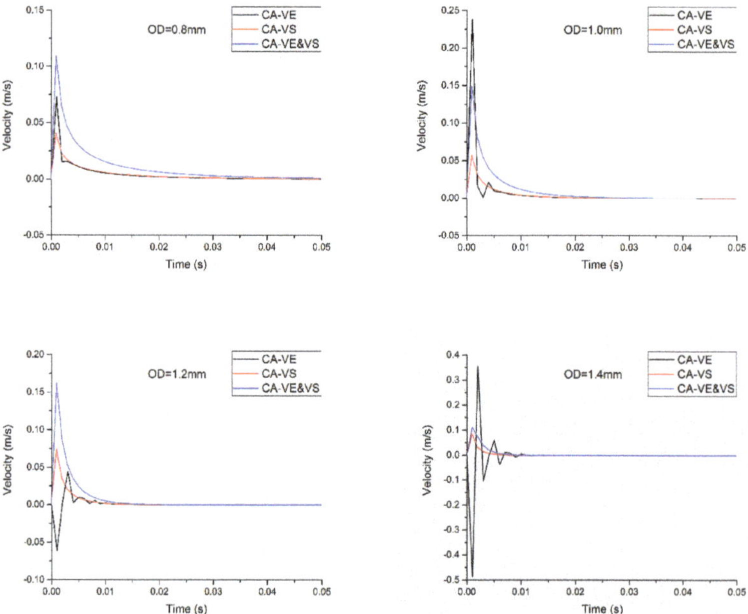

Figure 26. The velocity response of the three kinds of CTDARV: OD = 0.8 mm–1.4 mm.

5. Conclusions and Future Work

Based on the working principles of the three kinds of CTDARV, the simulation models of the three kinds of CTDARV are established by utilizing AMESIM. The numerical experiments on the three kinds of CTDARV are conducted and the performance comparisons of the three kinds of CTDARV are obtained, and the following conclusions are obtained:

(1) When the values of VED, VSD, VEM, SS, CAVE&CAVS, and OD are the same, the three kinds of CTDARV will eventually reach their respective stable pressure and stable displacement and will eventually reach the same stable flowrate and velocity. When all parameters have the same value, CA-VE has the highest stable pressure, CA-VE&VS has the highest stable displacement, CA-VS has the lowest stable pressure, and CA-VE has the lowest stable displacement. The stable pressure of CA-VE is significantly higher than that of CA-VS and CA-VE&VS. The stable displacement of CA-VE&VS is significantly higher than that of CA-VE and CA-VS, and the stable displacement of CA-VE and CA-VS has little difference.

(2) With the increase of the VED from 13 mm to 16 mm, the stable pressure of CA-VE remains constant, while the stable pressure of CA-VS and CA-VE&VS both decreases. As the VSD increases from 3 mm to 6 mm, the stable pressure of CA-VE and CA-VE&VS decreases, and the stable pressure of CA-VE decreases significantly. The pressure of CA-VS and CA-VE&VS does not fluctuate, and the pressure oscillation of CA-VE decreases gradually. With the increase of VEM from 0.01 kg to 0.04 kg, the stable pressure of CA-VE, CA-VS and CA-VE&VS remained unchanged, the pressure of CA-VS and CA-VE&VS did not fluctuate, and the pressure oscillation of CA-VE gradually increased. With the increase of SS from 5 N/mm to 20 N/mm, the stable pressure of CA-VE, CA-VS, and CA-VE&VS increased, but the increase was very small. With the increase of CAVE&CAVS from 15 degrees to 60 degrees, the stable pressure of CA-VE, CA-VS, and CA-VE&VS decreased, but the decrease was not significant. With the OD increasing from 0.8 mm to 1.4 mm, the stable pressure of CA-VE, CA-VS, and CA-VE&VS remained unchanged, the pressure of CA-VS and CA-VE&VS did not fluctuate, and the pressure oscillation of CA-VE gradually increased.

(3) With the increase of the VED from 13 mm to 16 mm, the stable displacement of CA-VE does not change, while the stable displacement of CA-VS and CA-VE&VS increases. As the VSD increases from 3 mm to 6 mm, the stable displacement of CA-VE increases (though the increase is small), while the stable displacement of CA-VE&VS decreases. When the VSD is 4 mm–6 mm, the stable displacement of CA-VS remains unchanged. With the increase of the VEM from 0.01 kg to 0.04 kg, the stable displacement of CA-VE, CA-VS, and CA-VE&VS remained unchanged. As the SS increases from 5 N/mm to 20 N/mm, the stable displacement of CA-VE, CA-VS, and CA-VE&VS decreases. As CAVE&CAVS increases from 15 degrees to 60 degrees, the stable displacement of CA-VE, CA-VS, and CA-VE&VS decreases. With the increase of OD from 0.8 mm to 1.4 mm, the stable displacement of CA-VE, CA-VS, and CA-VE&VS remained unchanged.

(4) With the increase of the VED from 13 mm to 16 mm, the velocity of CA-VE remained unchanged, while the velocity of CA-VS and CA-VE&VS increased. The velocity of CA-VE, CA-VS, and CA-VE&VS reached the maximum value in 0.001 s. As the VSD increases from 4 mm to 6 mm, the velocity of CA-VS remains unchanged, the velocity of CA-VE and CA-VE&VS decreases, and the velocity of CA-VE, CA-VS, and CA-VE&VS reaches the maximum value at 0.001 s. With the increase of VEM from 0.01 kg to 0.04 kg, the velocity oscillation of CA-VE gradually increases, and the velocity of CA-VS and CA-VE&VS shows little change. As SS increases from 5 N/mm to 20 N/mm, the velocity of CA-VE increases, while the velocity of CA-VS and CA-VE&VS decreases, but the increase and decrease are not significant. When CAVE&CAVS is 15 degrees and 30 degrees, the velocity of CA-VE is lower than that of CA-VS and CA-VE&VS. At 0.001 s, when CAVE&CAVS is 45 degrees and 60 degrees, the velocity of CA-VE is higher than that of CA-VS and CA-VE&VS. With the increase of OD from 0.8 mm to 1.4 mm, the velocity oscillation of CA-VE increases gradually, and the velocity of CA-VS and CA-VE&VS changes little.

In the future, we will manufacture several sets of different series of the three kinds of CTDARV and build a hydraulic test platform of the three kinds of CTDARV to verify the research results on the three kinds of CTDARV achieved in this paper.

Author Contributions: Conceptualization, H.L.; methodology, H.L.; validation, Q.Z.; formal analysis, Q.Z.; investigation, Q.Z.; data curation, H.L.; writing—original draft preparation H.L.; writing—review and editing, Q.Z.; supervision, Q.Z.; funding acquisition, H.L. All authors have read and agreed to the published version of the manuscript.

Funding: This work is supported by the National Natural Science Foundation of China [grant number 51365008], the Joint Foundation of Department of Science and Technology of Guizhou Province [grant number Qiankehe LH Zi [2015]7658], the Research Foundation of Guizhou Panjiang Coal Power Group Co., Ltd [grant number 701/700878212201], the Science and Technology Special Foundation of Department of Water Resources of Guizhou Province [grant number KT202113]. The authors are grateful to them for their support.

Data Availability Statement: The data presented in this study are available on request from the corresponding author.

Conflicts of Interest: The authors declare no conflict of interest.

References

1. Yuan, C.; Zhu, L.S.; Liu, S.Q.; Zunling, D.; Li, H. Numerical study on the cavitating flow through poppet valves concerning the influence of flow instability on cavitation dynamics. *J. Mech. Sci. Technol.* **2022**, *36*, 761–773. [CrossRef]
2. Min, W.; Li, C.; Wang, H.Y.; Zheng, Z.; Zhao, J.F.; Ji, H. Discharge coefficient of pilot poppet valve at low Reynolds number. *Flow Meas. Instrum.* **2022**, *85*, 10. [CrossRef]
3. Hao, Q.H.; Wu, W.R.; Tian, G.T. Study on reducing both flow force and cavitation in poppet valves. *Proc. Inst. Mech. Eng. Part C-J. Mech. Eng. Sci.* **2022**, *236*, 11160–11179. [CrossRef]
4. Yuan, C.; Zhu, L.S.; Du, Z.L.; Liu, S.Q. Numerical investigation into the cavitating jet inside water poppet valves with varied valve seat structures. *Eng. Appl. Comp. Fluid Mech.* **2021**, *15*, 391–412. [CrossRef]
5. Fornaciari, A.; Zardin, B.; Borghi, M.; Ceriola, M. Analysis of the flow force compensation in relief valves with conical poppet. In Proceedings of the 33rd Bath/ASME International Symposium on Fluid Power and Motion Control (FPMC), Online, 9–11 September 2020.
6. Min, W.; Wang, H.Y.; Zheng, Z.; Wang, D.; Ji, H.; Wang, Y.B. Visual experimental investigation on the stability of pressure regulating poppet valve. *Proc. Inst. Mech. Eng. Part C-J. Mech. Eng. Sci.* **2020**, *234*, 2329–2348. [CrossRef]
7. Filo, G.; Lisowski, E.; Rajda, J. Pressure Loss Reduction in an Innovative Directional Poppet Control Valve. *Energies* **2020**, *13*, 13. [CrossRef]
8. Hui, W.; Wenhua, J.; Yanru, Z.; Changli, S.; Chao, W. Dynamic Characteristics and Stability Analysis of Poppet Relief Valve. *Mach. Hydraul.* **2019**, *47*, 144–147.
9. Gomez, I.; Gonzalez-Mancera, A.; Newell, B.; Garcia-Bravo, J. Analysis of the Design of a Poppet Valve by Transitory Simulation. *Energies* **2019**, *12*, 18. [CrossRef]
10. Chiavola, O.; Frattini, E.; Palmieri, F.; Possenti, G. Poppet Valve Performance under Cavitating Conditions. In Proceedings of the 74th Conference of the Italian-Thermal-Machines-Engineering-Association (ATI), Modena, Italy, 11–13 September 2019.
11. Lei, J.B.; Tao, J.F.; Liu, C.L.; Wu, Y.J. Flow model and dynamic characteristics of a direct spring loaded poppet relief valve. *Proc. Inst. Mech. Eng. Part C-J. Mech. Eng. Sci.* **2018**, *232*, 1657–1664. [CrossRef]
12. Liu, J.B.; Xie, H.B.; Hu, L.; Yang, H.Y.; Fu, X. Flow force regulation of the main poppet in a large flow load control valve. *Proc. Inst. Mech. Eng. Part A-J. Power Energy* **2017**, *231*, 706–720. [CrossRef]
13. Ji, C.; Wang, J.R.; Mo, G.Y.; Zou, J.; Yang, H.Y. Instability of a poppet valve: Interaction of axial vibration and lateral vibration. *Int. J. Adv. Manuf. Technol.* **2018**, *94*, 3065–3074. [CrossRef]
14. Han, M.X.; Liu, Y.S.; Wu, D.F.; Zhao, X.F.; Tan, H.J. A numerical investigation in characteristics of flow force under cavitation state inside the water hydraulic poppet valves. *Int. J. Heat Mass Transf.* **2017**, *111*, 1–16. [CrossRef]
15. Bo, J.K. Multidisciplinary Design Optimization of a Hydraulic Poppet Valve Considering Fluid-solid Coupling. In Proceedings of the International Conference on Advanced Design and Manufacturing Engineering (ADME 2011), Guangzhou, China, 16–18 September 2011.
16. Wang, Q.; Luan, J.; He, X.H.; Zhou, D.J.; He, J. Transient characteristic analysis of a hydraulic poppet valve based on the dynamic mesh technique. In Proceedings of the International Conference on Mechanics and Mechatronics (ICMM), Changsha, China, 13–15 March 2015.
17. Yi, D.Y.; Lu, L.; Zou, J.; Fu, X. Interactions between poppet vibration and cavitation in relief valve. *Proc. Inst. Mech. Eng. Part C-J. Mech. Eng. Sci.* **2015**, *229*, 1447–1461. [CrossRef]
18. Min, W.; Ji, H.; Yang, L.F. Axial vibration in a poppet valve based on fluid-structure interaction. *Proc. Inst. Mech. Eng. Part C-J. Mech. Eng. Sci.* **2015**, *229*, 3266–3273. [CrossRef]
19. Zheng, S.J.; Quan, L. Optimization of Structure and Calculation of Flow-force on Poppet Valve under Converging Flow. In Proceedings of the International Conference on Materials and Products Manufacturing Technology (ICMPMT 2011), Chengdu, China, 28–30 October 2011.

20. Shi, X.; Chen, J. Numerical simulation based on CFD for the movement field of the hydraulic poppet valve. In Proceedings of the 1st International Conference on Intelligent System and Applied Material (GSAM 2012), Taiyuan, China, 13–15 January 2012.
21. Shi, J.Y. Study on the Effect of Cone Diameter on the Cone Resistance and Steady Flow Force of Hydraulic Poppet Valve. In Proceedings of the 3rd International Conference on Materials and Products Manufacturing Technology (ICMPMT 2013), Guangzhou, China, 25–26 September 2013.
22. Yi, D.Y.; Lu, L.; Zou, J.; Fu, X. Squeal noise in hydraulic poppet valves. *J. Zhejiang Univ. Sci. A* **2016**, *17*, 317–324. [CrossRef]
23. Chen, X.L.; Zhu, D.; Bao, G. Study on Static Characteristic of a Regulator Valve with Large Flow Capacity. In Proceedings of the International Conference on Fluid Power and Mechatronics FPM, Harbin, China, 5–7 August 2015.
24. Luan, J.; Wang, Q.; He, X.H.; Zeng, F.Q. Fluid-solid coupling analysis of hydraulic poppet valves based on CFD. In Proceedings of the International Conference on Mechanics and Mechatronics (ICMM), Changsha, China, 13–15 March 2015.
25. Rundo, M.; Altare, G. Comparison of Analytical and Numerical Methods for the Evaluation of the Flow Forces in Conical Poppet Valves with Direct and Reverse Flow. In Proceedings of the 72nd Conference of the Italian-Thermal-Machines-Engineering-Association (ATI), Lecce, Italy, 6–8 September 2017.
26. Guo, X.X.; Huang, J.H.; Quan, L.; Wang, S.G. Transient Flow Field Characteristic Analysis of Poppet Valve Based on Dynamic Mesh 6DOF Technique. In Proceedings of the International Conference on Fluid Power and Mechatronics FPM, Harbin, China, 5–7 August 2015.

Article

Low-Carbon and Energy-Saving Path Optimization Scheduling of Material Distribution in Machining Shop Based on Business Compass Model

Yongmao Xiao [1,2,3], Hao Zhang [4] and Ruping Wang [5,*]

[1] School of Computer and Information, Qiannan Normal University for Nationalities, Duyun 558000, China; xym198302@163.com
[2] Key Laboratory of Complex Systems and Intelligent Optimization of Guizhou Province, Duyun 558000, China
[3] Key Laboratory of Complex Systems and Intelligent Optimization of Qiannan, Duyun 558000, China
[4] College of Mechanical and Electrical Engineering, Xinjiang Agricultural University, Wulumuqi 830052, China; zhanghao_2022@126.com
[5] College of Management, China West Normal University, Nanchong 637002, China
* Correspondence: wrpqyi@163.com; Tel.: +86-13-922-899-161

Abstract: In order to reduce carbon emission and energy consumption in the process of raw material distribution, the workshop material distribution management model was established based on the business compass model; it can help guide enterprises to manage workshop production. Based on the raw material distribution equipment, a path calculation model considering the carbon emission and energy consumption in the process of raw material distribution was established. The dung beetle optimizer was selected for the optimization calculation. The dung beetle optimizer has the characteristics of fast convergence and high solution accuracy. The material distribution of an engine assembly workshop was taken as an example; the results showed that the optimized scheduling model could effectively optimize the material distribution route and reduce energy consumption and carbon emission in the distribution process on the basis of meeting the distribution demand.

Keywords: material distribution; path planning; low carbon; business compass; dung beetle optimizer

Citation: Xiao, Y.; Zhang, H.; Wang, R. Low-Carbon and Energy-Saving Path Optimization Scheduling of Material Distribution in Machining Shop Based on Business Compass Model. *Processes* **2023**, *11*, 1960. https://doi.org/10.3390/pr11071960

Academic Editor: Sergey Y. Yurish

Received: 23 May 2023
Revised: 12 June 2023
Accepted: 25 June 2023
Published: 28 June 2023

1. Introduction

The development of industrial enterprises has promoted the social economy but, at the same time, brought troubling environmental problems. Environmental and energy problems have become the focus of social development [1–3]. For manufacturing enterprises, each link has a large amount of carbon emissions, among which transportation is a link of concentrated energy consumption, and the transportation process generates a large amount of carbon dioxide [4]. An optimized transportation route plan can greatly reduce the transportation distance of production materials and reduce carbon emissions and energy consumption [5], which is of great significance for the optimization of the workshop material distribution path.

Workshop material distribution route optimization problem refers to that under certain constraints, according to the material requirements of the station, the distribution path of the car is planned so that it can carry out material distribution according to the planned path and achieve certain optimization objectives [6,7]. At present, due to the limited space in the line inventory area of some enterprises, the production of different products leads to the doubling of the types and quantities of parts, and the distribution staff distributes materials based on their experience, which leads to the unreasonable arrangement of the number of distribution cars and routes, which affects the delivery punctuality and increases the distribution cost and carbon emissions in the distribution process [8,9].

Many scholars have studied it. Yan et al. [10] took a pickup truck production assembly workshop as an example and established a material distribution optimization model with a fuzzy soft time window based on the average satisfaction of stations with the material arrival time, which improved the timeliness and accuracy of the material distribution in the workshop. Tong et al. [11] proposed a bidirectional material distribution strategy that could reflect the urgency of material demand at stations and proposed a hybrid genetic taboo search algorithm. The simulation is verified by an example of a machine parts processing workshop. Goel et al. [12] proposed a new algorithm combining firefly optimization and ant colony system algorithm for vehicle routing problems. The algorithm has been tested against two well-known routing problems, namely, the capacitated vehicle routing problem and the vehicle routing problem with a time window. Zhu et al. [13] established a distribution route optimization model with time window constraints and adopted a genetic algorithm based on tabu search to solve it. Taking a refrigerator assembly line as an example, the feasibility and effectiveness of the proposed method were verified. Erdodu et al. [14] studied the green vehicle routing problem, aiming to minimize the total distance and the total fuel consumption of all vehicle routes. The adaptive large neighborhood search was hybridized with two new local search heuristic methods, which were verified by comparison with other similar literature. Jie et al. [15] established a vehicle routing problem model considering the addition of a soft time window and random factors and proposed a hybrid algorithm combining a scanning algorithm and an improved particle swarm optimization algorithm to solve the problem. Tarhini et al. [16] proposed and tested a super heuristic algorithm based on evolutionary cuckoo search considering customer priority and required vehicle routing. Ren et al. [17] studied the green vehicle routing problem of a mixed energy fleet with a time window, aiming to reduce the total emissions of air pollutants and the total delay time of all vehicles, and developed different emission functions for each air pollutant.

The above studies provide a large number of solving ideas for workshop material distribution problems from distribution methods, distribution models, and algorithm models, and most of them use mathematical models combined with intelligent algorithms to optimize distribution. Although the above research has made some progress, there are still some problems, such as poor solution quality, easy falls into local optimal, and no consideration of environmental factors. Therefore, this paper combined the management of business compass philosophy and analysis of workshop path planning of various elements. The workshop material distribution management model based on the business compass was established to provide guidance for workshop production management. Considering the environmental issues of material distribution, carbon emissions, and energy consumption in the transportation process were taken as objectives, a multi-objective workshop material distribution route optimization model was established. The dung beetle algorithm was used to solve the problem, which has higher accuracy and stronger searchability. An engine assembly workshop was selected as an example for verification, three algorithms DBO, non-dominated sorting genetic algorithm-II (NSGA-II), and genetic algorithm (GA), were used to solve the problem, and the optimal path was obtained by comparison. The results showed that the model could reduce carbon emissions and energy consumption in the process of material transportation more effectively.

2. Workshop Material Distribution Based on Business Compass

2.1. Business Compass

Business compass is an enterprise sustainable development top-level design model that integrates Chinese and Western management thinking [18]. Management compass integrates Chinese philosophy and Western management science organically. Based on the people-oriented development concept and the five elements of "wood, fire, earth, gold and water" in ancient Chinese philosophy [19], it puts forward five core elements of "trend, path, skill, tool and profit" for the sustainable development of enterprises, as shown in Figure 1, and progressive construction of policy, industry, market, mission, vision, values, strategy, tactics, organization, technology, products, services, internal profit, external profit, society, and other 15 plates.

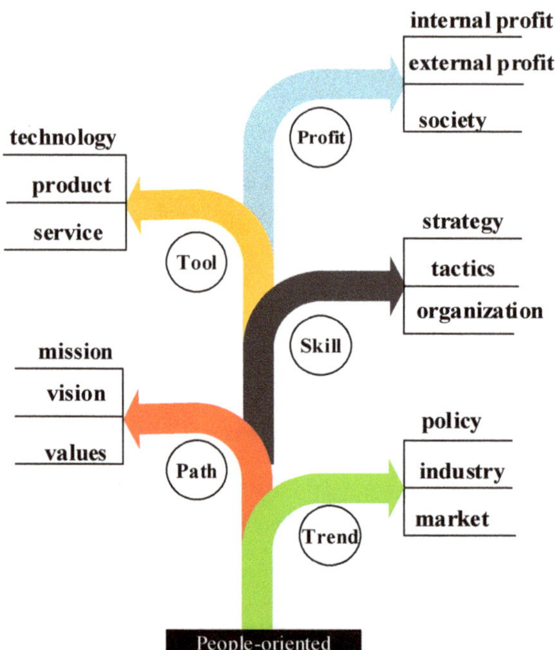

Figure 1. Business compass diagram.

Business compass consists of five elements of "trend, path, skill, tool and profit". The "trend" of the business compass refers to the situation faced by the enterprise, which mainly includes the relevant laws and regulations, economic development trends, and market demand changes. "Trend" has an important impact on the development of enterprises. Only by conforming to the changes in the situation can enterprises seize new development opportunities. The "path" of the business compass refers to the development direction of the enterprise. According to the changes in the situation, the enterprise timely adjusts the development direction and carries out internal reform. The "skill" of the business compass is a series of tactics and strategies put forward by enterprises according to the direction of development. The "tool" of the business compass means the new products and technologies developed by enterprises. Innovation is the primary productive force, and only by mastering core technologies can we achieve further development. The "profit" of the business compass refers to the profit distribution and responsibility distribution of the enterprise. A good profit distribution system can stimulate the enthusiasm of employees and better serve the enterprise [20–22].

Because the continuous distribution and circulation of "profit" will form a game between those with vested interests and those with new needs of interests and then produce a new external environment, that is, the formation of new potential. Just like the relationship between "wood, fire, earth, metal and water" among the five elements of Chinese philosophy, the five elements of "trend, path, skill, tool and profit" of business compass also form a circular ecological relationship with each other, influencing and promoting each other.

2.2. Workshop Material Distribution Management Based on Business Compass

It is of great significance for enterprises to reasonably plan the quantity and path of material transportation, ensure the timely supply of materials, and reduce transportation costs and carbon emissions. As the guiding concept of enterprise management, a business compass can provide guidance for the material distribution route planning of enterprises. Combined with the concept of the business compass, this paper proposes an optimization management model of workshop material distribution, as shown in Figure 2. The optimal management model of workshop material distribution takes enterprise management as the core and pays attention to the production management of enterprise production. The model consists of five elements, "business compass, operation management, workshop production, material distribution model, and distribution path plan". As a top-level design model of enterprise sustainable development, the business compass can provide a guiding ideology for enterprise management. The operation department is responsible for the operation management and production arrangement of the enterprise. Workshop production includes production plans, production equipment, and distribution equipment. The workshop arranges workshop work according to production plan requirements, reasonably plans various resources, and arranges related work of workshop equipment management. For material distribution, it is necessary to design the material distribution model according to the production situation of the workshop, determine the optimization objective, and use the optimization method to optimize the calculation of the material distribution model. Finally, the optimal route plan for material distribution in the workshop is obtained, which includes distribution vehicle, distribution time, distribution path.

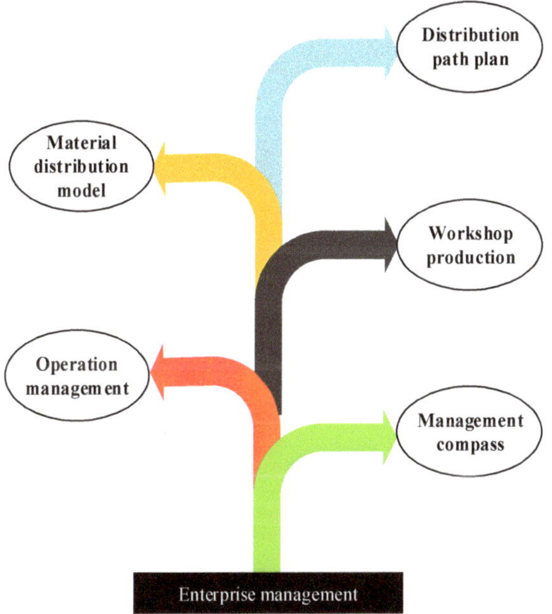

Figure 2. Workshop material distribution management model based on business compass.

3. Material Distribution Calculation Model

In this chapter, the energy consumption and loading performance of vehicles in distribution are considered comprehensively, and the comprehensive emission model is used to calculate related factors such as energy consumption and carbon emission of vehicles.

3.1. Calculation Model of Energy Consumption

In this paper, the comprehensive emission model is used to calculate vehicle energy consumption O_F from the micro point of view. On a path (i, j) with a length of d and a vehicle speed of V, the energy consumption of the vehicle driving on the road is calculated [23,24].

$$O_F(v) = \xi k N_Z V_l d/v + \xi p_c \gamma d/v \tag{1}$$

where ζ represents the mass ratio of fuel and air, k represents the friction coefficient of the engine, N_z represents the rotational speed of the vehicle engine, V_l represents the vehicle displacement, γ is a constant, p_c represents the total power required per vehicle mileage (kw), which is calculated as shown in the formula without considering the loss of power.

$$p_c = p_q/n_{tf} \tag{2}$$

where p_q is the total traction power of the vehicle per unit driving time. n_{tf} represents the efficiency of the vehicle's drivetrain. The simplification is shown below,

$$p_q = \left(Qv\alpha + \beta v^3\right)/1000 \tag{3}$$

where v is the actual driving speed of the vehicle (m/s), and Q is the total weight of the current vehicle (kg, dead weight, loaded weight). Under the condition of considering the path angle of 0, $\alpha = a + g\sin\theta + gCr\cos\theta$, among them, a is the acceleration (m/s^2), $\beta = 50C_dA\rho$, θ is the road angle, A is the vehicle in front of the surface area (m^2), g is the gravitational constant (9.8 m/s^2), C_d is the air drag coefficient, C_r is rolling resistance coefficient, and ρ is air density (kg/m^3). The energy consumption (unit, J) of the vehicle in the path (i,j) is obtained, and the specific calculation is shown as follows

$$Q_{F,ij}(Q_c + q_{ij}, v_{ij}, d_{ij}) = \lambda(kN_zV_l + (Q_c + q_{ij})\gamma\alpha_{ij}v + \beta\gamma v^3)d_{ij}/v_{ij} \tag{4}$$

where $\lambda = \zeta/\kappa\psi$, $\gamma = 1/1000n_{tf}\eta$ are constants, ψ represents the conversion factor of energy from g/s to 1/s. d_{ij} is the path length, v_{ij} is the average speed, and the load of the vehicle is $Q = Q_c + q_{ij}$, where Q_c is the dead weight of the vehicle and q_{ij} is the load of the vehicle on the path (i, j).

It can be seen from Formula (4) that vehicle energy consumption consists of three parts, the first part is cylinder energy consumption, the second part is mass energy consumption, which is influenced by the driving speed and load, and the third part is air resistance energy consumption. The first and third parts are affected by running time and speed, while the second part is affected by load and running time. The relevant data on diesel engines quoted in this paper are shown in Table 1 [25].

Table 1. Datasheet related to diesel-powered vehicles.

Symbol	Description	Reference Value
ζ	fuel–air mass ratio	1
k	engine friction coefficient	0.2
g	weight constant	9.8
C_d	air drag coefficient	0.7
ρ	air density	1.2041
C_r	rolling resistance coefficient	0.01
n_{tf}	vehicle drivetrain efficiency	0.4
η	diesel engine efficiency parameters	0.9
θ	road angle	0
f_c	unit fuel and CO_2 emission costs	7.99
κ	diesel fuel calorific value	44
ψ	conversion coefficient	737

3.2. Carbon Emissions Calculation Model

Vehicle carbon emissions are generally directly related to vehicle energy consumption. In addition to CO_2, methyl burning (CH_4) and nitrogen oxide (N_2O) also have a great impact on climate. In order to accurately calculate the number of greenhouse gases emitted by vehicles, all greenhouse gases generated in the process of vehicle speech should be converted into CO_2 equivalent according to a certain conversion factor for calculation [26,27].

In this paper, emissions are calculated by fuel consumption. Direct conversion of fuel to CO_2 equivalent only requires multiplying the amount of fuel consumed by the vehicle with the corresponding conversion factor.

$$E_c = \delta_c \cdot O_F \tag{5}$$

where δ_c is the corresponding emission factor of fuel, O_F is the amount of fuel used. All the vehicles used in this paper are diesel-powered cars, and the carbon emission is CO_2 equivalent calculated comprehensively, including concentrated greenhouse gases. The conversion coefficient of diesel based on IPCC2006 is 2.73 $KgCO_2/L$.

3.3. Model Assumptions and Constraints

When establishing the multi-objective balance model, it is necessary to list some assumptions that make the model valid: (1) there is only one material distribution center; (2) the location of the material distribution center and each working place is known; (3) the demand for materials in each working place is known; (4) the vehicle is of one type, and the loading capacity is known; the vehicle shall not exceed its loading capacity when distributed to the working place; (5) the needs of each workplace must be met.

According to the above assumptions, the decision variables for multi-objective optimization of the material distribution route are set as follows: $x_{ijm} = 1$, which means that vehicle (deliverer) m arrives from the required work place i to the required work place j; $x_{ijm} = 0$, indicating that the vehicle (distribution personnel) m did not arrive from the demand work place i to the demand work place j. The constraints of multi-objective optimization of the material distribution path are as follows:

(1) There is only one vehicle for distribution in each required working place.

$$\sum_{m=1}^{M} y_{im} = 1 \tag{6}$$

where i is the number of working places, $i = 1, 2 \ldots N$; m is the number of distribution vehicles (distribution personnel), $m = 1, 2 \ldots M$; M is the number of operators responsible for the task of material distribution. Each work place can be assigned to any one operator, and each distribution personnel has a material truck with the same capacity to distribute materials to N work places. y_{im} distributes the number of vehicles (distribution personnel) for each required work location.

(2) The material demand on each path should not exceed the carrying weight of the vehicle.

$$\sum_{i=1}^{N} g_i y_{im} \leq Q \tag{7}$$

where g_i is the demand of each required working place, and Q is the maximum load of the vehicle.

(3) Distribution vehicles (distribution personnel) leave from the distribution center.

$$\sum_{m=1}^{M} y_{0m} = 1 \tag{8}$$

(4) The distribution vehicles (distribution personnel) return to the distribution center after finishing the distribution.

$$\sum_{m=1}^{M} \sum_{i=1}^{N} x_{i0m} = M \tag{9}$$

4. Dung Beetle Optimizer

Dung beetles feed on the dung of animals. The dung beetle likes to roll feces into balls, roll them to reliable places to hide, and then slowly eat them. Dung beetles can roll a ball of dung much larger than themselves and use celestial bodies (the sun, moon, etc.) to navigate, making the ball roll in a straight line. When left in total darkness, however, the beetle's path is no longer straight but curved and sometimes rounded. Many other factors, such as uneven ground and wind, can cause the beetles to deviate from their original direction. In addition, dung beetles are likely to encounter obstacles as they roll, preventing them from moving forward. To do this, dung beetles usually climb on top of the dung ball and dance (involving a series of spins and pauses), which determines their direction of travel [28,29].

Dung beetles can be observed in the way of life for dung has the following two main purposes: (1) some dung is used to lay their eggs and raise the next generation; (2) the rest is used for food. Specifically, dung beetles bury balls of dung in which females lay their eggs. It is important to note that the dung ball is not only a place for the larvae to grow but also provides the food necessary for the larvae to survive. Therefore, dung balls play an irreplaceable role in the survival of dung beetles. In addition, some dung beetles steal dung balls from other dung beetles, called thieves, which is also a common phenomenon in nature. Researchers mainly proposed a dung beetle optimizer (DBO) for global search and local utilization based on the above living habits of dung beetles and inspired by their rolling, dancing, foraging, stealing, and reproduction behaviors [30].

4.1. Dung Beetles Roll a Ball

(1) When there is no obstacle in front of dung beetles, dung beetles will navigate according to the light source during dung ball rolling. Assuming that the intensity of the light source will affect the position of dung beetles, the position of dung beetles during dung ball rolling will be updated as follows:

$$x_i(t+1) = x_i(t) + \alpha \times k \times x_i(t-1) + b \times \Delta x$$
$$\Delta x = |x_i(t) - X^w| \tag{10}$$

where t represents the current iteration number, $x_i(t)$ represents the position information of the ith dung beetle in the t iteration, $k \in (0, 0.2]$ is the disturbance coefficient, b is the random number between $(0, 1)$, α is -1 or 1, X^w represents the global worst position and Δx is used to simulate the change of light intensity.

(2) When a dung beetle has an obstacle in front of it, it needs to dance to realign itself to get a new route. The tangent function is used to simulate the dancing behavior so as to obtain a new rolling direction, which is only considered between $[0, \pi]$. Once the beetle has successfully determined its new direction, it should continue to roll the ball backward. The beetle's position is updated as follows:

$$x_i(t+1) = x_i(t) + tan(\theta)|x_i(t) - x_i(t-1)| \tag{11}$$

where θ is the deflection angle, and its value is $[0, \pi]$.

4.2. Dung Beetle Reproduction

Female dung beetles provide a safe environment for their offspring by rolling the ball to a safe place to lay eggs and hiding it. Inspired by this, we proposed a boundary selection strategy that mimics the area where females lay their eggs:

$$Lb^* = max(X^* \times (1 - R), Lb)$$
$$Ub^* = min(X^* \times (1 + R), Ub) \tag{12}$$

where X^* represents the current optimal position, Lb^* and Ub^* represent the lower limit and upper limit of the spawning area, respectively, T_{max} represents the maximum number of iterations, $R = 1 - t/T_{max}$, Lb, and Ub represent the lower limit and upper limit of the optimization problem, respectively.

Once female dung beetles have identified a spawning area, they choose that area to raise their young and lay eggs. Each female dung beetle produced only one egg in each iteration. It can be seen that the boundary range of the spawning area was dynamic and mainly determined by the R-value. Therefore, the position of the brood ball is also dynamic during iteration, which is defined as follows:

$$B_i\,(t+1) = X^* + b_1 \times (B_i\,(t) - Lb^*) + b_2 \times (B_i\,(t) - Ub^*) \tag{13}$$

where $B_i(t)$ represents the position information of the ith brood ball in the t iteration, b_1 and b_2 are $1 \times D$ random vectors, and D represents the dimension of the optimization problem.

4.3. Dung Beetle Foraging

The eggs laid by females will gradually grow, and some mature young dung beetles will come out of the ground in search of food. The optimal feeding area of young dung beetles is modeled as follows:

$$\begin{aligned} Lb^b &= max(X^b \times (1 - R), Lb) \\ Ub^b &= min(X^b \times (1 + R), Ub) \end{aligned} \tag{14}$$

where X^b represents the global optimal location, Lb^b and Ub^b represent the lower limit and upper limit of the optimal foraging area, respectively.

The position of the young beetle is updated as follows:

$$x_i\,(t+1) = x_i\,(t) + C_1 \times (x_i\,(t) - Lb^b) + C_2 \times (x_i\,(t) - Ub^b) \tag{15}$$

where $x_i(t)$ represents the position of the ith young dung beetle in the t iteration, C_1 is a random number obeying normal distribution, and C_2 is a random vector $(0, 1)$.

4.4. Dung Beetle Stealing

Some dung beetles steal dung balls from other dung beetles. The location of the thief dung beetle is updated as follows:

$$x_i\,(t+1) = X^b + S \times g \times \left(\left|xi\,(t) - X^*\right| + \left|x_i\,(t) - X^b\right|\right) \tag{16}$$

where represents the position of the ith thief dung beetle in the t iteration, g is a $1 \times D$ random vector obeying normal distribution, and S is a constant.

The population was divided into different roles according to the ratio of 6:6:7:11. The proportion of rolling dung beetles is 6, that of breeding dung beetles is 6, that of young dung beetles is 7, and that of thieves is 11.

As a novel swarm intelligent optimization algorithm, DBO mainly has six steps:

(1) Initialize the parameters of the dung beetle population and DBO;
(2) Calculate the fitness values of all target agents according to the objective function;
(3) Update the position of all dung beetles;
(4) Determine whether each target agent exceeds the boundary;
(5) Update the current optimal solution and its fitness value;
(6) Repeat the above steps until t meets the termination criteria, and output the optimal global solution and its fitness value.

DBO is based on subpopulations, each of which performs a different search. Unlike snake optimization (SO) and Ephemera algorithm (MA), the DBO algorithm is not based on two populations but on multiple subpopulations, four of which are divided by the authors. Different regional search strategies (including spawning areas and optimal foraging areas) can promote the utilization behavior of the DBO algorithm. The algorithm pursues a stronger search ability to avoid falling into local optimal. The DBO flow chart is shown in Figure 3.

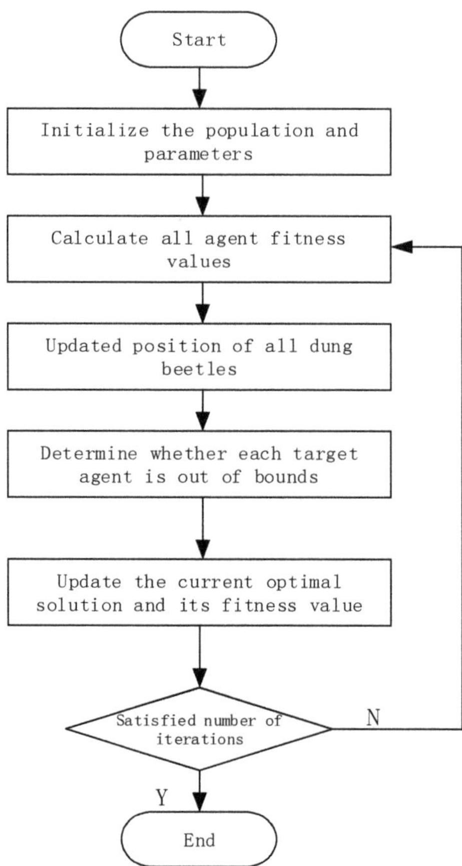

Figure 3. DBO flow chart.

The process of DBO includes initializing the dung beetle population, calculating the fitness value, and updating the position of all dung beetles. The update of dung beetle location is the focus of the algorithm, and DB is divided into four subgroups based on the dung beetle's ball-rolling, foraging, stealing, and reproduction behaviors, each of which performs a different search method based on their behavior. Because of its unique population division and search method, the DBO algorithm takes into account both global exploration and local development and has the characteristics of fast convergence and high solution accuracy.

5. Case Study

5.1. Material Distribution in Workshop

An enterprise is an engine assembly manufacturing enterprise. The materials needed for the production of the workshop are stored in a warehouse center, with a total of eight stations responsible for production. Table 2 shows the distance from the material storage warehouse to each station and the distance between each station. Table 3 shows the material demand list of each station, including the service time of vehicles at each station in the process of material distribution, the time waiting for loading and unloading of materials, and the time window limit of each station. It is stipulated that the maximum load of the vehicle responsible for material distribution is 70, and the maximum speed is not more than 50. Matlab programming is used to solve the above model.

Table 2. Distance between stations and between stations and material supermarket.

Station Number	0	1	2	3	4	5	6	7	8
0	0	40	60	75	90	200	100	160	80
1	40	0	65	40	100	50	75	110	100
2	60	65	0	75	100	100	75	75	75
3	75	40	75	0	100	50	90	90	150
4	90	100	100	100	0	100	75	75	100
5	200	50	100	50	100	0	70	90	75
6	100	75	75	90	75	70	0	70	100
7	160	110	75	90	75	90	70	0	100
8	80	100	75	150	100	75	100	100	0

Table 3. Station material requirement list.

Distribution Requirement	Station Number							
	1	2	3	4	5	6	7	8
Material requirement	20	25	15	23	16	32	20	10
Service time	1	0.5	1	1	1	1.5	1	0.8
Waiting time	0.8	0.9	1	0.5	0.5	1	0.8	0.9
Time window	[6, 7]	[5, 7]	[1, 3]	[4, 7]	[3, 5]	[2, 5]	[4, 6]	[1.5, 4]

5.2. Optimization Results and Analysis

This paper takes the workshop of discrete manufacturing enterprises as the research object, takes the minimum carbon emission and energy consumption of material distribution as the optimization goal, and establishes the optimization model of the workshop logistics distribution route. DBO is used to solve the model. The algorithm parameters are set as follows: the initial population size N is 30, the number of iterations T_{max} is 300, and the control parameters $K = 0.1$, $b = 0.3$, and $S = 0.5$. In order to verify the effectiveness of DBO, three algorithms, namely DBO, non-dominated sorting genetic algorithm-II (NSGA-II), and genetic algorithm (GA), were used to solve the distribution route, respectively, and the solution results of three different algorithms were compared and analyzed. Among them, the population number of GA and NSGA-II is 60, the crossover rate is 0.9, the variation rate is 0.1, and the iteration times of the algorithm are 300 [31–33]. Figures 4–9 show the algorithm iteration process.

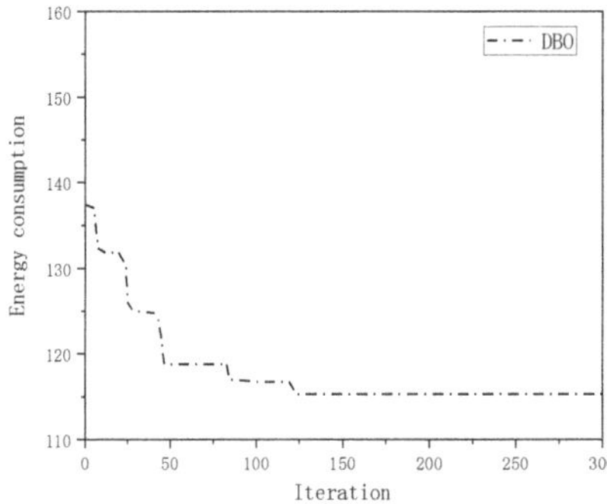

Figure 4. Energy consumption iteration diagram of DBO.

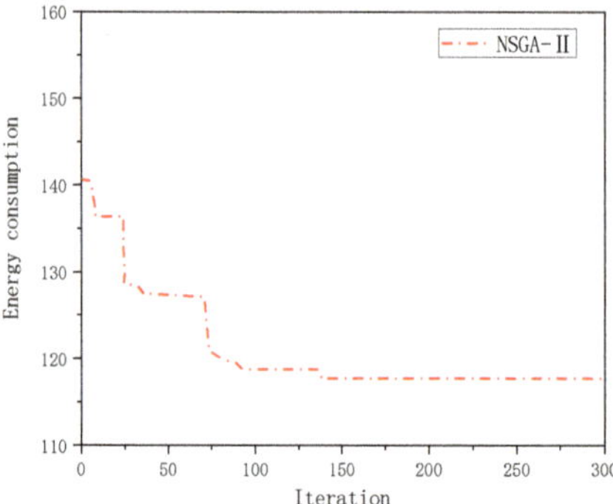

Figure 5. Energy consumption iteration diagram of NSGA-II.

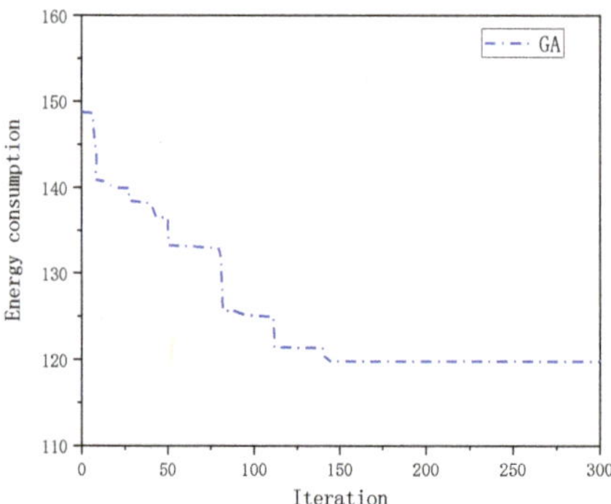

Figure 6. Energy consumption iteration diagram of GA.

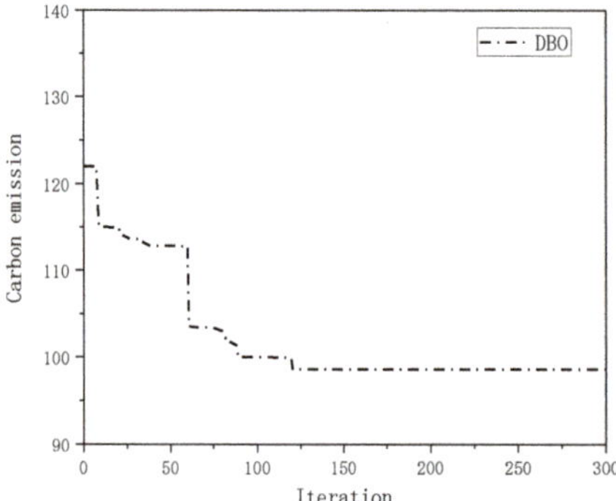

Figure 7. Carbon emission iteration diagram of DBO.

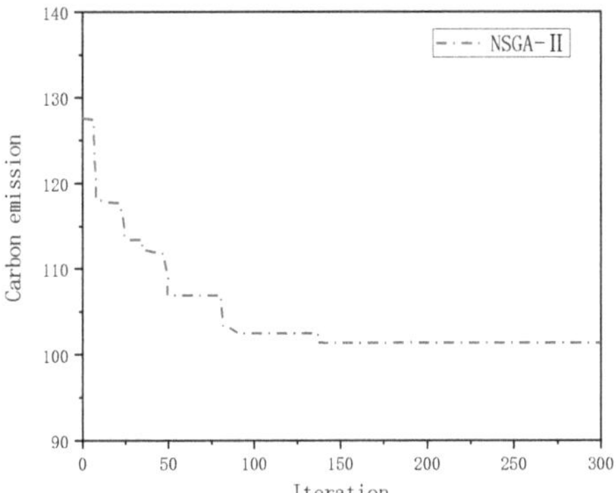

Figure 8. Carbon emission iteration diagram of NSGA-II.

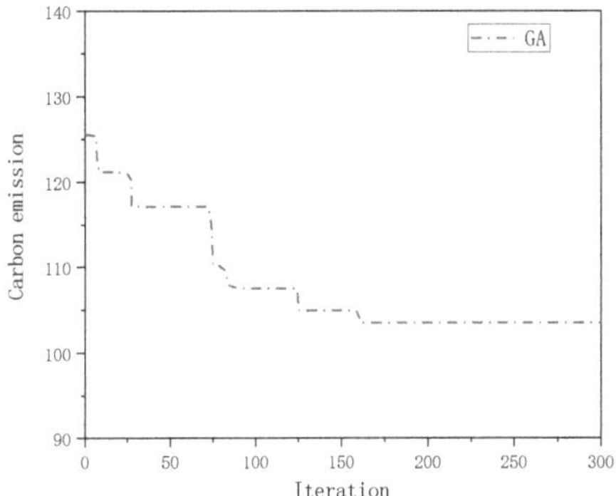

Figure 9. Carbon emission iteration diagram of GA.

Figures 4–6 show the iterative process of energy consumption of the three algorithms. As can be seen from the figure, as the number of iterations increases, the algorithm begins to optimize the path, and the energy consumption in the distribution process gradually decreases until it becomes stable. In the simulation results, energy consumption after DBO optimization is 115.32 J; energy consumption after NSGA-II optimization is 117.78 J; energy consumption after GA optimization is 119.86 J. In the iterative process, the energy consumption of the objective function value obtained by the DBO solution is the lowest, the second is NSGA-II, and the last is GA. The number of iterations used by the DBO algorithm to achieve stable optimal energy consumption is about 125 times, which is significantly improved compared with NSGA and GA, and DBO has a faster iteration speed.

Figures 7–9 show the iterative process of carbon emission of the three algorithms. As can be seen from the figure, as the number of iterations increases, the algorithm begins to optimize the path, and the carbon emission iteration in the distribution process gradually decreases until it becomes stable. In the simulation results, the carbon emission after DBO optimization is 98.62 kg; the carbon emission after NSGA-II optimization is 101.36 kg; and the carbon emission after GA optimization is 103.58 kg. In the iterative process, the objective function value obtained by the solution of DBO has the lowest carbon emission result, the second is NSGA-II, and the last is GA. The number of iterations used by the DBO algorithm to achieve stable optimal carbon emission is about 120 times, which is significantly improved compared with NSGA and GA, and DBO has a faster iteration speed. The above results show the effectiveness of the DBO algorithm proposed in this paper. The algorithm can expand the search range of solutions, improve the search speed of solutions, and improve search accuracy.

The optimized distribution path is shown in Table 4; the optimal distribution route calculated by DBO includes three distribution vehicles. The material distribution route of vehicle 1 is 0-3-52-5, the load factor is 86%, and the distance to complete one distribution is 5.8 km. The material distribution route of vehicle 2 is 0-6-4-0, the load rate is 53%, and the distance to complete distribution is 3.9 km. The material distribution route of vehicle 3 is 0-8-7-1-0, the load rate is 78%, and the distance to complete one distribution is 7.3 km. After the optimization of DBO, the material distribution distance is 17, the energy consumption in the distribution process is 115.32, and the carbon emission is 98.62. From the above simulation results and analysis, it can be seen that the optimized path of the DBO algorithm covers a shorter distance, and the distribution generates less energy consumption and carbon emissions, which is conducive to improving the production efficiency of enterprises.

Table 4. Results of vehicle route planning using DBO.

Distribution Vehicles	Distribution Path	Loading Rate	Delivery Distance (km)	Is It within the Time Window?
1	0-3-2-5-0	86%	5.8	Yes
2	0-6-4-0	53%	3.9	Yes
3	0-8-7-1-0	78%	7.3	Yes

5.3. Comparison with Previous Literature

The literature [10–17] provides a large number of solution ideas for workshop material distribution problems from the distribution method, distribution model, and algorithm model. Although the above research has made some progress, there are still problems, such as poor solution quality, easy falls into local optimal, and no consideration of environmental factors. Therefore, based on the enterprise management philosophy of management compass, this paper analyzes the elements of enterprise logistics distribution and takes distribution carbon emission and energy consumption as targets to establish a multi-objective optimization model of workshop material distribution route. DBO was used to solve the model with higher accuracy and stronger searching ability.

6. Conclusions

Workshop material distribution route planning is an important part of workshop product production decision-making, which is of great significance to workshop production management. Aiming at the environmental problems caused by the material distribution path in the manufacturing workshop, this paper makes the following research on the workshop material distribution path.

1. Based on the model of the business compass, the workshop material distribution management model was established, and the enterprise management process to realize the material distribution was analyzed.
2. A material distribution calculation model based on energy consumption and carbon emission was established.
3. Using DBO to solve this problem model, the comparison of DBO, NSGA- II, and GA shows the advantages of DBO; the algorithm has good global search ability and high search accuracy and can effectively optimize the material distribution route of manufacturing workshops. The effectiveness of the algorithm was verified by a case of material distribution in an engine assembly workshop.

The simulation results show this method can obtain the low carbon and low energy distribution path of workshop materials, which realizes the material delivery on time with the least energy consumption and carbon emission. In order to further adapt to the increasingly complex machine shop production environment, the author will study the dynamic material demand and the workshop material distribution path planning under different material types.

Author Contributions: Conceptualization, Y.X. and H.Z.; methodology, Y.X. and R.W.; software, Y.X.; validation, R.W.; formal analysis, Y.X. and R.W.; investigation, Y.X. and H.Z.; writing—original draft preparation, Y.X. and H.Z.; writing—review and editing, Y.X. and H.Z.; visualization, Y.X. All authors have read and agreed to the published version of the manuscript.

Funding: This research was funded by the reform project of teaching content and curriculum system of higher education in Guizhou Province in 2022, grant number: GZJG20220331.

Institutional Review Board Statement: Not applicable.

Informed Consent Statement: Informed consent was obtained from all subjects involved in the study.

Data Availability Statement: Not applicable.

Conflicts of Interest: The authors declare no conflict of interest.

References

1. Ai, X.F.; Jiang, Z.G.; Zhang, H.; Wang, Y. Low-carbon product conceptual design from the perspectives of technical system and human use. *J. Clean. Prod.* **2020**, *244*, 118819. [CrossRef]
2. Xia, T.; An, X.; Yang, H.; Jiang, Y.; Xu, Y.; Zheng, M.; Pan, E. Efficient Energy Use in Manufacturing Systems—Modeling, Assessment, and Management Strategy. *Energies* **2023**, *16*, 1095. [CrossRef]
3. Liu, H.; Fan, L.; Shao, Z. Threshold effects of energy consumption, technological innovation, and supply chain management on enterprise performance in China's manufacturing industry. *J. Environ. Manag.* **2021**, *300*, 113687. [CrossRef] [PubMed]
4. Fatemeh, F.; Afshar, N.B. A bi-objective green location-routing model and solving problem using a hybrid metaheuristic algorithm. *Int. J. Logist. Syst. Manag.* **2018**, *30*, 366–385.
5. Xu, L.A.; Wang, N.A.; Ling, X.B. Study on Conflict-free AGVs Path Planning Strategy for Workshop Material Distribution Systems. *Procedia CIRP* **2021**, *104*, 1071–1076.
6. Li, S.; Zhao, H. Optimization of Material Distribution Path of Manufacturing Workshop Based on IPSO Algorithm. *Mach. Des. Manuf.* **2019**, *10*, 209–212.
7. Farooq, B.; Bao, J.; Raza, H.; Sun, Y.; Ma, Q. Flow-shop path planning for multi-automated guided vehicles in intelligent textile spinning cyber-physical production systems dynamic environment. *J. Manuf. Syst.* **2021**, *59*, 98–116. [CrossRef]
8. Wang, C.L.; Wang, Y.; Zeng, Z.Y.; Lin, C.Y.; Yu, Q.L. Research on Logistics Distribution Vehicle Scheduling Based on Heuristic Genetic Algorithm. *Complexity* **2021**, *2021*, 8275714. [CrossRef]
9. Liu, S.; Zhang, Y.; Liu, Y.; Wang, L.; Wang, X.V. An 'Internet of Things' enabled dynamic optimization method for smart vehicles and logistics tasks. *J. Clean. Prod.* **2019**, *215*, 806–820. [CrossRef]
10. Yan, Z.; Meu, F.; Ge, M. Path optimization method of workshop logistics based on fuzzy soft time windows. *Comput. Integr. Manuf. Syst.* **2015**, *21*, 2760–2767.
11. Tong, F.; Xu, J. Two-way material distribution path planning for intelligent workshops considering workstation priority. *J. Mech. Electr. Eng.* **2021**, *38*, 1465–1471.
12. Goel, R.; Maini, R. A hybrid of Ant Colony and firefly algorithms (HAFA) for solving vehicle routing problems. *J. Comput. Sci.* **2018**, *25*, 28–37. [CrossRef]
13. Zhu, F.; Cao, T. Optimization Method of Material Delivery Path in Workshop Based on Time Window Constraints. *Mach. Des. Manuf.* **2023**, *1*, 136–139.
14. Erdodu, K.; Karabulut, K. Bi-objective green vehicle routing problem. *Int. Trans. Oper. Res.* **2022**, *29*, 1602–1626. [CrossRef]
15. Jie, K.W.; Liu, S.Y.; Sun, X.J. A hybrid algorithm for time-dependent vehicle routing problem with soft time windows and stochastic factors. *Eng. Appl. Artif. Intell.* **2022**, *109*, 104606. [CrossRef]
16. Tarhini, A.; Danach, K.; Harfouche, A. Swarm intelligence-based hyper-heuristic for the vehicle routing problem with prioritized customers. *Ann. Oper. Res.* **2022**, *308*, 549–570. [CrossRef]
17. Ren, X.; Huang, H.; Feng, S.; Liang, G. An improved variable neighborhood search for bi-objective mixed-energy fleet vehicle routing problem. *J. Clean. Prod.* **2020**, *275*, 124155. [CrossRef]
18. Wang, R.P. *Business Compass*; Science Press China: Beijing, China, 2020.
19. Sun, X.P. Taoism Naturalistic Technology Views. *Stud. Dialectics Nat.* **2022**, *38*, 42–47.
20. Xiao, Y.; Wang, R.; Yan, W. Optimum Design of Blank Dimensions Guided by a Business Compass in the Machining Process. *Processes* **2021**, *9*, 1286. [CrossRef]
21. Wang, R.P.; Yi, J. Characteristics and Mission of Management Thinking with Chinese Characteristics Based on the Business Compass Perspective. In Proceedings of the 8th International Symposium on Project Management (ISPM 2020), Beijing, China, 4–5 July 2020.
22. Xiao, Y.; Zhou, J.; Wang, R.; Zhu, X.; Zhang, H. Energy-Saving and Efficient Equipment Selection for Machining Process Based on Business Compass Model. *Processes* **2022**, *10*, 1846. [CrossRef]
23. Xiao, Y.; Zhao, R.; Yan, W. Analysis and Evaluation of Energy Consumption and Carbon Emission Levels of Products Produced by Different Kinds of Equipment Based on Green Development Concept. *Sustainability* **2022**, *14*, 7631. [CrossRef]

24. Zhang, X.; Wang, L. Optimization of apparel material distribution route based on carbon emission. *J. Text. Res.* **2020**, *41*, 5.
25. Demir, E. An adaptive large neighborhood search heuristic for the Pollution-Routing Problem. *Eur. J. Oper. Res.* **2012**, *223*, 346–359. [CrossRef]
26. Zhan, X.L.; Zhang, C.Y.; Meng, L. Low Carbon modeling and optimization of milling parameters based on improved gravity search algorithm. *China Mech. Eng.* **2020**, *31*, 11.
27. Chen, J.; Liao, W.; Yu, C. Route optimization for cold chain logistics of front warehouses based on traffic congestion and carbon emission. *Comput. Ind. Eng.* **2021**, *161*, 107663. [CrossRef]
28. Dacke, M.; Baird, E.; El, J.B.; Warrant, E.J.; Byrne, M. How dung beetles steer straight. *Annu Rev Entomol.* **2021**, *66*, 243–256. [CrossRef]
29. Yin, Z.; Zinn-Björkman, L. Simulating rolling paths and reorientation behavior of ball-rolling dung beetles. *J Theor Biol.* **2020**, *486*, 110106. [CrossRef]
30. Xue, J.; Shen, B. Dung beetle optimizer: A new meta-heuristic algorithm for global optimization. *J. Supercomput.* **2022**, *79*, 7305–7336. [CrossRef]
31. Su, Y.; Sun, W. Analyzing a Closed-Loop Supply Chain Considering Environmental Pollution Using the NSGA-II. *IEEE Trans Fuzzy Syst.* **2018**, *27*, 1066–1074. [CrossRef]
32. Xu, T.; Yao, L.; Xu, L.; Chen, Q.; Yang, Z. Image Segmentation of Cucumber Seedlings Based on Genetic Algorithm. *Sustainability* **2023**, *15*, 3089. [CrossRef]
33. Guo, W.; Lei, Q.; Song, Y.; Lyu, X. A learning interactive genetic algorithm based on edge selection encoding for assembly job shop scheduling problem. *Comput. Ind. Eng.* **2021**, *109*, 107455. [CrossRef]

Article

IOTA Data Preservation Implementation for Industrial Automation and Control Systems

Iuon-Chang Lin [1,*], Pai-Ching Tseng [2], Yu-Sung Chang [1] and Tzu-Ching Weng [2]

[1] Department of Management Information Systems, National Chung Hsing University, 145 Xingda Road, Taichung 40227, Taiwan; yorkisinhere@gmail.com

[2] Ph.D. Program of Business, Feng Chia University, Taichung 40724, Taiwan; tpcp630@gmail.com (P.-C.T.); tcweng@fcu.edu.tw (T.-C.W.)

* Correspondence: iclin@nchu.edu.tw; Tel.: +886-422840864; Fax: +886-422857173

Abstract: Blockchain 3.0, an advanced iteration of blockchain technology, has emerged with diverse applications encompassing various sectors such as identity authentication, logistics, medical care, and Industry 4.0/5.0. Notably, the integration of blockchain with industrial automation and control systems (IACS) holds immense potential in this evolving landscape. As industrial automation and control systems gain popularity alongside the widespread adoption of 5G networks, Internet of Things (IoT) devices are transforming into integral nodes within the blockchain network. This facilitates decentralized communication and verification, paving the way for a fully decentralized network. This paper focuses on showcasing the implementation and execution results of data preservation from industrial automation and control systems to IOTA, a prominent distributed ledger technology. The findings demonstrate the practical application of IOTA in securely preserving data within the context of industrial automation and control systems. The presented numerical results validate the effectiveness and feasibility of leveraging IOTA for seamless data preservation, ensuring data integrity, confidentiality, and transparency. By adopting IOTA's innovative approach based on Directed Acyclic Graph (DAG), the paper contributes to the advancement of blockchain technology in the domain of Industry 4.0/5.0.

Keywords: IOTA; Industry 4.0/5.0; blockchain; data preservation; industrial automation and control systems

Citation: Lin, I.-C.; Tseng, P.-C.; Chang, Y.-S.; Weng, T.-C. IOTA Data Preservation Implementation for Industrial Automation and Control Systems. *Processes* **2023**, *11*, 2160. https://doi.org/10.3390/pr11072160

Academic Editor: Sergey Y. Yurish

Received: 13 June 2023
Revised: 15 July 2023
Accepted: 16 July 2023
Published: 19 July 2023

1. Introduction

The application of Industry 4.0/5.0 is experiencing exponential growth across various fields. As Internet of Things (IoT) devices are deployed, a substantial amount of data is generated, which is traditionally centralized on client servers or cloud servers. However, such a centralized infrastructure poses challenges such as a single point of failure and a lack of trust between devices [1]. One solution to address these challenges is the implementation of a distributed system based on blockchain technology. The application of blockchain in Industry 4.0/5.0 has seen diverse use cases [2].

For instance, Datta et al. [3] proposed a security scheme that enables encrypted record storage for information and energy transactions between vehicles. Hang et al. [4] utilized Hyperledger Fabric to establish a fish farm platform, ensuring data integrity. Guan et al. [5] designed a two-tier distributed energy transaction system that safeguards transaction information through blockchain consensus mechanisms. Grecuccio et al. [6] leveraged IoT devices to record the food supply chain process and established an Ethereum node as a gateway to broadcast the data to smart contracts on the blockchain network for storage.

Despite the progress made, there are still several challenges in the application of blockchain in Industry 4.0/5.0. For instance, sensor data monitoring over an extended period is limited, and the costs associated with integrating blockchain into IoT systems, due to handling fees and scalability issues, can be substantial. Furthermore, while smart contracts have shown promise in supply chain management, the historical data from IoT devices cannot be recorded on the blockchain, and the data transmission process from the IoT perception layer to the network layer lacks openness and transparency. To address these challenges of scalability, efficiency, and handling fees, Popov introduced a distributed ledger called Tangle, which is based on the Directed Acyclic Graph (DAG) structure [7] Tangle differs from traditional blockchain ledgers as it utilizes the DAG structure as the foundation, enabling faster transactions and comprehensive recording of the history of Industry 4.0/5.0.

According to a report by Ericsson [8], it is projected that by 2021, there will be approximately 28 billion globally connected smart devices. Additionally, more than 15 billion devices have already adopted machine-to-machine (M2M) communication [8]. With the widespread adoption of the Internet of Things in various aspects of life, Wireless Sensor Networks (WSNs) have emerged as a crucial component for the application of Industry 4.0/5.0. WSNs are characterized by their low energy consumption, small size, and ease of deployment.

In the realm of IoT environments, ensuring real-time data reception and transmission with data integrity is crucial. To address the challenges related to massive data transmission and efficiency, Alsboui et al. [9] proposed the Mobile-Agent Distributed Intelligence Tangle-Based approach (MADIT). They utilized the IOTA Masked Authenticated Messaging (MAM) protocol to ensure data privacy on the Tangle. While ledger data are publicly transparent, sensitive data may need to be uploaded to the ledger. Therefore, Zhang et al. [10] introduced LDP, which preserves data confidentiality while uploading it to the distributed ledger technology (DLT). In scenarios where data sizes exceed the single transaction limit of the ledger, J. Jayabalan and N. Jeyanthi [11] proposed a model that encrypts medical data and stores it on IPFS, while storing the IPFS-generated index on the DLT, ensuring both data integrity and confidentiality.

Moving on to Docker, it is an extensively adopted open-source platform for containerization, enabling developers to package applications and their dependencies into lightweight, portable containers. Since its release in 2013, Docker has become an indispensable tool for building, shipping, and running applications. Containerization technology has gained significant popularity due to its ability to provide lightweight, portable, and isolated execution environments for applications. Researchers have explored various aspects of containerization, including performance, security, and usability. In a study by Divya and Sri [12], the potential of containerization technology and edge-fog cloud infrastructure is showcased in enabling efficient and scalable fall detection systems in healthcare. The paper emphasizes the importance of considering specific application workload requirements when selecting a containerization platform and infrastructure architecture.

Another research by Singh et al. [13] proposes an innovative solution to address the challenges of secure and efficient task containerization in IoT systems. The paper highlights the significance of considering both security and efficiency requirements in designing containerization solutions and demonstrates the effectiveness of game-theoretic approaches in tackling these challenges.

In the context of industrial automation and control systems, the existing architecture relies on a centralized server or cluster head to manage, identify, and encrypt connections among a large number of deployed sensors in the sensing area. However, this centralized approach presents challenges in terms of scalability, security, and privacy. Additionally, ensuring the confidentiality and integrity of data during transmission from the sensor collection end to the user end is crucial.

Furthermore, the application of blockchain in Industry 4.0/5.0 brings forth several challenges. These challenges encompass the inability to monitor sensor data over an extended period, the high costs associated with importing blockchain into the Internet of Things (IoT) due to handling fees and scalability issues, the lack of IoT data recording in the blockchain, and the lack of transparency in the data transmission process from the IoT perception layer to the network layer.

These challenges underscore the significance of addressing the limitations in current architectures and exploring the potential of blockchain technology to overcome them. By developing decentralized and secure solutions, achieving scalable, cost-effective, and transparent data management in industrial automation and control systems becomes feasible. The resolution of these challenges can enhance the reliability, efficiency, and trustworthiness of Industry 4.0/5.0 applications, facilitating their widespread adoption and unlocking their full potential in various domains.

2. Related Works

2.1. IOTA

IOTA is an open-source decentralized ledger technology operating on a peer-to-peer network. It utilizes the Directed Acyclic Graph (DAG) method to store each transaction and was officially launched in around 2018. The development of IOTA is primarily driven by the Berlin-based non-profit organization, IOTA Foundation. The name "IOTA" originates from the ninth letter of the ancient Greek alphabet, symbolizing tiny things and representing the smallest unit of currency issued by IOTA.

In the era of the Internet of Everything, IOTA enables devices to conduct micropayments and exchange data without incurring additional costs. The underlying structure of IOTA, known as Tangle, employs a net-like ledger structure. Unlike traditional blockchains, Tangle allows for multiple forks and does not require transactions to be specified behind specific blocks. Instead, new transactions are randomly selected for verification by referencing two existing transactions. As a result, transactions can be generated synchronously, leading to significantly improved speed and scalability compared to conventional blockchains.

An IOTA account serves as a means to prove ownership of transactions within the Tangle. Similar to a bank account, it is created using a seed, which differentiates it from traditional name- and password-based accounts. Importantly, the seed remains solely accessible to the account owner and represents their identity on the internet. This approach ensures decentralization and maintains anonymity within the IOTA network. To generate addresses, IOTA employs Winternitz One Time Signature (WOTS) for which the seed acts as the master key. Multiple private keys can be derived from the seed, and each private key generates a unique address. It is worth noting that each address can only be used once and can hold any amount of IOTA currency. The total balance of an account is determined by the cumulative sum of currencies across all addresses within that account.

The seed serves as the sole master key for proving ownership of IOTA currency within messages or addresses. It consists of an 81-character Tryte string comprising 26 uppercase English letters and the number 9. The number of possible seeds is immense, reaching approximately 8.7×10^{115}, making the likelihood of two identical seeds extremely low. A seed can generate different private keys by varying the index and security levels. There are three security levels available, each corresponding to different private key lengths. Higher security levels feature longer private key lengths, which reduce the risk of theft. Security level 1 is suitable for storing low-value IoT device data, while security level 2 is utilized for wallet transactions and high-value IoT devices. For transactions requiring the utmost security, such as exchanges, security level 3 is employed.

There are five main steps in the transaction sending process, which are described in detail as follows:

1. Generate transaction information:

The first step is to create a single transaction. You must specify an Address and a Value for each transaction. You can also define a Tag to classify different transactions. Message is the message content of the transaction. In zero-value transactions, the Message is used to trace the root address, and a Timestamp is automatically generated by IOTA. To send a valuable transaction, at least two IOTA transaction messages are required; one with a positive value to allow the receiver to obtain encrypted currency, and one with a negative value to allow the sender to deduct the transferred amount.

2. Package into Bundle:

A Bundle is a collection of transactions. After the transaction information is generated, all transactions will be packaged into Bundles. The sum of the Value of all transactions in the Bundle must be zero, and the Bundle also has a unique address for querying transactions. After the packaging is complete, use the Seed to generate the private key, and use the private key to sign the Bundle.

3. Choose two Tips:

The POW calculation must be performed before the transaction is attached to the ledger. The object of performing POW is in the Tangle, using Markov chain Monte Carlo (MCMC) to randomly select two Tips, called Branch and Trunk, and then add the Hash of the two Tips to the Bundle.

4. Do proof of work:

Perform POW (proof of work) operation on the selected two Tips, and then append the calculated Nonce value to the Bundle.

5. Broadcast to the Tangle network:

After verifying other transactions, the Bundle is broadcast to the Tangle for storage, and the new transaction is Tips, waiting to be verified by others.

2.2. Signature and Verification

As depicted in Figure 1, the signature is generated during the Bundle generation process, where the Bundle is signed to validate ownership, and the resulting signature is appended to the Signature Fragment within the Bundle. To ensure the security and integrity of the signature, the Hash value of the Bundle is first obtained, and the private key is derived from the Seed using WOTS. The length of the private key varies depending on the chosen security level. For the lowest security level, the private key length is 27×81 Trytes. The private key is divided into segments, with each segment consisting of 81 Trytes. Unequal hash operations are performed on these segments based on the Hash value of the Bundle. By converting the Bundle Hash into its decimal representation, an integer between -13 and 13 is obtained. This integer is then subtracted from 13, and the resulting values are hashed onto the segments. The combination of these hashed values constitutes the signature. It is worth noting that higher security levels result in longer private keys and correspondingly longer signatures. As illustrated in Figure 1, the first and second Trytes of the Signed Data (Normalized Bundle Hash) are represented by the letters L and W, respectively, with their corresponding decimal values being 12 and -4. Subtracting these values from 13 yields 1 and 17, respectively, indicating that Segment 1 undergoes one sponge function calculation and Segment 2 undergoes seventeen sponge function calculations. Once all the values are integrated, the resulting combination forms the signature.

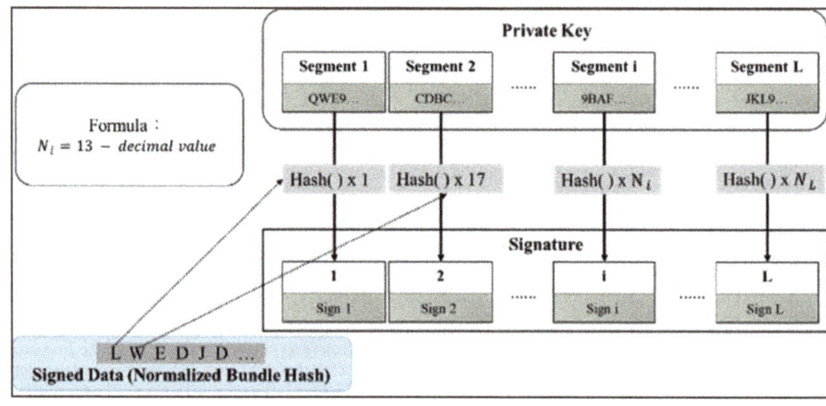

Figure 1. Transaction signature.

The node verifies the signature in the transaction through the Bundle Hash and address. The method is similar to the signature. First, obtain the value of the Bundle Hash, and then convert it to a decimal. Then add 13 to each Tryte value, and then each signature fragment in the signature corresponds to the result of the Bundle Hash operation, and the number of hash operations is different. Finally, combine the hash values of the signature fragments and perform two hash operations to obtain the transaction address. Verify that the address of the Bundle is consistent with this address to know whether the signature is correct. Because the address generation method is also WOTS, in the step of the private key, the segment is subjected to 26 sponge function operations, and the $13 - d$ (decimal value) operation is performed when signing, and $13 + d$ (decimal value) is performed again during verification. The result of $13 - d + 13 + d$ is 26 sponge operations. This method can achieve the purpose of signature verification without leaking the private key and seed. As shown in Figure 2, the decimal places of the first two Trytes of Signed Data are 12 and -4, and the values after adding 13 are 25 and 9. After 25 and 9 operations, the private values of Segment 1' and Segment 2' are obtained. Key hash value. After all the fragments are merged and hashed, the address can be obtained. If the addresses match, the transaction is valid.

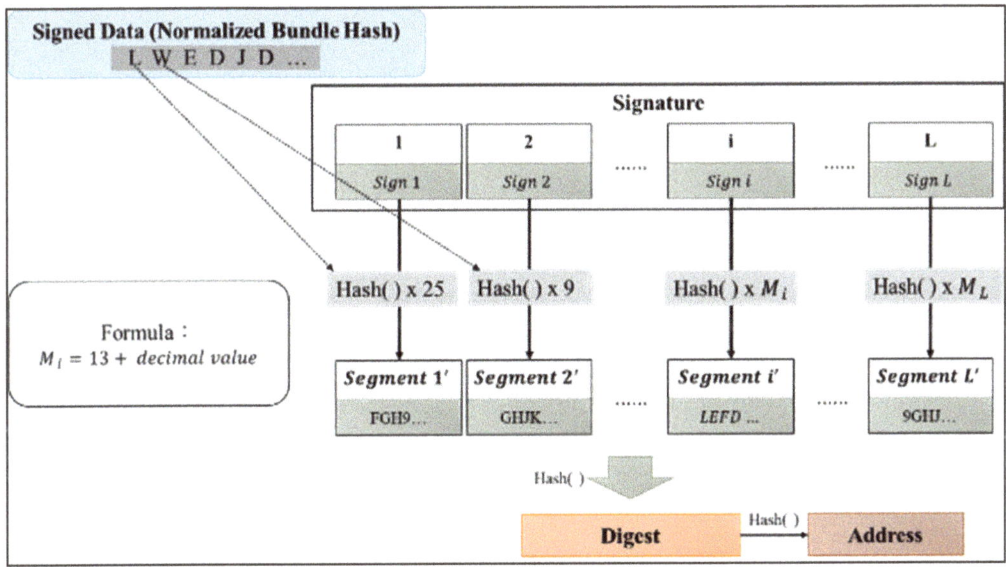

Figure 2. Signature verification.

2.3. Proof of Work

Proof of Work (PoW) is a cryptographic process used to demonstrate that a problem has been solved, granting the right to write data into the ledger. These problems are challenging to solve, but their correctness can be easily verified. PoW is employed to safeguard the network against spam attacks, as the associated computational effort incurs a certain cost. In IOTA, PoW is not performed by miners but rather delegated to nodes for execution by individuals initiating transactions. Prior to writing data into the ledger, every node must execute PoW. As illustrated in Figure 3, the Nonce value consists of 81 Trytes within the transaction message. To carry out PoW, the Nonce value must be randomly filled in, and the transaction message, including the Nonce, is converted into a Trits format for hashing operations. The resulting Minimum Weight Magnitude (MWM) of the final Trit sequence must be zero. The difficulty of achieving MWM varies based on the network. In the main network, the MWM is set to 14, meaning that the next 14 Trits after PoW execution must be zero. In the test network, the MWM is nine. When the transaction initiator node completes PoW and broadcasts the transaction to other nodes for storage, these nodes independently perform the calculation using the provided Nonce value. If the calculation result satisfies the MWM requirement, the transaction can be written into the Tangle.

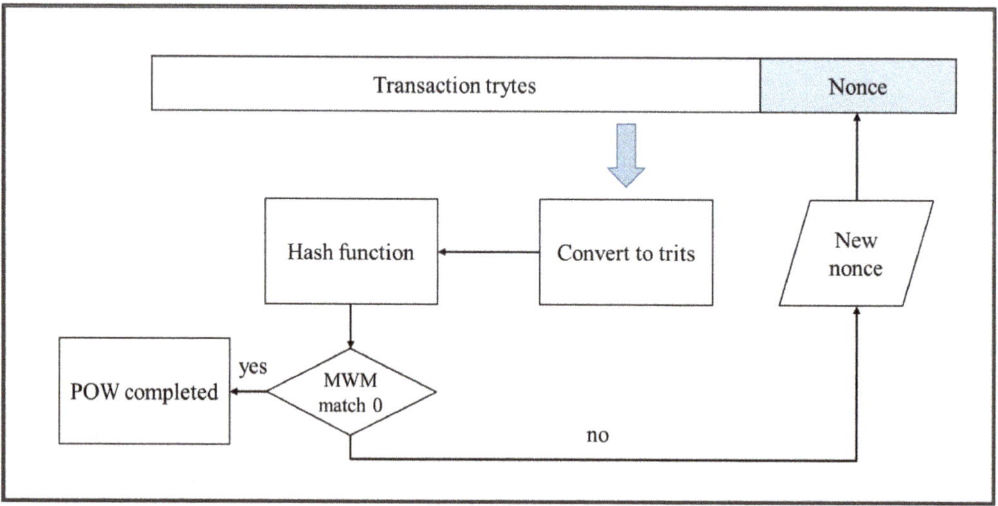

Figure 3. Proof of work.

2.4. Channel

MAM's Channel is like streaming media. Publishers can publish messages regularly, viewers can receive messages through subscriptions, and only the owner can publish messages. In IOTA, this owner is the seed owner. If the Seed is stolen, the attacker can publish messages at will. There are three modes in the Channel to control the message flow, namely Public, Private, and Restricted. In the Public mode, everyone can directly obtain the message content while in the Private message content, only the owner of the Seed can unlock the encrypted content. In the Restricted mode, encrypted content can be unlocked by using a key called Sidekey. By sharing the Sidekey, confidential information can be easily opened by specific people for viewing. The address generation methods of the three modes are not the same. Although they all use the Merkle-tree Signature Scheme, they are different in the last Root. The detailed address is generated as follows:

1. Public: Address = Root.
2. Private: Address = hash (Root), hidden messages are decrypted using Root.
3. Restricted: Address = hash (Root), hidden messages are decrypted using Sidekey.

2.5. Sponge Function

The sponge function is a cryptographic algorithm. It uses a limited state to receive input bit streams of any length. After the data are "absorbed" into the sponge, the desired result is "squeezed out", and then it can satisfy any Length of the output. Sponge function can be divided into two stages, namely Absorbing and Squeezing. The information is first input and compressed repeatedly, and then the result is repeatedly extruded. As shown in Figure 4, the values M0~M3 need to be input; after inputting M0, go through the calculation of XOR and function, then input the value of M1, and repeat the calculation until all the values are counted. Then when the value is taken out, it will go through the function calculation again to obtain the value of Z0. If the length is not enough, continue the calculation until the required length is obtained.

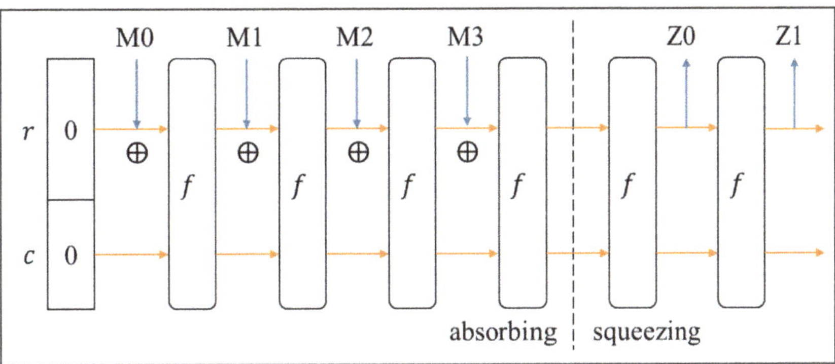

Figure 4. Sponge function.

3. Implementation

3.1. Method

The proposed study will employ a research design focused on evaluating the implementation and execution of data preservation from industrial automation and control systems to the IOTA DAG technology within the context of Industry 4.0/5.0.

1. Research Design: The research design will be based on a practical implementation and execution approach, aiming to assess the feasibility and effectiveness of utilizing the IOTA DAG technology for data preservation in industrial automation and control systems. The design will involve setting up a testbed or simulation environment to simulate real-world scenarios and evaluate the performance of the proposed solution.

2. Data Collection: Data collection will involve capturing sensor data from industrial automation and control systems. The specific environmental parameters and performance metrics to be collected will be determined based on the objectives of the study. The collected data will be securely transferred and stored using the IOTA DAG technology, ensuring data integrity throughout the process.

3. Implementation and Execution:

The proposed solution for data preservation using the IOTA DAG technology will be implemented and executed within a testbed or simulation environment. This will involve designing and deploying the necessary infrastructure, configuring the sensor networks, and integrating the IOTA DAG technology. The execution phase will focus on monitoring and evaluating the performance of the data preservation process, assessing factors such as scalability, efficiency, and cost-effectiveness.

By adopting a practical implementation approach and focusing on the evaluation of the IOTA DAG technology for data preservation, the study aims to provide insights into the feasibility and effectiveness of this solution in industrial automation and control systems within the Industry 4.0/5.0 context. The system designed in this paper is divided into three main members: Base Station, Cluster Head, and Sensor Node. The hardware configuration of the system is as follows.

3.2. Base Station

The base station is divided into two parts, namely Gateway and IOTA nodes. The Base Station runs on the Windows 10 operating system and is equipped with Intel's i7 processor and 16G of memory. The Proxy Server run by Gateway is written in node.js language, while IOTA nodes are set up using Hornet, and only the 15,600 Port is open for external node synchronization. In the data storage part, use Chronicle to store the data in Scylla's NoSQL database.

Because MAM data upload must choose a website with SSL/TLS standards, the Base Station must apply for a Domain Name and generate a certificate. To set up a node, you must first download the public account snapshot from the IOTA website, then select a fixed node, and wait for the node to complete synchronization before uploading data to the node.

3.3. Cluster Head

Cluster Head uses a Raspberry Pi 4/8G device with a network card, as shown in Figure 5. The operating system uses a Linux system and uses the version of ubuntu-20.04.2 preinstalled-server-arm64+raspi. Cluster Head also uses node.js to write, uses mosca's suite to run MQTT Server, and uses IOTA's node.js Client Library to encrypt and sign data.

Figure 5. Raspberry pi 4.

3.4. Sensor Node

Sensor Node uses the development version of Arduino Nano 33 IoT with 256 KB of CPU Flash Memory and 32 KB of SRAM. And the part of the sensor that uses the RFID Reader of RC522, as shown in Figure 6, uses the C/C++ language to write the program.

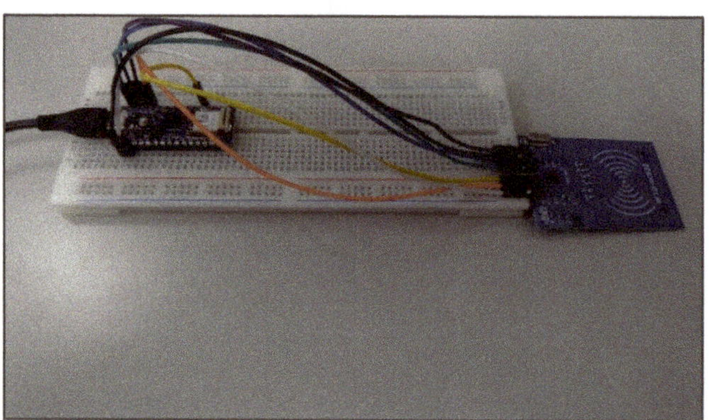

Figure 6. Arduino nano 33 IoT.

3.5. Gateway

The Gateway in the Base Station is responsible for configuring new identity data for the new device, as shown in Algorithm 1. When a new device wants to join the network, it will first indicate whether it is Cluster Head or Sensor Node, and then obtain the latest data from Channel 3, and determine whether the new device is in the blacklist. If everything is correct, Gateway will configure a new AES key and ID for the new device and obtain the whitelist of the network from Channel 2, and then append the ID of the new device to the whitelist and upload it to Channel 2.

Algorithm 1: New Device.

Input: field, ch, information
Output: device_id, encryption key
count = 2
tag = SECONDCHANNEL
mac ← *get address from information*
blacklist ← *fetch mam data from channel 3*
If *mac in the blacklist* **then**
    return failed
else
    fetchdata ← *fetch mam data from channel 2*
    ID ← *generate random number for new device*   **If** *field is* cluster **then**
        generate cluster key by hash(BSkey $||$ *ID*
    else if *field is sensor* **then**
    generate sensor key by hash(BSkey $||$ *ch* $||$ *ID)*
    end if
    channelState ← *read channel detail from channel2.json*
    channelState.count = *count*
    newdata ← *pack ID to fetchdata*
    message ← *createmessage(channelState, newdata)*
    mamAttach(message, 3, 14, tag)
    Strore the mam root to channel2.json
end if
return ID, key

Assuming that the wireless sensor network has a detection tool for malicious devices, when a malicious device is detected, the Blacklist program can be called, as shown in Algorithm 2. Pass in the ID and MAC address of the malicious device, then obtain the latest blacklist from Channel 3 and obtain the latest network identity list from Channel 2. Then record the ID, MAC, and time of the malicious device on the blacklist and upload it to Channel 3. Finally, the malicious device is removed from the list of Channel 2 and then re-uploaded to Channel 2, and the latest root is stored in channel2.json.

Algorithm 2: Blacklist.

Input: id, mac
Output: root
count = 3
tag = THIRDCHANNEL
blacklist ← *fetch mam data from channel 3*
fetchdata ← *fetch mam data from channel 2*
newlist ← *pack id and mac to blacklist*
channelState ← *read channel detail from channel3.json*
channelState.count = *count*
message ← *createmessage(channelState, newlist)*
mamAttach(message, 3, 14, tag)
Strore the mam root to channel3.json
If *id is in fetchdata* **then**
    newdata ← *remove id from fetchdata*
    channel2 ← *read channel detail from channel2.json*
    channel2.count = 2
    msg ← *createmessage(channelState, newdata)*
    mamAttach(msg, 3, 14, "SECONDCHANNEL")
    Strore the mam root to channel2.json
end if
return mam root

As shown in Algorithm 3, when the Cluster Head wants to register the connection identity data, it will send the Cluster Head ID and encrypted data to the Gateway. Then Gateway will use the Cluster Head ID and its own key to perform MD5 operations and use the first 16 Digest as the decryption key and the last 16 Digest as the IV of the AES CBC. Then use the Key and IV to unlock the encrypted content to obtain the timestamp and the identity ID. After the Gateway verifies that the Cluster Head identity is correct and the Channel 2 list exists, a temporary certificate and the validity of the certificate will be generated. Then use the key and IV of the Cluster Head to encrypt the certificate data and send it back to the Cluster Head.

Algorithm 3: Register.

Input: CHID, encryptdata
Output: BSID, return_data
$return\_data = ""$
$concat\_id \leftarrow BSkey$ *concatenate with* CHID
$SKey \leftarrow MD5$ *hash*$(concat\_id)$
$key \leftarrow get\ first\ 16\ digest\ from\ SKey$
$iv \leftarrow get\ last\ 16\ digest\ from\ SKey$
$content \leftarrow AES\ CBC\ decrypt($encryptdata$,\ key,\ iv)$
$TS1 \leftarrow content\ XOR\ CHID$
If $TS1$ *is timeout* **then**
    return failed
else
    $fetchdata \leftarrow fetch\ mam\ data\ from\ channel\ 2$
    If $CHID$ *is not in* $fetchdata$ **then**
        return failed
    else
        $TE \leftarrow compute\ the$ overdue time
        $Pi \leftarrow MD5\ hash\ the\ value\ of\ (CHID||BSID||TE)$
        $TC \leftarrow MD5\ hash\ the\ value\ of\ (Pi||BSkey)$
        $phrace \leftarrow package\ TC, Pi, TE,\ and\ the\ timestamp\ TS2$
        $return\_data \leftarrow AES\ encrypt(phrase, key, iv)$
    end if
end if
$return\ return\_data$

The logged-in Pseudocode is shown in Algorithm 4. The Cluster Head will send the certificate-related data to the Gateway for identity verification and at the same time send the first root data of the MAM and the decrypted sidekey to the Gateway. When the Gateway receives the message, it will verify whether the certificate has expired and whether the Cluster Head has the correct certificate data. Then the Gateway will generate the AES CBC Key and IV and decrypt the encrypted content and, at the same time, verify whether the timestamp is within a reasonable transmission time. Finally, the MAM information of the Cluster Head is recorded in Channel 1, which is used to unlock the MAM data stream uploaded by the Cluster Head. Then Gateway stores the latest root of Channel 1 in channel1.json and records the IP, *sidekey*, and TE of the Cluster Head in whitelist.json.

In the part of secure communication, because the Cluster Head has been authenticated during the previous registration and login, the secure communication does not verify too much information; it only confirms whether the IP of the Cluster Head is in the whitelist and whether the validity of the certificate of the Cluster Head has expired. If it is correct, it will be transferred to the IOTA node of the Base Station, as shown in Algorithm 5.

3.6. Raspberry pi

Algorithm 6 is the Pseudocode of the Cluster Head. The Cluster Head will receive the data passed by the Sensor Node and use the ID of the Sensor Node and its own Key to perform the MD5 hash function calculation. Then the first 16 Digests of the generated value are used as the Key, and the last 16 Digests are used as the IV. Then the Key and IV are calculated by the AES CBC to obtain the encrypted content of the Sensor Node. When it is confirmed that the encrypted content of the Sensor Node can be correctly unlocked and the timestamp is in line with the upload delay, the Cluster Head will package the sensor data and send it to the 3000 Port of the Base Station so that the Gateway will forward the packet to the IOTA node for upload. Finally, the Cluster Head will store the latest Root data in channel.json.

Algorithm 4: Login.

Input: CHID, encryptdata, C, TE, R, P
Output: BSID, confirm
IP $\leftarrow$ get ip address from TCP/IP
If *TE is timeout* **then**
    return failed
end if
P $\leftarrow$ concatenate BSID, CHID, TE*
If $P \neq P^*$ **then**
    return failed
end if
Compute TC by MD5 hash(P $\|$BSkey)
C $\leftarrow$ MD5 hash(CHID $\|$TC $\|$R)*
If $C \neq C^*$ **then**
    return failed
else
    concat_id $\leftarrow$ BSkey concatenate with CHID
    SKey $\leftarrow$ MD5 hash(concat_id)
    key $\leftarrow$ get first 16 digest from SKey
    iv $\leftarrow$ get last 16 digest from SKey
    content $\leftarrow$ AES CBC decrypt(encryptdata, key, iv)
    TS3 $\leftarrow$ get timestamp from content
    If *TS3 is timeout then*
        return failed
    else
        sidekey $\leftarrow$ get sidekey from content
        root $\leftarrow$ get first root from content
        Store IP·sidekey·TE to whitelist.json
        rootKey $\leftarrow$ MD5 hash(CHID $\|$sidekey)
        mam publish(root, sidekey)
        Store mam publish root to channel1.json
    end if
end if
return confirm $\leftarrow$ true

Algorithm 5: MAM publish.

Input: mam transaction data
Output: forward transaction to hornet node
IP $\leftarrow$ get ip address from TCP/IP
If *IP is not in the whitelist.json* **then**
    return failed
else
    TE $\leftarrow$ get TE from the whitelist.json
    If *the TE is timeout* **then**
        return failed
    else
        Forward to 14265 port where the hornet is located
    end if
end if

Algorithm 6: Cluster Head.

Input: SID, mqtt_data
seed ← get mam seed from config
sidekey ← get mam key from config
TC ← get temporary confirm from storage
TE ← get time expired from storage
If *not register* **then**
    call base station port 3001 *to register*
end if
If *not login* **then**
    create new channel by using seed, sidekey
    create mamMessage by channelState, TC, TE
    call base station port 3002 *to login*
else
    If *TE is timeout* **then**
        login again
    else
        concat_id ← CHkey concatenate with SID
        SKey ← MD5 hash(concat_id)
        key ← get first 16 *digest from SKey*
        iv ← get last 16 *digest from SKey*
        content ← AES CBC decrypt(mqtt_data, *key, iv*)
        channel ← read channel detail from channel.json
        msg ← createmessage(channelState, content)
        mamAttach(msg, 3, 14, "CLUSTERHEAD")
        Strore the mam root to channel.json
    end if
end if

3.7. Arduino Nano

Algorithm 7 is the upload process of the Sensor Node. The Sensor Node can collect data on a regular basis, but because the experimental device is an RFID Reader, the Sensor Node will passively obtain the data. When the Sensor Node obtains the data, it will obtain the Key and IV from the configuration file and convert the sensor data to hexadecimal and then encrypt it with AES CBC. Then it is sent to the Broker of the Cluster Head through the MQTT protocol. After the Subscriber of the Cluster Head obtains the Sensor Node data from the Broker, the data will be packaged and uploaded to the Tangle ledger.

Algorithm 7: Sensor Node.

Input: sensor data
Output: SID, mqtt_data
iv ← get AES iv from config
key ← get AES encryption key from config
If *WiFi is not connected* **then**
    reconnect()
end if
If *MQTT broker is not connected* **then**
    reconnect()
end if
While *read the RFID tag from reader* **do**
    hexdata ← change data to hex string
    TS4 ← get timestamp of sensor node
    encdata ← AES CBC encrypt(TS4, hexdata, iv, key)
    publish(SID, encdata) to cluster head
end while

---

4. Execution Results

The execution results depicted in Figures 7–12 provide valuable insights into the implementation and functionality of the proposed system. Figure 7 showcases the verification screen for Gateway registration and login, displaying the successfully logged-in IP of the Cluster Head. Subsequently, the MAM message is forwarded to the IOTA node. In Figure 8, the Cluster Head execution and login screen illustrate the encrypted message generated during registration and login, along with the screen displaying the first MAM message stream sent after a successful login. Figure 9 demonstrates the process in which the Cluster Head receives the message from the sensor node. Upon receiving the sensor node message, the Cluster Head repackages it and uploads it to the Tangle.

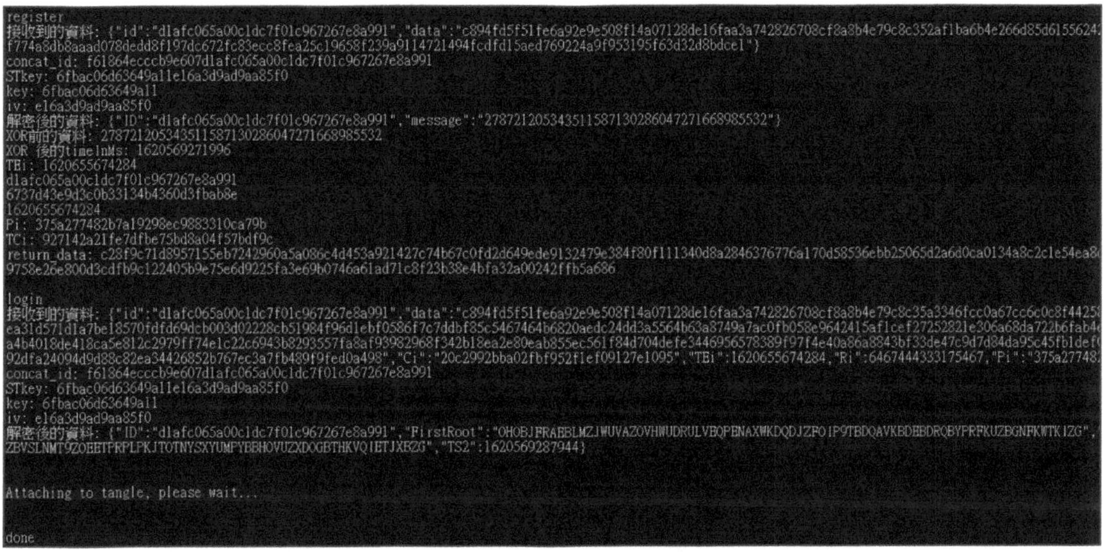

register
接收到的資料: {"id":"d1afc065a00c1dc7f01c967267e8a991","data":"c894fd5f51fe6a92e9e508f14a07128de16faa3a742826708cf8a8b4e79c8c352af1ba6b4e266d85d6155624
f774a8db8aaad078dedd8f197dc672fc83ecc8fea25c19658f239a9114721494fcdfd15aed769224a9f953195f63d32d8bdce1"}
concat_id: f61864ecccb9e607d1afc065a00c1dc7f01c967267e8a991
STkey: 6fbac06d63649a11e16a3d9ad9aa85f0
key: 6fbac06d63649a11
iv: e16a3d9ad9aa85f0
解密後的資料: {"ID":"d1afc065a00c1dc7f01c967267e8a991","message":"2787212053434511587130286047271668985532"}
XOR前的資料: 2787212053434511587130286047271668985532
XOR 後的timeInMs: 1620569271996
TEi: 1620655674284
d1afc065a00c1dc7f01c967267e8a991
6737d43e9d3c0b33134b4360d3fbab8e
1620655674284
Pi: 375a277482b7a19298ec9883310ca79b
TCi: 927142a21fe7dfbe75bd8a04f57bdf9c
return_data: c28f9c71d8957155eb7242960a5a086c4d453a921427c74b67c0fd2d649ede9132479e384f80f111340d8a2846376776a170d58536ebb25065d2a6d0ca0134a8c2c1e54ea8
9758e26e800d3cdfb9c122405b9e75e6d9225fa3e69b0746a61ad71c8f23b38e4bfa32a00242ffb5a686

login
接收到的資料: {"id":"d1afc065a00c1dc7f01c967267e8a991","data":"c894fd5f51fe6a92e9e508f14a07128de16faa3a742826708cf8a8b4e79c8c35a3346fcc0a67cc6c0c8f44258
ea31d571d1a7be18570fdfd69dcb003d02228cb51984f96d1ebf0586f7c7ddbf85c5467464b6820aedc24dd3a5564b63a8749a7ac0fb058e9642415af1cef27252821e306a68da722b6fab4e
a4b4018de418ca5e812c2979ff74e1c22c6943b8293557fa8af93982968f342b18ea2e80eab855ec561f84d704defe3446956578389f97f4e40a86a8843bf33de47c9d7d84da95c45fb1def0
92dfa24094d9d88c82ea34426852b767ec3a7fb489f9fed0a498","Ci":"20c2992bba02fbf952f1ef09127e1095","TEi":1620655674284,"Ri":6467444333175467,"Pi":"375a277482
concat_id: f61864ecccb9e607d1afc065a00c1dc7f01c967267e8a991
STkey: 6fbac06d63649a11e16a3d9ad9aa85f0
key: 6fbac06d63649a11
iv: e16a3d9ad9aa85f0
解密後的資料: {"ID":"d1afc065a00c1dc7f01c967267e8a991","FirstRoot":"OHOBJFRAEBLMZJWUVAZOVHWUDRULVEQPENAXWKDQDJZFOIP9TBDQAVKBDEBDRQBYPRFKUZBGNFKWTKIZG",
ZBVSLNMT9ZOEETPRPLPKJTOTNYSXYUMFYBBHOVUZXDOGBTHKVQIETJXEZG","TS2":1620569287944}

Attaching to tangle, please wait...

done

Figure 7. Base station operation screen.

ready
registering...
時戳: 1620894390085
加密後資料: c894fd5f51fe6a92e9e508f14a07128de16faa3a742826708cf8a8b4e79c8c352af1ba6b4e266d85d615562426f12
dedd8f197dc672fc83ecc8acdb22d9f30c70c6aa3f08aa1ad93aedca8835fff190336b0a44930d6b1ef659
已註冊
logging...
loginin firstroot: OHOBJFRAEBLMZJWUVAZOVHWUDRULVEQPENAXWKDQDJZFOIP9TBDQAVKBDEBDRQBYPRFKUZBGNFKWTKIZG
loginin sidekey: LJNQJJELTBVLVHHNUBIJDLVBWAJQPWSKHIDPUMHFNZQKJTYAFPQSGAXAWKIDALAUYDCMXXJXNLPXRDUFE
loginin Ci: 8ab80fe719a0eceeeeb0b1868565e53a
loginin TEi: 1620980760865
loginin Ri: 3784181989429487
loginin Pi: e6a9673f47e1f4d3456284c84e99a4d9
{ return_data: '' }
Seed: MFQZFWXMSIIUK9JTWRE9BVOJSEPISXBCUUEZPSJDDZZTUV9HPVRQUTIZLQKMTIJOOMPWGJKNFCZGKIISF
Address: UBMCJOQLVZJFRJHIXDRFBHRVA9ZRGJLWC9WHLMDB9TRAQVKFHOBQNPXSUQEGBQVONTDQECDUKOWJRTAOH
Root: OHOBJFRAEBLMZJWUVAZOVHWUDRULVEQPENAXWKDQDJZFOIP9TBDQAVKBDEBDRQBYPRFKUZBGNFKWTKIZG
NextRoot: VXKOVYWBHLSWLLOPPVPRZKXJ9BIHQPOGFCUSORYUCTJMWHLLNALUZPWZXDZIHRNH9USXOOPDMA9BFVUTK
sidekey: LJNQJJELTBVLVHHNUBIJDLVBWAJQPWSKHIDPUMHFNZQKJTYAFPQSGAXAWKIDALAUYDCMXXJXNLPXRDUFE
Attaching to tangle, please wait...
done

Figure 8. Cluster head login and registration screen.

Figure 10 portrays the sensor node reading the sensor data. After receiving the sensor data, the Arduino Nano encrypts the packet and forwards it to the Raspberry Pi via MQTT. Once the sensing data are successfully uploaded to the Tangle, users can utilize the IOTA Tangle Explorer to query the recorded data. Furthermore, Figure 11 presents the First Root, certificate information, certificate validity, and hashed sidekey of the Cluster Head in WSNs. Figure 12 showcases the Dashboard of the node in the Base Station, enabling users to view the current running status. The "Synced" indication on the node signifies that it has been operating synchronously with the IOTA network.

These experimental results provide visual representations of the system's functionality and data flow, highlighting the successful execution of key processes such as registration, login, message transmission, encryption, and data recording. The figures offer a comprehensive understanding of the system's operation, further supporting the proposed methodology's effectiveness and demonstrating the feasibility of preserving data from industrial automation and control systems using IOTA technology.

Figure 9. The cluster head receives the packet screen.

Figure 10. Sensor node upload package screen.

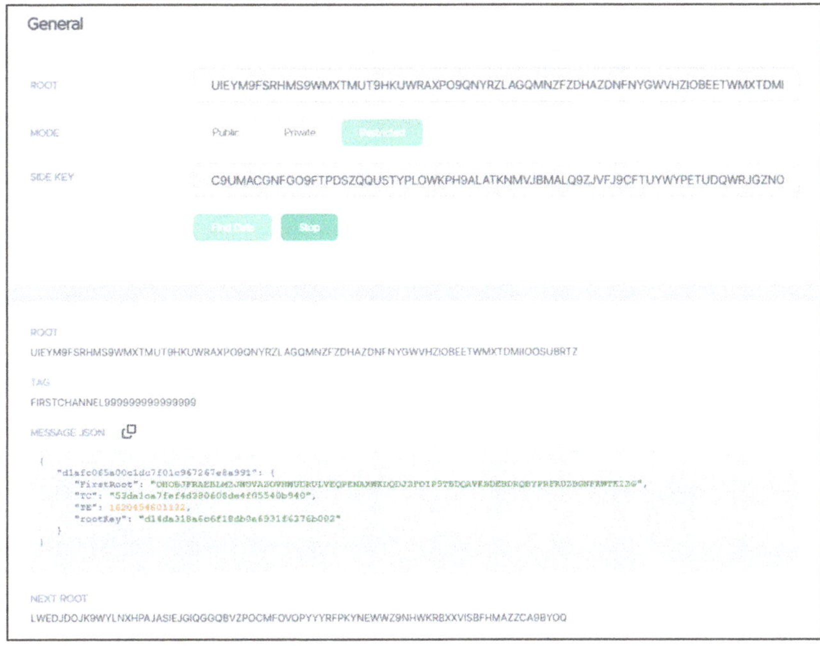

Figure 11. The content of the message stored in the base station channel 1.

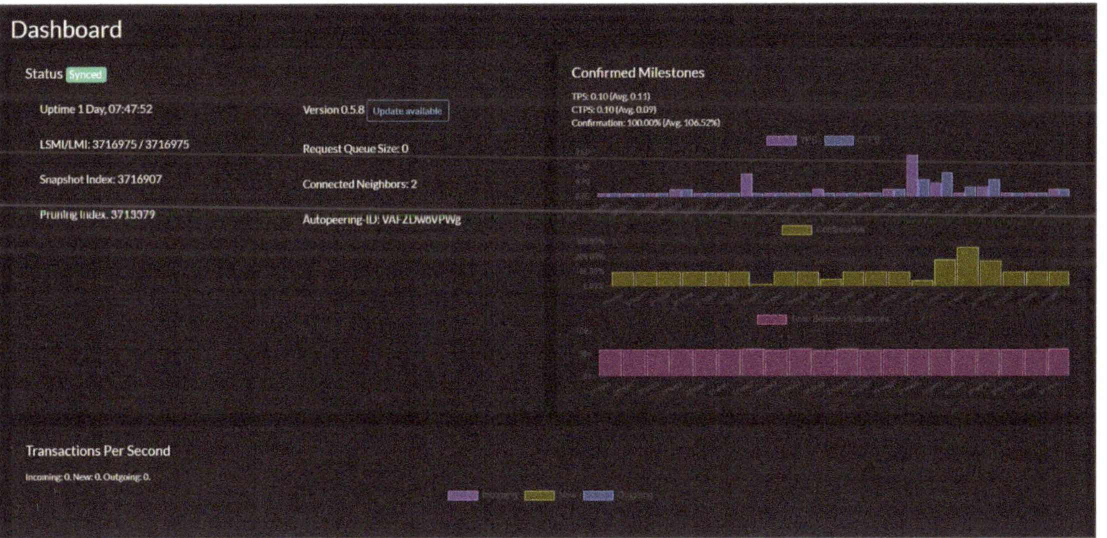

Figure 12. Node dashboard screen.

5. Conclusions

Numerous achievements have been made in the application of IOTA in industrial automation and control systems, with several papers proposing solutions to integrate IOTA into these systems. However, these solutions often overlook the constraints faced by wireless sensor networks, where many sensor nodes have limited computing power and storage capacity. Particularly in industrial control networks, sensing data must traverse multiple network layers before being uploaded to the distributed network, necessitating the protection of security and privacy in the intermediate stages. The IOTA Foundation has proposed solutions for IoT devices, including the design of hardware devices and firmware such as CryptoCore and STM X-Cube-IOTA1. Nevertheless, the lack of hardware support poses a hindrance to the development of industrial automation and control systems as these devices still require substantial computing power. Although lightweight Bee nodes have been enhanced by IOTA for data upload actions, not all hardware is capable of supporting these nodes. Therefore, this paper focuses on conducting identity authentication for resource-constrained sensing devices and securely uploading their sensing data to the Tangle for storage. The implementation of IOTA in industrial automation and control systems is demonstrated in this paper, with satisfactory execution results.

By addressing the challenges of scalability, security, and privacy in existing centralized architectures, the proposed solution holds promising implications for various stakeholders in this domain.

1. Enhanced Scalability: The utilization of the IOTA DAG technology allows for improved scalability in managing a large number of sensors deployed in industrial automation and control systems. The decentralized nature of the DAG structure enables efficient communication and verification among the sensors, fostering a more scalable and resilient network.

2. Strengthened Security and Privacy: The integration of the IOTA DAG technology provides enhanced security and privacy measures for data preservation. The utilization of a distributed ledger ensures data integrity and confidentiality during the transmission process, mitigating vulnerabilities associated with centralized approaches. This has significant implications for ensuring the confidentiality and integrity of sensitive data in industrial automation and control systems.

3. Transparent and Immutable Data Record: The adoption of the IOTA DAG technology enables the establishment of a transparent and immutable record of the collected data. This has implications for auditability, compliance, and trustworthiness in various domains, such as supply chain management and regulatory compliance, within the Industry 4.0/5.0 context.

4. Potential Cost Reduction: The implementation and execution of data preservation using the IOTA DAG technology have the potential to reduce costs associated with importing blockchain technology into the Internet of Things (IoT). The decentralized nature of the DAG structure eliminates the need for intermediaries, resulting in potential cost savings for the stakeholders involved.

Ultimately, these implications can drive the widespread adoption of decentralized and secure approaches in the industrial automation and control systems domain, unlocking their full potential and benefits.

Author Contributions: Methodology, P.-C.T.; Software, Y.-S.C.; Validation, T.-C.W.; Project administration, I.-C.L. All authors have read and agreed to the published version of the manuscript.

Funding: This research was funded by the Ministry of Science and Technology, grant number 112-2218-E-005-007.

Data Availability Statement: Not applicable.

Conflicts of Interest: The authors declare no conflict of interest.

References

1. Agrawal, R.; Verma, P.; Sonanis, R.; Goel, U.; De, A.; Kondaveeti, S.A.; Shekhar, S. Continuous security in IoT using blockchain. In Proceedings of the 2018 IEEE International Conference on Acoustics, Speech and Signal Processing (ICASSP), Calgary, AB, Canada, 15–20 April 2018; pp. 6423–6427.
2. Wang, Q.; Zhu, X.; Ni, Y.; Gu, L.; Zhu, H. Blockchain for the IoT and industrial IoT: A review. *Internet Things* **2020**, *10*, 100081. [CrossRef]
3. Datta, S.K.; Haerri, J.; Bonnet, C.; Da Costa, R.F. Vehicles as connected resources: Opportunities and challenges for the future. *IEEE Veh. Technol. Mag.* **2017**, *12*, 26–35. [CrossRef]

4. Hang, L.; Ullah, I.; Kim, D.-H. A secure fish farm platform based on blockchain for agriculture data integrity. *Comput. Electron. Agric.* **2020**, *170*, 105251. [CrossRef]
5. Guan, Z.; Lu, X.; Wang, N.; Wu, J.; Du, X.; Guizani, M.J. Towards secure and efficient energy trading in IIoT-enabled energy internet: A blockchain approach. *Future Gener. Comput. Syst.* **2020**, *110*, 686–695. [CrossRef]
6. Grecuccio, J.; Giusto, E.; Fiori, F.; Rebaudengo, M. Combining Blockchain and IoT: Food-Chain Traceability and Beyond. *Energies* **2020**, *13*, 3820. [CrossRef]
7. Popov, S.J. The Tangle. White Paper 2018; Version 1.4.3. Available online: http://cryptoverze.s3.us-east-2.amazonaws.com/wp-content/uploads/2018/11/10012054/IOTA-MIOTA-Whitepaper.pdf (accessed on 18 June 2023).
8. Ericsson. White Paper, Cellular Networks for Massive IoT, Uen 284 23-3278. January 2020. Available online: https://www.ericsson.com/en/reports-and-papers/white-papers/cellular-networks-for-massive-iot{-}{-}enabling-low-power-wide-area-applications (accessed on 18 July 2023).
9. Alsboui, T.; Qin, Y.; Hill, R.; Al-Aqrabi, H. Enabling distributed intelligence for the Internet of Things with IOTA and mobile agents. *Computing* **2020**, *102*, 1345–1363. [CrossRef]
10. Zhang, K.; Tian, J.; Xiao, H.; Zhao, Y.; Zhao, W.; Chen, J. A Numerical Splitting and Adaptive Privacy Budget-Allocation-Based LDP Mechanism for Privacy Preservation in Blockchain-Powered IoT. *IEEE Internet Things J.* **2023**, *10*, 6733–6741. [CrossRef]
11. Jayabalan, J.; Jeyanthi, N. Scalable Blockchain Model Using Off-Chain IPFS Storage for Healthcare Data Security and Privacy. *J. Parallel Distrib. Comput.* **2022**, *164*, 152–167. [CrossRef]
12. Divya, V.; Sri, R.L. Docker-Based Intelligent Fall Detection Using Edge-Fog Cloud Infrastructure. *IEEE Internet Things J.* **2020**, *8*, 8133–8144. [CrossRef]
13. Singh, C.; Kumari, P.; Mishra, R.; Gupta, H.P.; Dutta, T. Secure Industrial IoT Task Containerization with Deadline Constraint: A Stackelberg Game Approach. *IEEE Trans. Ind. Inform.* **2022**, *18*, 8674–8681. [CrossRef]

Communication

The Industrial Digital Energy Twin as a Tool for the Comprehensive Optimization of Industrial Processes

Alejandro Rubio-Rico [1,*], **Fernando Mengod-Bautista** [1], **Andrés Lluna-Arriaga** [1,*], **Belén Arroyo-Torres** [1] and **Vicente Fuster-Roig** [2,*]

[1] Instituto Tecnológico de la Energía (ITE), Avda. Juan de la Cierva, 24, 46980 Valencia, Spain; fernando.mengod@ite.es (F.M.-B.); belen.arroyo@ite.es (B.A.-T.)

[2] Instituto de Tecnología Eléctrica, Universitat Politècnica de Valencia, Camino de Vera s/n Edificio 6C, 46022 Valencia, Spain

* Correspondence: alejandro.rubio@ite.es (A.R.-R.); andres.lluna@ite.es (A.L.-A.); vicente.fuster@ite.es (V.F.-R.); Tel.: +34-961366670 (A.R.-R.)

Abstract: Industrial manufacturing processes have evolved and improved since the disruption of the Industry 4.0 paradigm, while energy has progressively become a strategic resource required to maintain industrial competitiveness while maximizing quality and minimizing environmental impacts. In this context of global changes leading to social and economic impact in the short term and an unprecedented climate crisis, Digital Twins for Energy Efficiency in manufacturing processes provide companies with a tool to address this complex situation. Nevertheless, already existing Digital Twins applied for energy efficiency in a manufacturing process lack a flexible structure that easily replicates the real behavior of consuming machines while integrating it in complex upper-level environments. This paper presents a combined multi-paradigm approach to industrial process modeling developed and applied during the GENERTWIN project. The tool allows users to predict energy consumption and costs and, at the same time, evaluates the behavior of the process under certain productive changes to maximize consumption optimization, production efficiency and process flexibility.

Keywords: digital twin; smart manufacturing; energy; energy efficiency; productive flexibility

Citation: Rubio-Rico, A.; Mengod-Bautista, F.; Lluna-Arriaga, A.; Arroyo-Torres, B.; Fuster-Roig, V. The Industrial Digital Energy Twin as a Tool for the Comprehensive Optimization of Industrial Processes. *Processes* **2023**, *11*, 2353. https://doi.org/10.3390/pr11082353

Academic Editor: Sergey Y. Yurish

Received: 5 June 2023
Revised: 31 July 2023
Accepted: 1 August 2023
Published: 4 August 2023

1. Introduction

In the context of evolution from "climate change" to "climate crisis", the industrial sector must face a series of changes and evolve to new, more efficient manufacturing models in the context of the revolution of the Industry 4.0 paradigm and digitalization. Different tools are sufficiently developed and tested to be applied with success and help maximize productivity and minimize energy consumption and costs, both concepts highly related. The energy efficiency of manufacturing systems arises again as a key domain on which there is still a high degree of improvement to be achieved. Nevertheless, manufacturing companies present different levels of development and application of energy efficiency measures, and this situation gets even worse in cases where a minimum degree of digitization is required. In fact, a dynamic and continuous evaluation of the energy consumption of the process and its relationship with the internal and external environment seems to be necessary to effectively face previously cited challenges in a globally connected and competitive world.

Digital Twins (DT) are one of the main tools developed in this context. Appearing in the early 2000s [1], the concept represents the idea of developing a digital copy of a physical object. This copy, technically known as a twin, allows for the implementation of process improvements and optimization of its operation in a fully controlled and risk-free environment. However, different approaches can be examined when Digital Twins are developed and deployed— some of them more linked to the Internet of Things (IoT) concept than others, but in any case, are focused on replicating the actual performance of the manufacturing process. In this context, Digital Twins prove to be a valuable tool to predict and improve energy consumption as a specific use case. This document will examine the current context of DT developed and implemented in project GENERTWIN for improving the energy efficiency of manufacturing processes, their potential as energy efficiency improvement tools, and their consequent characteristics and particularities. The paper focuses on highlighting the advantages of multi-method integration of energy-production models in complex environments with the main objective of replicating energy consumption, energy cost and productive performance.

2. Digital Twins for Energy Efficiency Improvement of the Industrial Process

2.1. The Challenge of Improving Energy Efficiency in the Manufacturing Industry

The way to reach a good level of energy efficiency in manufacturing processes has traditionally been based on two trends; (1) the introduction of new process technologies to improve efficiency and (2) the introduction of specific improvements in plant operations [2]. However, this second approach has never been sufficient to understand the causality inherent in the operation of manufacturing processes in sufficient detail to propose improvements that dynamically adapt to the process context and changes. Some energy efficiency analysis techniques, such as performance measurement and verification protocols, come close to the level of detail needed to achieve such intuition. These kinds of tools, complemented by internationally recognized implementation support protocols such as ISO 50001 [3] referred to Energy Management Systems or energy audits (in Spain, according to Royal Decree 56/2016 following Energy Efficiency Directives), provide a general development framework that is explicitly adapted to the particularities of each application case [4]. As a consequence, the state of the art of energy efficiency analysis of industrial processes is usually quite far from the implementation of more advanced techniques than those described, seeking, in any case, high-impact technological changes but not so much an optimization methodology based on dynamically updated measures. Figure 1 summarizes the state of the art of tools for analyzing energy efficiency on manufacturing processes.

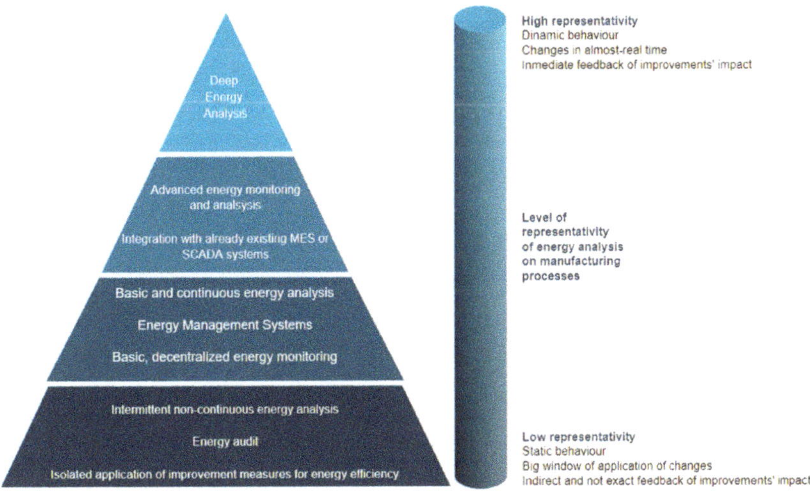

Figure 1. Different levels of representativity and commonly used tools for analyzing and optimizing the energy efficiency of the process. Source: ITE.

Even though it is not simple to accurately evaluate the current situation of the different manufacturing sectors in terms of energy efficiency techniques' degree of adoption (And, specifically, digitalization degree as a key dimension for energy efficiency optimization in time), systematic evaluations of some sub-sectors reveal a large potential for energy efficiency improvements in energy-intensive sectors [5]. This conclusion can be extended to other manufacturing sectors and can be affected by the economic environment and other complex factors, resulting in high discrepancies in energy efficiency among sub-sectors [6]. Another important consideration in scenarios of a medium-high degree of adoption of energy efficiency measures is the digital barrier of the traditionally conceived architecture of processes, which grow according to economic or legal requirements or needs, rather than evolving and growing in a more organic way and based on a structure that supports the integration of the basic mechanisms for calculating and evaluating energy efficiency. A representative example is the ceramics sector in Spain, which has achieved significant reductions in CO_2 emissions over the past two decades but still faces a great challenge of emissions reduction, most of which are directly related to the improvement of energy efficiency of the manufacturing process [7].

2.2. Digital Twins Usage for Energy Efficiency Optimization in Manufacturing Environments

The present and future context of manufacturing companies requires rethinking the strategy for maximizing process energy efficiency while meeting the required production and quality standards. Consequently, it is necessary to have tools that allow flexible adaptation to every change in productive and non-productive environments. In this context, multiple authors describe the DT as a key enabler in accomplishing this challenge despite not being sufficiently implemented in Smart Manufacturing Systems (SMS) [8]. It is important to note that the design of DT has not been developed at the same pace as their interest has grown [9]. In fact, the concept proves to be sufficiently flexible to be applied to a heterogeneous set of industries, such as healthcare [10], farming [11] or construction [12]. This last example of application leads to highlight the importance of energy optimization in almost all use cases involving the usage of building and/or equipment. In the particular case of the manufacturing industry, this leads to considering energy consumption optimization as a key and valuable application of DT to industrial applications closely linked to operational efficiency [12].

Different approaches of DT focused on improving the energy efficiency of the manufacturing process arise when examining the state-of-the-art, most of them related to the concept of making existing and future manufacturing systems more sustainable. Energy consumption depends on different factors (it gets even more complex depending on the sector and manufacturing activities), but in any case, the manufacturing system operation, including equipment, human and material resources management, turns out to be the dimension most directly linked to the dynamic energy consumption of the industrial plant [13]. The concept of dynamic energy consumption is meant to reflect the importance of considering energy consumption as something variable in time, which leads to the formulation of how to integrate the energy consumption dimension of the plant with the inherent and future production and operational flexibility in an organic and coherent way. Manufacturing system reconfiguration and optimization is one of the references DT architectures allowing increasing productivity while reducing energy consumption, to the point of including energy consumption and energy cost as two of the most important Key Performance Indicators (KPI) when reconfiguring the optimized productive scenario. However, this reference architecture may cover different phases of the plant's use from the design and engineering phase, construction phase and, in the context of the construction sector, even the demolition phase, in addition to the already referenced maintenance and operation phase, making it possible to apply it to the concept of lifecycle of the project [14]. Even though energy consumption optimization has high importance in each and every one of the phases described above, the approach of DT presented in this paper is mainly focused on the operation phase of the industrial site.

In this application scenario, DT can potentially provide the company with the ability to predict the behavior of its process in different scenarios. These scenarios can be oriented to analyze different dimensions of the process, such as predictive maintenance, performance monitoring and control, better resource planning, or even facilitating renewable energy integration, to mention a few examples [15]. The DT can, in fact, be designed bearing one or more of these functionalities in mind, and it may even evolve to include new ones at a later stage. Nevertheless, the existence of Digital Twin manufacturing systems reflects how high levels of fidelity and complexity can be reached when proposing a bottom-up approach from the design phase of the SMS [16]. This approach seems highly interesting and even essential to the development of new manufacturing systems aligned with the Industry 4.0 paradigm, along with dynamic and advanced mechanisms to evaluate and promote energy efficiency in the complex context of manufacturing processes. As a consequence, it seems important to consider a flexible and adaptable architecture of DT highly focused on energy consumption prediction and evaluation considering process parameters, the productive environment and external factors following the requirements to provide energy efficiency improvements.

Digital Twins, as a tool for effectively integrating energy consumption with operational performance in industrial plants, is currently not sufficiently developed, at least considering the need to calculate a dynamic energy consumption value susceptible to being affected by every important productive parameter and integrated into their environmental context. When descending to the machine level, some models consider mathematical approaches to reproduce power consumption along with carbon footprint from the basis of the job scheduling of the simplified application case [17]. This approach proposes a correct focus on the problem of predicting energy consumption but also poses the problem of the replication of the model to other types of machines or scopes. In addition, the approach has the disadvantage of not considering a broader application and contextual environment in the model, as previously mentioned. This model should be integrated into a Digital Twin and would also cover part of the specifications needed to consider a Digital Twin for energy efficiency. Another alternative method also involves nonlinear planning models, to mention an example [18]. In any case, the results provided by the mathematical methods are not as important as the fact they depend on the quantity, quality and representativity of data and to which extent it can be extrapolated to other machines or processes.

Alternative approaches integrate both energy and production data through cyber-physical systems, focused on sustainable production [19] but basically framed within the traditional concept of an energy monitoring system. In fact, most approaches employ discrete modeling of the process or production chain to introduce discretized energy consumption concepts [20]. These approaches present the same problem of a purely mathematical approach as the previously described, that is, the lack of a dynamic design of calculation criteria intrinsic to the operation of machinery in which possible internal and external changes are considered. Those models are focused on the site supply chain logistics and optimization of material resources in most cases in which energy consumption shows a characteristic dynamic behavior that this kind of model cannot process. On the other hand, this kind of model can be adapted to different machines and environments in an easier way. These models aligned with the concept of DT for energy efficiency of the process are supposed to be designed to allow for the study of the final and disaggregated impact of the variation of specific operational control parameters while improving technical service and maintenance—all of which should be performed in a digital environment that is controlled, connectable and with a much lower risk.

Further approaches involve working with sufficiently representative datasets, including energy consumption and the rest of the related productive and non-productive parameters, to apply purely mathematical algorithms or other advanced enablers, such as machine learning algorithms. In fact, Artificial Intelligence (AI) as a technology is a very valuable key enabler to boost and utilize the capabilities a DT can provide in the best possible way [14]. However, again, in this case, we start from a difficult point since this approach imposes the need for a high quantity of well-structured and significant registers of energy and related data to train the corresponding algorithms. Some examples show good results of a Neural Network (NN) trained to predict energy consumption in specific applications at a low abstraction level of a machine, and considering all the necessary data for training, the algorithm is available. Nevertheless, this approach also poses the problem of the difficulty of the task of developing energy consumption models based on the interrelated parameters that influence energy consumption [21].

As a conclusion of the previous approach, it can be deduced that the real challenge lies in developing Digital Twins with a specific focus on the detailed calculation of the energy consumption of the machinery, which, in turn, must consider operational characteristics of the models, but, at the same time, not excessively depending of data availability since, in most cases, and particularly related to energy digitalization, the data are not easily accessible and complete datasets including energy, productive and non-productive data are not available.

2.3. Project GENERTWIN

The objective of the GENERTWIN "Sistema Digital de Análisis de procesos industriales para GENERación de escenarios alternativos bajo consideraciones productivas y de eficiencia energética" project is the development and implementation of a DT applicable to different manufacturing sectors considering the optimization needs and energy dependence of these processes. The Digital Twin (DT) is developed following the identified needs after a deep situation analysis, including the identification of the use cases and the functionalities that this twin must provide to the company.

The DT is developed to a low level of detail and considers the internal dynamics of the machinery involved in the process, its operation, the relationship with the necessary resources and the impact of the different possible variations on the quality and type of products obtained. By considering it as a DT, the model consequently scales in complexity and capabilities, allowing not only to analyze the operation and interactions of the aforementioned elements that are part of the process but also to add value to the company's decision-making process. The DT presents an architecture specifically designed to maximize the representativeness of energy consumption and cost prediction and its relationship with productive and non-productive contexts.

2.4. Constraints and Challenges

The development and implementation of the proposed architecture of DT present a series of constraints and challenges. The first of them is the identification of the real expectations and needs of the companies on which the development of the tool is focused. On many occasions, a bad design of the DT can lead to an excess of work in the development that entails a higher final cost. On the other hand, insufficient development of the model can result in the DT losing the representativeness of its results. Therefore, an optimal balance between both points must be found to achieve a balanced, accessible system that meets the needs of the company.

In order to mitigate the potential design problem as much as possible, a previous design phase has been developed in which the company using the process, participates by receiving the appropriate feedback. In addition, a multisectoral analysis of requirements and possible interests of other follower companies is carried out in the project to maximize the representativeness and conduct the final solution to the most effective development possible. Another major challenge to solve when developing a model is data accessibility, as previously explained. It is common to find a lack of information, even among process specialists, about the details of its operation and internal mechanisms. In addition, the relatively low penetration rate of digital monitoring systems collecting energy and production data is also a handicap that hinders development, as cited above.

3. Digital Twin Development

3.1. Basic Architecture

The Digital Twin is structured in different layers, as shown in Figure 2. The lower layer refers to the Energy Productive Model (EPM), which represents the simulation core considering the dynamic models of the processes together with the simulation of discrete events to reproduce the manufacturing activity. The Digital Twin environment, which gathers the interaction and analysis functionalities in a digital environment, is developed using a specialized hybrid and multi-paradigm simulation software, including multi-agent modeling, which also allows integrating and testing models and interfaces. Finally, there is an upper layer of the system itself, which is designed to be integrated with other types of plant systems with specific interconnectivity capabilities and connect the simulation models as analytical advanced digital tools with the Internet of Thing layers and other automation architectures [22].

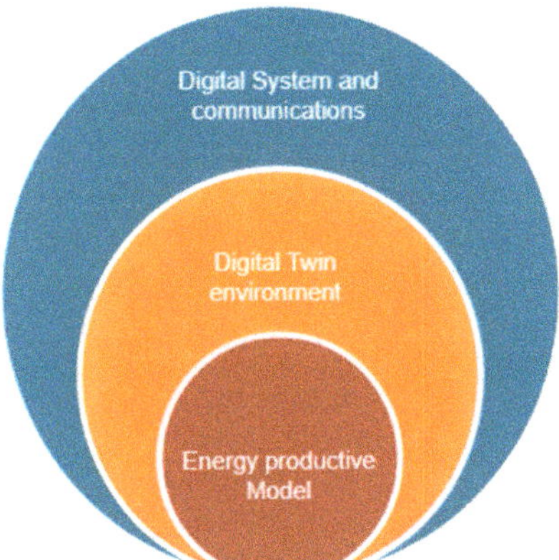

Figure 2. Structure of the DT architecture in basic high-level layers. Source: ITE.

The lower layer of the DT contains all simulation methods and behaviors at the machine and resource levels. This layer combines the necessary technologies and calculation mechanisms to solve every energy prediction problem in the most standardized and replicable way. Every part of this layer is developed to maximize the ratio of the representativity of the predictions against the complexity of the programmed behavioral functions. A representative example developed in the project is the implemented mechanism for calculating the energy consumption of an industrial furnace. Every piece of this simulation element has been developed, combining the expert knowledge of this kind of thermal-transfer process with some simplifications with the goal to be potentially adapted to similar industrial furnaces of similar characteristics. It has also been developed, guaranteeing the exact correlation between energy consumption and production parameters, such as the dimensions of the product, type of material and other characteristics (ambient conditions, quantity of products per productive line, initial humidity of the sub-product entering the furnace, etc.). Including this kind of parameter also justifies the importance of the proposed architecture of DT connected to the production schedule. It is important to note that every problem has been divided into sub-problems to be solved by different means, including machine state characterization and implementing a variety of calculation algorithms with the common framework of replicating the energy behavior of the machine, its power consumption curve and all the necessary calculations of energy consumed in every case.

The use of System Dynamics (SD) calculation methods is highlighted as a powerful approach to replicating individual behaviors of different agents, especially for transient states. In fact, the representativeness of the process is guaranteed in this layer as a solution for the referred challenge in Section 2.4 since the proposed architecture allows for separate testings of every element and module, while the general requirements, such as comparing the integration of the power curves of different consuming elements with general consumption (which is the most commonly available energy data in the manufacturing industry), can be met.

The intermediate layer of the DT includes all the environments in which machines develop their behavior and interact with human and material resources and the typical constraints of manufacturing scenarios, such as work schedules, scheduled stops, or fail ures of machinery. This layer is an important environment in which previously referred machines develop their behaviors and where the impact of external changes (not depending on the machine configuration itself or operational parameters) is calculated. This layer also integrates the impact of different machines to provide aggregated KPIs affecting production lines or even a whole factory, depending on the scope of the DT and the need for output information required. This layer has been developed following the needs and restrictions of companies participating in the project.

Finally, the higher layer of the DT is mainly focused on connecting and exchanging information with other systems and allowing interactions with users through User Inter faces (UI). The development of this layer is determined by the already existing data and the type and amount of data to be exchanged. It also includes the use of internal or external DataBases (DB) according to the characteristics imposed by such data. The project has considered non-relational databases. In particular, a time series database because of the structure of the registers. However, it also can connect to relational DB, such as those developed in SQL.

The proposed architecture aims to deal with some of the problems identified in Section 2.4. The communication layer is intended to solve the challenge of data availability, along with the simulation layer; when the data are available (in fact, when data are available with enough quality and can be effectively imported from already deployed energy meters) this layer implements all the necessary mechanism to integrate dataflow in the DT while, at the same time, simulation layer will include numerical or mathematical models to represent each DT element behavior, following a black-box approach. Alternatively, when the data are only partially available or not available at all, the burden of effectively replicating the behavior of the elements of which the DT is composed falls upon the Simulation layer implementing energy models depending on a first-principles approach in a higher degree rather than excessively depending on data. The following Figure 3 represents these layers structure and the related element, tools and dimensions with which each layer is related.

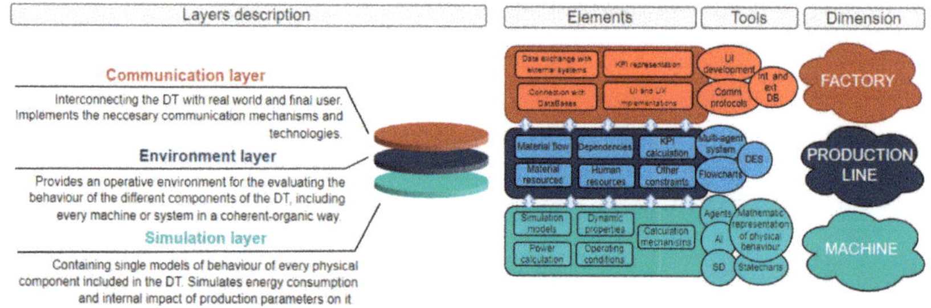

Figure 3. Description and elements of each layer of the DT. Source: ITE.

3.2. The DT Implementation Process

The process of implementing a DT is complex and, sometimes, not straightforward particularly during model design and development. In fact, it is difficult to find research documents about DT implementation methodologies with formal applications to the man ufacturing industry. Similar approaches, in this case concerning the development and application of a DT to a ship, show how the manufacturing process of the object is followed by corresponding steps of development of the DT, from design/development of the "digital mockup", to operation/application of the DT to production phase [23]. Other references current implementation procedures based on defining individual DT to be integrated into an aggregated DT as a hierarchical composition of the rest [24]. This kind of methodol ogy is designed to consider the deployment of multi-purpose DT using a standardized architecture whose data model is, in this case, based on AutomationML. Nevertheless the scope of this kind of DT is much greater than the proposed in this paper, focused on the energy efficiency of the process. The DT for Energy Efficiency developed and applied in GENERTWIN could potentially serve to be integrated into higher-level DT like those described in the preferred source. At a low level, referred methodology splits DT into different components such as storage, method, access control, communication interface, etc. [24]. This approach is not compatible with the architecture presented in the project since the interdependence of each of the three layers containing all the mentioned elements prevents the development of each of the elements independently if this development path is intended to be followed.

Therefore, it is determined that the implemented methodology should include a design phase focused on identifying key functionalities. Also, due to the importance of developed models and their performance and degree of accuracy during operation, the development phase is divided into different steps to iteratively develop the model. Finally, the operation phase is not considered since the proposed methodology just aims to be applied to the deployment procedure, not to the operation procedure. In consequence, a standardized methodology was developed and followed to guide the implementation process based on stages, in which clear objectives are defined with end users. It is important to agree on the functionalities the system must achieve before launching the development and to continue working on these functionalities during the development of the model. Figure 4 represents all the referred steps and the methodology structure.

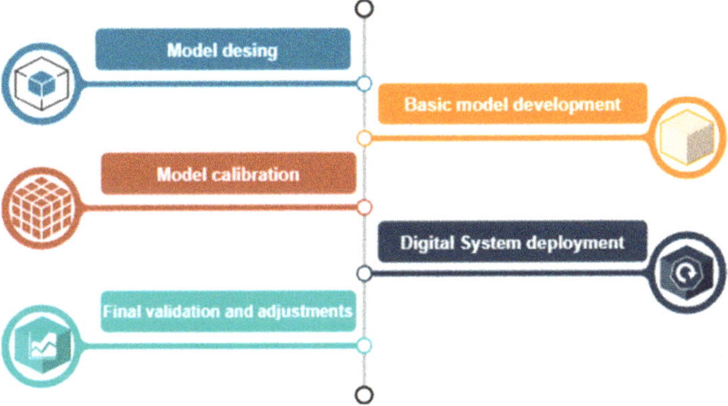

Figure 4. Necessary steps to accomplish in the project. Source: ITE.

It should be noted that the developed methodology in the project aims to structure a highly iterative process, such as a development that includes a Digital Twin. For this reason, the model must be constantly reexamined and reevaluated to ensure that it meets the required representativeness, allowing it to be sufficiently flexible to add necessary new elements or functionalities. However, the project seeks a balance between sufficient representativeness in terms of energy performance and the feasibility of a time-bound development. This approach seems the most adequate to deal with managing client expectations of the size and complexity of the implemented DT, a major challenge previously identified in Section 2.4.

3.3. Functionalities and Examples of Implementation

As mentioned above, DT for energy efficiency analysis and optimization offers a wide range of possibilities for simulations and improvements to the manufacturing process. The main functionalities provided by the developed DT cover energy consumption and cost prediction of the replicated environment as a core. Some practical application cases are described in the following point, but it is important to note how this approach replicates a detailed level of energy consumption and productive behavior of every machine implemented and how its impact is scaled, along with the impact of other machines, resources, and other agents in a complex environment, as represented in the architecture's description. This application is aligned with the potential of DT identified in previous research, which can provide valuable information in production phases evaluating an operating system under different conditions [24]. It is also noted that DT can help small and medium-sized enterprises to improve operational performance and data acquisition systems. In fact, the GENERTWIN project considers data will not always be easily available and integrable in the DT. This lack of data is compensated with the use of physical and mathematical models that are not excessively dependent on data availability.

An example of energy prediction is shown below in Figure 5. The following graphs show a reconstruction of a power load curve of a process under specific operating conditions of a thermal heating element during a transitory heating phase.

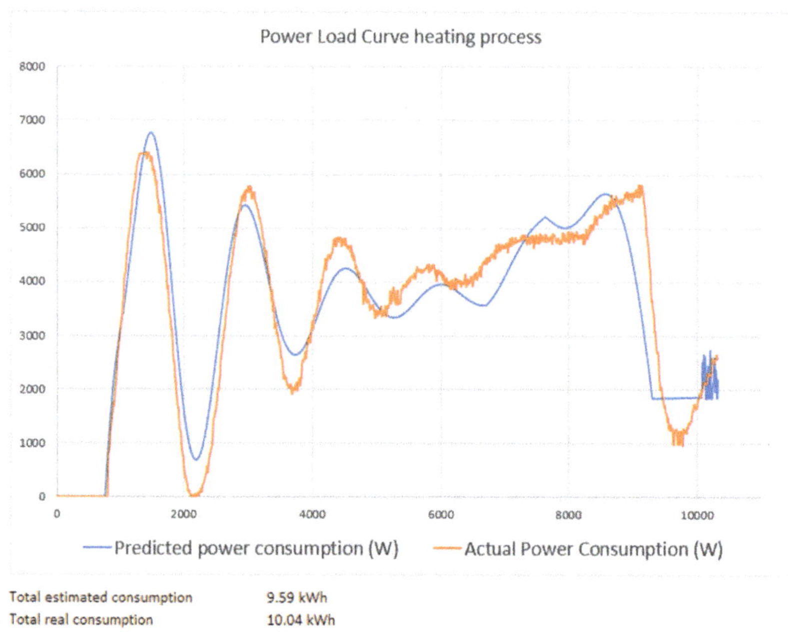

Figure 5. Power load prediction and actual prediction of heating element, in kW. Source: ITE.

The model, in this case, shows a good accuracy, calculated through the Mean Absolute Percentage Error (MAPE):

$$MAPE = \frac{100}{n} \times \sum_{i=1}^{n} \left| \frac{y_{actual} - y_{predicted}}{y_{actual}} \right|$$

Energy consumption and the economic cost of energy are also calculated. This model has been implemented by adapting and combining convection, radiation, and heat storage equations in a single model. An example of a convection heat transfer equation is possible, the simplest one of them:

$$\dot{Q}_{conv} = hA_s(T_s - T_\infty)$$

For this specific example, this combination of equations is used because the heating process is made without admitting biomass or other kind of materials. In the event of a material input entering the region heated by the element, its characteristics are integrated into the model to adapt simulations to it. This example shows the level of detail of simulations implemented since this element is integrated into bigger and more complex elements and modules, as explained before. This continuous nature of the energy variable must, however, be integrated into the discrete logic of the operation of the industrial machinery. The following Figure 6 represents a State Machine Diagram (SMD), or statechart, which implements the behavior of different elements integrated into a single logic agent, either machine or resource:

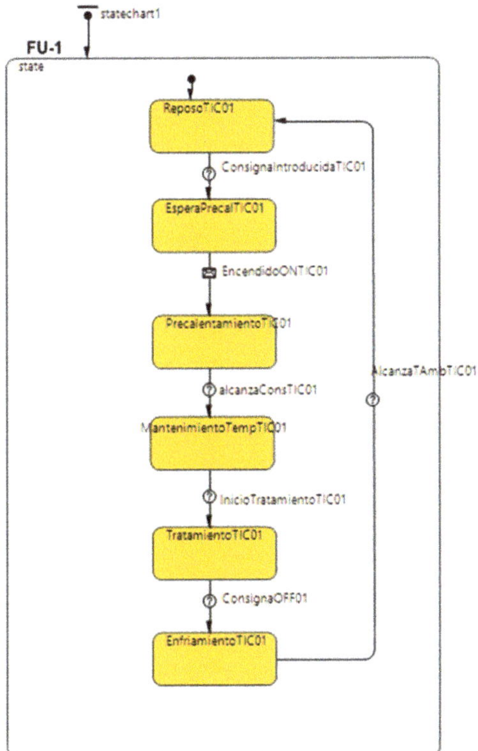

Figure 6. Example of statechart implementing machine's behaviors. Source: ITE.

This chart is implemented by splitting the real behavior of a component into different phases, which significantly helps to model this behavior in different parts while maintaining its continuous nature. Although being a powerful combination of both continuous and discrete behaviors, individual models of energy consumption and statecharts are not the single tools implemented in the proposed architecture of DT for energy efficiency of the process. In fact, some complex structures are created to integrate data and scale up KPI calculation. The following Figure 7 shows an example of the implementation of a mass flow balance.

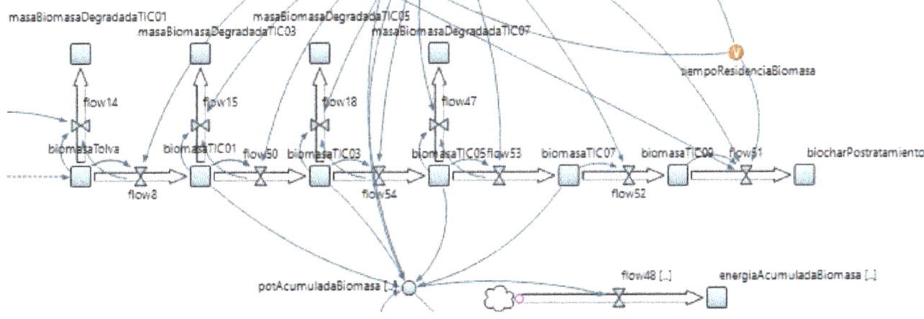

Figure 7. Example of mass flow balance for a horizontal element. Source: ITE.

The development of this mass flow includes the use of differential equations to dynamically evaluate changes, in this case, related to a biomass' flow through each horizontal section, considering degraded and not-degraded fraction change:

$$\frac{dm_t}{dt} = \frac{dm_{nd}}{dt} + \frac{dm_d}{dt}$$

To conclude, the following final example shows in Figure 8 a more complex system of burners of a machine, including process parameters, energy consumption and data interactions of the DT, with the final objective of calculating the energy consumption of different integrated elements of a single machine. This structure of variables, equations and data connections considers the behavior of each burner of a furnace along with interactions with the characteristics of the burning zone, the air, the gas, and the rest of the burners and their influence on the heart of the furnace.

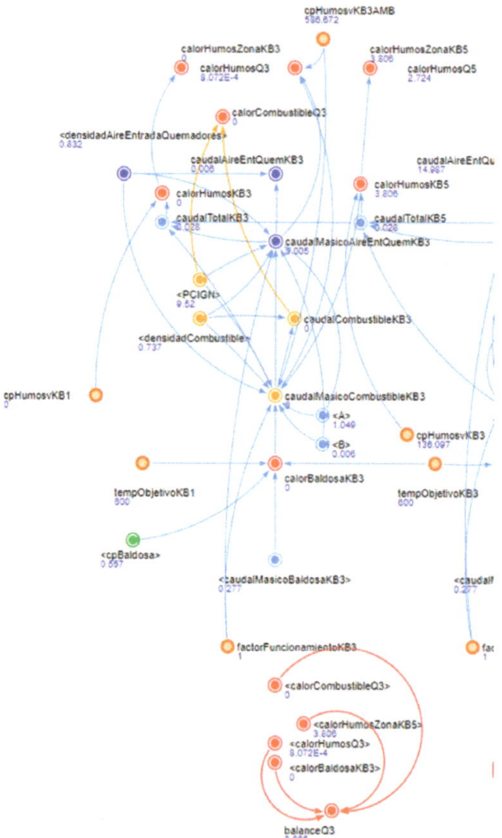

Figure 8. Implementation of correlation between parameters and energy consumption calculation of a complex environment of a DT of the thermal process. Source: ITE.

In the following Figure 9, a standard example of a dashboard for energy consumption is presented, designed to dynamically show a representation of energy consumption and cost divided by period, machines and other criteria aligned with final user requirements.

Contents, UI design, and navigation are adapted to meet final client requirements in this case, to predict energy consumption per hour, day, week and month, energy consumption per machine, and the power load curve of the whole process and every machine. It is important to emphasize the complexity underlying showing all this information, in this case, including a complete cost-calculation mechanism applied to energy consumption forecasting to give much more representative results of the economic impact of every simulated scenario.

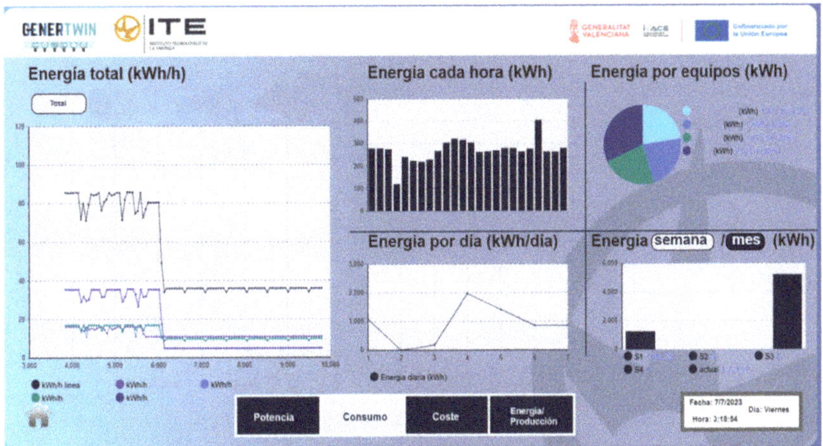

Figure 9. Example of User Interface implemented for visualizing energy consumption and cost of each DT scenario generated. Source: ITE.

3.4. Application and Impact

The DT is designed with the clear functionality of predicting process behavior in the face of substantial production changes, especially focused on the impact on energy consumption of both productive and non-productive variables. In this sense, it simulates different states of the system, displaying predictions and integrating them into the data environment in which the process is developed. This allows optimization adjustments to be made and even optimal production sequencing algorithms to be executed according to the energy cost. At this point, a definition of Energy Efficiency must be provided as a core concept of the proposed definition of DT. The International Performance Measurement and Verification Protocol (IPMVP) [25] proposes defining a baseline situation and calculating differences before and after applicating measures for improving energy consumption, as represented in Figure 10. This protocol includes different options for evaluating the impacts of changes affecting energy consumption (such as temperature for HVAC systems, for example) as adjustments. A final form of adjustment is, in fact, using Simulation Models to assess the exact impact of Energy Efficiency measures in a reliable way when measuring is not possible.

Consequently, it can be stated that the proposed DT simulation environment is completely aligned with this approach if all the representative parameters affecting the energy consumption of simulated systems are included. The energy efficiency of a system will be considered as energy savings depending on the defined context as follows:

$$EE(\%) = \frac{EC_{baseline} - EC_{change}}{EC_{baseline}} \times 100$$

This formula includes a calculation of energy consumption (EC) in both the baseline simulation scenario (Business As Usual) and the analyzed scenario, in which changes to models have been made to evaluate the response of the DT. EC will be calculated in energy consumption units (kWh, for example) or in another conveniently chosen KPI, which assures to assess the pursued representation of impacts. Some examples of KPI could be kWh/produced units. Nevertheless, the selection of a specific KPI will depend entirely on the defined simulation. For example, for a scenario of a fixed quantity of produced units, the assessment of consumption in kWh may be sufficient. With this in mind, the following Table 1 shows the potential for improving the economic cost of energy by shifting the batches according to two scenarios of different industrial flexibility.

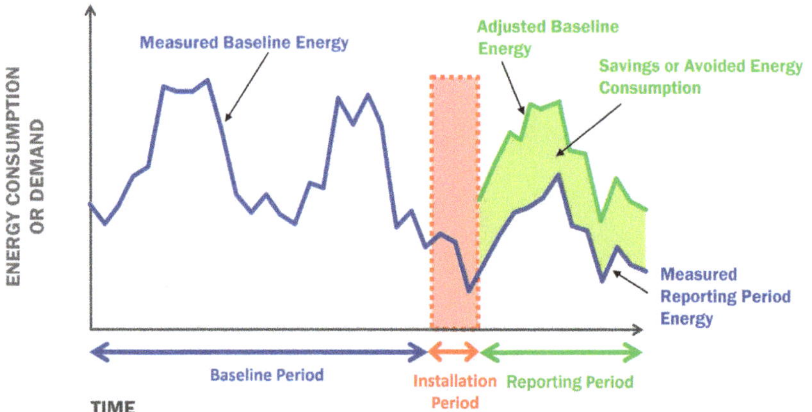

Figure 10. Energy savings adjustments defined by IPMVP. Source: EVO.

Table 1. Impact of scheduling optimization based on previous results. Source: ITE.

Application Case	Energy Cost Improvement (%)	Considerations
Ceramic industry process	4	Low flexibility
Glass industry process	37	High flexibility

This scenario of optimization applies the concept of DT developed to estimate energy cost and re-organize the sequence based on a minimum energy cost criterion, considering energy estimation and consequent cost estimation following the energy cost structure of each company. The first application case considers the impact on the energy cost of modifying the production schedule in an industrial furnace (ceramic case). In this case, the flexibility admitted due to productive constraints is low because of the demand-oriented production model of the company, and in consequence, some restrictions have been imposed on the calculation algorithm. In the second case (glass production process), each production batch is quite independent of the rest, and the machinery has a high degree of availability. These results are obtained by comparing two scenarios of minimum energy cost and maximum energy cost, respectively, implementing an optimization tool based on a solver using an approach of mixed integer linear programming (MILP).

In a second application case summarized below in Table 2, two kinds of simulations were launched along with a calculation of a specific KPI of energy consumption per produced unit (In this case, in an application case of the automotive industry) following variation of different process parameters. As a result, in both a constant period and constant production set of simulations, a similar impact of increasing the medium time during which an operator manually manipulates machinery is detected. On the other hand, other what-if scenarios were analyzed, such as an increase in losses (break of products in some production steps), concluding a relatively low impact on energy consumption compared to baseline simulation.

Finally, another possible use of the DT involves modifying the design or the use of the resources of the process. This approach may seem to be similar to previous works as described in Section 2.2, but GENERTWIN includes all of these approaches in a single multi-paradigm development that combines discrete and continuous dynamics as the main strength, which constitutes the project's primary strength. In this way, the model can be easily adapted to include new functionalities once the base knowledge of the process is strongly replicated.

Table 2. Impact of different simulation cases on energy consumption. Source: ITE.

Simulation N°	Type of Simulation	Description	KPI (kWh/Unit)	Impact *
S1	Constant period—3 working days	Baseline	0.526	0%
S2	Constant period—3 working days	Increasing procedure time + 40 s	0.542	3.1%
S3	Constant period—3 working days	Increasing losses by 4.5%	0.527	0.2%
S4	Constant period—3 working days	Variation in operator speed	0.53	0.8%
S5	Constant production—10,000 units	Baseline	0.527	0%
S6	Constant production—10,000 units	Increasing procedure time + 40 s	0.542	3.0%
S7	Constant production—10,000 units	Increasing losses by 4.5%	0.528	0.3%
S8	Constant production—10,000 units	Variation in operator speed	0.529	0.6%

* Impact related to baseline case in each type of simulation.

4. Discussion

Finally, we would like to highlight that the functionalities and impacts shown can be obtained in the productive environment with the implementation of the energy analysis models developed in the form of Digital Twins that integrate energy data with production data. These models specialize in analyzing the optimization of energy and its cost with reference to production, and in this case, are adjusted to two concrete industrial processes, such as the ceramic tiles furnace process and break friction materials manufacturing. Although it is considered that the methodology and approach presented can be extrapolated to other production process lines, this will be the future line of work to be developed by ITE in the following research and development projects. One of the most important objectives to be achieved is to obtain, characterize and determine industrial scenarios in which energy cost improvements can be achieved by considering situations of energy production flexibility in their production planning. In this sense, this approach of the proposed DT is sufficiently developed to acquire its own entity as the Digital Energy Twin (DET) of a manufacturing process.

So, the tool allows users to predict energy consumption and costs and, at the same time, evaluate the behavior of the process under certain productive changes in order to maximize consumption optimization, production efficiency and process flexibility. And in this respect, with the intention of obtaining the best results and total interoperability, another of the main challenge on which future development must work is the effective connection of the DT to the reality of the plant by means of key automation industry systems like Manufacturing Execution System (MES) or Supervisory Control and Data Acquisition (SCADA) systems, as appropriate. The connectivity of the application with direct Machine to Machine (M2M) field elements should also be addressed by analyzing the options of developing customized communication drivers such as Modbus TCP or making use of OPC UA as a reference protocol with a high level of application in the industry.

5. Conclusions

The proposed development and application framework of DT for energy efficiency implemented in project GENERTWIN aims to simulate the energy consumption behavior of the process under different productive and contextual changes. For this purpose, a multi-paradigmatic model is developed that encompasses the complexity of the production environment and the impact of the energy consumption of the process to offer advanced analysis possibilities such as scenario generation, energy cost optimization or energy impact assessment of production actions. The proposed Digital System model of energy—production advanced analysis is implemented in two demonstration experimental pilots: (a) ceramic tiles furnace process and (b) break friction materials manufacturing process in the automotive industry.

Author Contributions: Conceptualization, V.F.-R. and A.L.-A.; methodology, V.F.-R., A.L.-A. and A.R.-R.; software, F.M.-B. and B.A.-T.; validation, A.R.-R., F.M.-B. and B.A.-T.; formal analysis, A.R.-R. and A.L.-A.; investigation, A.R.-R., F.M.-B. and B.A.-T.; resources, A.R.-R.; data curation, F.M.-B. and B.A.-T.; writing—original draft preparation, V.F.-R., A.L.-A. and A.R.-R.; writing—review and editing, A.L.-A. and A.R.-R.; visualization, F.M.-B.; supervision, V.F.-R.; project administration, A.R.-R. All authors have read and agreed to the published version of the manuscript.

Funding: GENERTWIN "Sistema Digital de Análisis de procesos industriales para GENERación de escenarios alternativos bajo consideraciones productivas y de eficiencia energética", IMDEEA/2022/16 has been co-financed by IVACE (Instituto Valenciano de Competitividad Empresarial) and ERDF funds (European Regional Development Fund).

Data Availability Statement: The data presented in this study are available on request from the corresponding author. The data are not publicly available due to confidentiality and trade secret of participating companies.

Conflicts of Interest: The authors declare no conflict of interest.

References

1. Grieves, M. *Digital Twin: Manufacturing Excellence through Virtual Factory Replication*; White Paper; Michael W. Grieves, LLC: Cocoa Beach, FL, USA, 2014; pp. 1–7.
2. May, G.; Stahl, B.; Taisch, M.; Kiritsis, D. Energy management in manufacturing: From literature review to a conceptual framework. *J. Clean. Prod.* **2017**, *167*, 1464–1489. [CrossRef]
3. International Organization for Standardization. Energy Management Systems-Requirements with Guidance for Use (ISO 50001); 2018. Available online: https://www.iso.org/standard/69426.html (accessed on 31 July 2023).
4. Dörr, M.; Wahren, S.; Bauernhansl, T. Methodology for energy efficiency on process level. *Procedia CIRP* **2013**, *7*, 652–657. [CrossRef]
5. Dolge, K.; Kubule, A.; Blumberga, D. Composite index for energy efficiency evaluation of industrial sector: Sub-sectoral comparison. *Environ. Sustain. Indic.* **2020**, *8*, 100062. [CrossRef]
6. Xiong, S.; Ma, X.; Ji, J. The impact of industrial structure efficiency on provincial industrial energy efficiency in China. *J. Clean. Prod.* **2019**, *215*, 952–962. [CrossRef]
7. Asociación Europea de la Industria Cerámica. Hoja de Ruta de la Industria Cerámica, 2050. Available online: https://www.ascer.es/verDocumento.ashx?documentoId=2714&tipo=pdf (accessed on 20 July 2023).
8. Leng, J.; Wang, D.; Shen, W.; Li, X.; Liu, Q.; Chen, X. Digital twins-based smart manufacturing system design in Industry 4.0: A review. *J. Manuf. Syst.* **2021**, *60*, 119–137. [CrossRef]
9. Tekinerdogan, B.; Verdouw, C. Systems architecture design pattern catalog for developing digital twins. *Sensors* **2020**, *20*, 5103. [CrossRef] [PubMed]
10. Maeyer, C.; Markopoulos, P. Future outlook on the materialisation, expectations and implementation of Digital Twins in healthcare. In Proceedings of the 34th British HCI Conference (HCI2021), London, UK, 20–21 July 2021. [CrossRef]
11. Nasirahmadi, A.; Hensel, O. Toward the Next Generation of Digitalization in Agriculture Based on Digital Twin Paradigm. *Sensors* **2022**, *22*, 498. [CrossRef] [PubMed]
12. Opoku, D.; Perera, S.; Osei-Kyei, R.; Rashidi, M. Digital twin application in the construction industry: A literature review. *J. Build. Eng.* **2021**, *40*, 102726. [CrossRef]
13. Schoonenberg, W.; Farid, A. A Dynamic Production Model for Industrial Systems Energy Management. In Proceedings of the 2015 IEEE International Conference on Systems, Man, and Cybernetics, Hong Kong, China, 9–12 October 2015; pp. 1–7. [CrossRef]
14. Mo, F.; Rehman, H.U.; Monetti, F.M.; Chaplin, J.C.; Sanderson, D.; Popov, A.; Maffei, A.; Ratchev, S. A framework for manufacturing system reconfiguration and optimisation utilising digital twins and modular artificial intelligence. *Robot. Comput.-Integr. Manuf.* **2023**, *82*, 102524. [CrossRef]
15. Yu, W.; Patros, P.; Young, B.; Klinac, E.; Walmsley, T.G. Energy digital twin technology for industrial energy management: Classification, challenges and future. *Renew. Sustain. Energy Rev.* **2022**, *161*, 112407. [CrossRef]
16. Zhang, C.; Xu, W.; Liu, J.; Liu, Z.; Zhou, Z.; Pham, D. A Reconfigurable Modeling Approach for Digital Twin-based Manufacturing System. *Procedia CIRP* **2019**, *83*, 118–125. [CrossRef]
17. Fang, K.; Uhan, N.; Zhao, F.; Sutherland, J.W. A new approach to scheduling in manufacturing for power consumption and carbon footprint reduction. *J. Manuf. Syst.* **2011**, *30*, 234–240. [CrossRef]
18. Zhang, Z.; Tang, R.; Peng, T.; Tao, L.; Jia, S. A method for minimizing the energy consumption of machining system: Integration of process planning and scheduling. *J. Clean. Prod.* **2016**, *137*, 1647–1662. [CrossRef]
19. Ma, S.; Zhang, Y.; Lv, J.; Yang, H.; Wu, J. Energy-cyber-physical system enabled management for energy-intensive manufacturing industries. *J. Clean. Prod.* **2019**, *226*, 892–903. [CrossRef]
20. Keshari, A.; Sonsale, A.N.; Sharma, B.K.; Pohekar, S.D. Discrete event simulation approach for energy efficient resource management in paper pulp industry. *Procedia CIRP* **2018**, *78*, 2–7. [CrossRef]

21. Kant, G.; Sangwan, K. Predictive Modelling for Energy Consumption in Machining Using Artificial Neural Network. *Procedia CIRP* **2015**, *37*, 205–210. [CrossRef]
22. Holler, J.; Tsiatsis, V.; Mulligan, C.; Karnouskos, S.; Avesand, S.; Boyle, D. *From Machine-to-Machine to the Internet of Things: Introduction to a New Age of Intelligence*; Elsevier Science: Amsterdam, The Netherlands, 2014.
23. Ferreno-González, S. Aproximación Metodológica a la Implantación del Gemelo Digital en Buques. Available online: https://ruc.udc.es/dspace/bitstream/handle/2183/30974/FerrenoGonzalez_Sara_TD_2022.pdf?sequence=2 (accessed on 31 July 2023).
24. Schroeder, G.N.; Steinmetz, C.; Rodrigues, R.N.; Henriques, R.V.B.; Rettberg, A.; Pereira, C.E. A methodology for digital twin modeling and deployment for industry 4.0. *Proc. IEEE* **2020**, *109*, 556–567. [CrossRef]
25. Efficiency Valuation Organization: International Performance Measurement and Verification Protocol (IPMVP). Available online: https://evo-world.org/en/products-services-mainmenu-en/protocols/ipmvp (accessed on 31 July 2023).

MDPI AG
Grosspeteranlage 5
4052 Basel
Switzerland
Tel.: +41 61 683 77 34

Processes Editorial Office
E-mail: processes@mdpi.com
www.mdpi.com/journal/processes

www.ingramcontent.com/pod-product-compliance
Lightning Source LLC
LaVergne TN
LVHW070054170726
843515LV00007B/1535